Lecture Notes in Mathematics 1548

Subseries: Mathematisches Institut der Universität
und Max-Planck-Institut für Mathematik, Bonn – vol. 17
Adviser: F. Hirzebruch

Tohsuke Urabe

Dynkin Graphs and Quadrilateral Singularities

Springer-Verlag Berlin Heidelberg GmbH

Author

Tohsuke Urabe
Department of Mathematics
Tokyo Metropolitan University
Minami-Ohsawa 1-1
Hachioji-shi
Tokyo, 192-03, Japan

Mathematics Subject Classification (1991):
Primary: 14J17
Secondary: 14E15, 14B05, 32S25, 32S30, 32G10

DOI 10.1007/978-3-540-47769-3

MyCopy version of the original edition 1993

46/3140-543210 - Printed on acid-free paper
www.springer.com/mycopy

Contents

Chapter 4

Chapter 0

0
Introduction

In this book we will study hypersurface quadrilateral singularities. Because the study of them can be reduced to the study of elliptic K3 surfaces with a singular fiber of type I_0^*, we will study such K3 surfaces, too. The combinations of rational double points that can occur on fibers in the semi-universal deformations of quadrilateral singularities are considered. We will show that the possible combinations can be described by a certain law from the viewpoint of Dynkin graphs. This is equivalent to saying that the possible combinations of singular fibers in elliptic K3 surfaces with a singular fiber of type I_0^* can be described by a certain law using classical Dynkin graphs appearing in the theory of semi-simple Lie groups.

In the appendix we explain that a similar description can be given for plane sextic curves. The theory developed in this book provides a long list of singular points that can occur on plane sextic curves. (Because of the complexity of the world of all plane sextic curves, the list is, however, probably not complete.)

In this book we always assume that the ground field is the complex field $\mathbf{C}$.

There are 6 types of hypersurface quadrilateral singularities (Arnold [1], [2]); each of them has the following normal form of the defining function and the Milnor number μ; all have modules number 2:

$J_{3,0}$: $x^3 + ax^2y^3 + y^9 + bxy^7 + z^2$, $\mu = 16$. $(4a^3 + 27 \neq 0)$,

$Z_{1,0}$: $x^3y + ax^2y^3 + bxy^6 + y^7 + z^2$, $\mu = 15$. $(4a^3 + 27 \neq 0)$,

$Q_{2,0}$: $x^3 + yz^2 + ax^2y^2 + bx^2y^3 + xy^4$, $\mu = 14$. $(a^2 \neq 4)$,

$W_{1,0}$: $x^4 + ax^2y^3 + bx^2y^4 + y^6 + z^2$, $\mu = 15$. $(a^2 \neq 4)$,

$S_{1,0}$: $x^2z + yz^2 + y^5 + ay^3z + by^4z$, $\mu = 14$. $(a^2 \neq 4)$,

$U_{1,0}$: $x^3 + xz^2 + xy^3 + ay^3z + by^4z$, $\mu = 14$. $(a(a^2 + 1) \neq 0)$,

Note that any connected Dynkin graph of type A, D or E corresponds to a singularity on a surface called a *rational double point* (Durfee [6]).

Let X be a class of quadrilateral singularities. Let $PC(X)$ denote the set of Dynkin graphs G with components of type A, D or E only such that there exists a fiber Y in the semi-universal deformation family of a singularity belonging to X satisfying the following two conditions depending on G:

(1) Fiber Y only has rational double points as singularities.
(2) The combination of rational double points on Y corresponds to graph G.

We will study set $PC(X)$ for $X = J_{3,0}$, $Z_{1,0}$, $Q_{2,0}$, $W_{1,0}$, $S_{1,0}$ and $U_{1,0}$.

For quadrilateral singularities we have important results due to Looijenga (Looijenga [9]). They enable us to reduce the study of the above $PC(X)$ to the study of lattice embeddings. A Dynkin graph G belongs to $PC(X)$ if, and only if, the root lattice $Q(G)$ has an embedding into Λ_3 satisfying certain conditions depending on X, where Λ_3 denotes the even unimodular lattice with signature (19, 3). (See Section 2.) On the other hand, for lattice embeddings we have excellent arithmetic results due to Nikulin (Nikulin [11]). With Nikulin's results we can always determine whether or not G belongs to $PC(X)$. This is the viewpoint of Mr. F.-J. Bilitewski, who made the list of $PC(X)$ for $X = J_{3,0}$, $Z_{1,0}$ and $Q_{2,0}$ in his dissertation. In this book we show that his vast list has simple description from the view-point of the graph theory of Dynkin graphs. Moreover, we generalize this result to the remaining 3 quadrilateral singularities $W_{1,0}$, $S_{1,0}$ and $U_{1,0}$. Also, some results on plane sextic curves can be obtained.

The theory developed in this book was derived from the theory of elementary and tie transformations. These transformations are certain operations for Dynkin graphs. Fundamental theories of these transformations have been developed in Urabe [16] and in Urabe [18]. We will not repeat the descriptions in [16] and [18]. However, we believe readers will be able to understand the theory put forth in this book.

Dynkin graphs of type A, B, C, D, E, F or G related to simple Lie groups are used in our theory. Dynkin graphs associated with non-reduced root systems of type BC are used as aids. Bourbaki [3] contains explanations of Dynkin graphs of type A, B, C, D, E, F and G and root systems of type BC.

Let G be a connected Dynkin graph with r vertices. Recall that vertices in G have one-to-one correspondence with members of the corresponding root basis. This root basis is a basis of a certain Euclid space E of dimension r and consists of special vectors called roots. The collection of all roots in E is called the root system R. We can choose the normal inner product on E such that the longest root in R has length $\sqrt{2}$. Then, any root in R has length $\sqrt{2}$, 1, $\sqrt{2/3}$ or $1/\sqrt{2}$.

Now, associated with a finite subset S of a Euclid space consisting of vectors with length $\sqrt{2}$, 1, $\sqrt{2/3}$ or $1/\sqrt{2}$, we can draw a graph Γ by the following rules:

(1) Vertices in Γ have one-to-one correspondence with vectors in S.

(2) Any vertex in Γ has one of 4 different expressions depending on the length of the corresponding vector in S.

length	$\sqrt{2}$	1	$\sqrt{2/3}$	$1/\sqrt{2}$
expression	○	●	◎	⊗

(3) If two vectors $\alpha, \beta \in S$ are orthogonal, then we do not connect the corresponding two vertices in Γ. α ∗   ∗ β. (∗ denotes one of ○, ●, ◎, and ⊗.)

(4) If α and $\beta \in S$ are not orthogonal, then the corresponding two vertices in Γ are connected by a single edge. α ∗——∗ β.

(5) If α and $\beta \in S$ are proportional, i.e., $\beta = t\alpha$ for some real number t, then the edge connecting the corresponding two vertices is replaced by a bold edge. α ∗━━∗ β.

If S is a root basis of a finite root system, then the graph Γ is the corresponding Dynkin graph in our theory. A Dynkin graph is a finite union of connected Dynkin graphs.

If S is an extended root basis, i.e., the union of a finite root basis Δ and (-1) times maximal roots associated with irreducible components of Δ, then the graph Γ is the corresponding extended Dynkin graph.

Note here that our Dynkin graphs are slightly different from the standard Dynkin graphs in Bourbaki [3]. Our graphs carry more information on the length of vectors than the standard ones. Let G be a connected Dynkin graph in our sense. We can produce the corresponding standard Dynkin graph by the following procedure:

(1) If G has a part like $\circ\!\!-\!\!-\!\!\bullet$ or $\bullet\!\!-\!\!-\!\!\otimes$, replace it by $\circ\!\!\Rightarrow\!\!\circ$.
(2) If G has a part like $\circ\!\!-\!\!-\!\!\circledcirc$, then replace it by $\circ\!\!\Rrightarrow\!\!\circ$.
(3) Finally, replace all vertices by a small white circle o.

Also, applying the same procedure to an extended Dynkin graph in our theory, we get the standard extended Dynkin graph.

A root having a length of $\sqrt{2}$ is called a *long root*, and a root having a length shorter than $\sqrt{2}$ is called a *short root*.

For any connected Dynkin graph with only vertices o corresponding to long roots, our graph in our definition coincides with the standard graph in Bourbaki [3].

See Section 5 for the exact explanation of Dynkin graphs.

To state theorems we need two definitions (Urabe [15], [16], [18]).

Definition 0.1. Elementary transformation: The following procedure is called an *elementary transformation* of a Dynkin graph:

(1) Replace each connected component by the corresponding extended Dynkin graph.
(2) Choose in an arbitrary manner at least one vertex from each component (of the extended Dynkin graph) and then remove these vertices together with the edges issuing from them.

Definition 0.2. Tie transformation: Assume that by applying the following procedure to a Dynkin graph G we have obtained the Dynkin graph $\overline{G}$. Then, we call the following procedure a *tie transformation* of a Dynkin graph:

(1) Attach an integer to each vertex of G by the following rule:
Let $\alpha_1, \alpha_2, \ldots, \alpha_k$ be the root basis associated with a connected component G' of G. Let $\sum_{i=1}^{k} n_i\alpha_i$ be the associated maximal root. Then, the attached integer to the vertex corresponding to α_i is n_i.
(2) Add one vertex and a few edges to each component of G and make it into the extended Dynkin graph of the corresponding type. Attach the integer 1 to each new vertex.
(3) Choose, in an arbitrary manner, subsets A, B of the set of vertices of the extended graph $\widetilde{G}$ satisfying the following conditions:
    ⟨a⟩ $A \cap B = \emptyset$.
    ⟨b⟩ Choose arbitrarily a component $\widetilde{G}''$ of the extended graph $\widetilde{G}$ and let V be the set of vertices in $\widetilde{G}''$. Let l be the number of elements in $A \cap V$. Let $n_1, n_2, \ldots, n_l$ be the numbers attached to $A \cap V$. Also, let N be the sum of the numbers attached to elements in $B \cap V$. (If $B \cap V = \emptyset$, $N = 0$.) Then, the greatest common divisor of the $l+1$ numbers $N, n_1, n_2, \ldots, n_l$ is 1.
(4) Erase all attached integers.
(5) Remove vertices belonging to A together with the edges issuing from them.

(6) Draw another new vertex called θ corresponding to a long root. Connect θ and each vertex in B by a single edge.

Remark. After following the above procedure (1)–(6) the resulting graph $\overline{G}$ is often *not* a Dynkin graph. We consider only the cases where the resulting graph $\overline{G}$ is a Dynkin graph and then we call the above procedure a *tie transformation*.

The number $\#(B)$ of elements in the set B satisfies $0 \leq \#(B) \leq 3$. $l = \#(A \cap V) \geq 1$.

When the Dynkin graph G contains a_k of connected components of type A_k, b_l of components of type D_l, c_m of components of type E_m, d_n of components of type B_n, ..., we identify the formal sum $G = \sum a_k A_k + \sum b_l D_l + \sum c_m E_m + \sum d_n B_n \cdots$ with graph G.

The following is the first part of our main results.

Theorem 0.3. *Consider one of $J_{3,0}$, $Z_{1,0}$, and $Q_{2,0}$ as the class X of hypersurface quadrilateral singularities. A Dynkin graph G belongs to $PC(X)$ if, and only if, either (1) or (2) of the following holds:*

(1) G is one of the following exceptions.

(2) G can be made from one of the following essential basic Dynkin graphs by elementary or tie transformations applied 2 times (Four kinds of combinations — i.e., "elementary" twice, "tie" twice, "elementary" after "tie", and "tie" after "elementary" — are all permitted.) and G contains no vertex corresponding to a short root.

	The essential basic Dynkin graphs:	*The exceptions:*
The case $X = J_{3,0}$:	$E_8 + F_4$	$3A_3 + 2A_2$
The case $X = Z_{1,0}$:	$E_7 + F_4, E_8 + BC_3$	*None*
The case $X = Q_{2,0}$:	$E_6 + F_4, E_8 + F_2$	$3A_3 + A_2$

Example. Let us show that $2E_7$ and $A_7 + D_6$ are members of $PC(J_{3,0})$.

Consider the Dynkin graph $E_8 + F_4$ first. This is the essential basic Dynkin graph for $J_{3,0}$. We apply a tie transformation to this graph. At the second step we have the following graph:

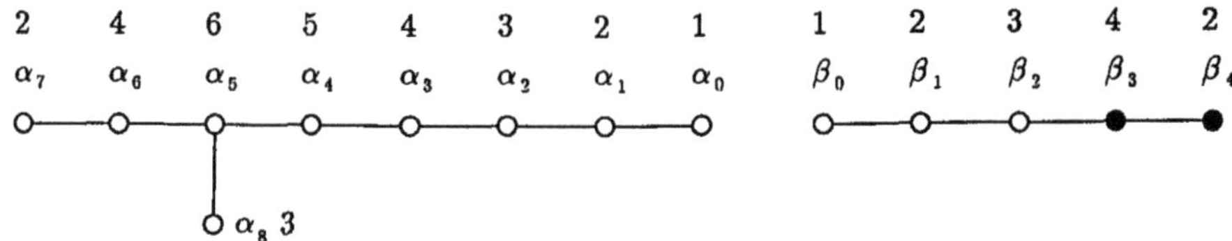

Set $A = \{\alpha_1, \beta_4\}$ and $B = \{\alpha_0, \beta_0\}$. We can check to determine if the condition on G.C.D. is satisfied for each component. Under this choice we get the graph $E_7 + B_6$ as the result of the tie transformation.

Now, we can apply a transformation to $E_7 + B_6$ again. At the start we have the following graph:

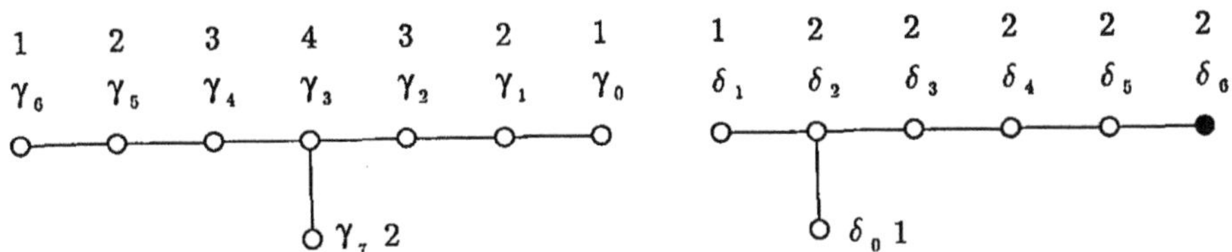

If an elementary transformation is applied and if we erase the vertices γ_7 and δ_6, we get graph $A_7 + D_6$. If a tie transformation is applied and if we choose $A = \{\gamma_0, \delta_6\}$ and $B = \{\delta_1\}$, we get graph $2E_7$.

By the above theorem one knows $A_7 + D_6$, $2E_7 \in PC(J_{3,0})$.

Remark. In the above theorem type BC_3 and type F_2 Dynkin graphs appear. We explan them briefly here:

Following is an F_2 type graph: •——•. Since it is a subgraph of F_4, we call it an F_2 type Dynkin graph. The extended F_2 type Dynkin graph is nothing more than a triangle with three black circles • at three angles. The 3 coefficients of the maximal root are 1, 1, and 1.

Let $L = \sum_{i=1}^{k} \mathbf{Z}v_i$ be a positive definite unimodular lattice equipped with the bilinear form (,) satisfying $v_i^2 = (v_i, v_i) = 1$ for $1 \le i \le k$ and $(v_i, v_j) = 0$ for $i \neq j$. Set $R_m = \{\alpha \in L \mid \alpha^2 = (\alpha, \alpha) = m\}$. The union $R = R_1 \cup R_2$ is a root system of type B_k, and $\Delta = \{v_1\} \cup \{-v_i + v_{i+1} \mid 1 \le i < k\}$ is the root basis of type B_k. The maximal root is $v_{k-1} + v_k$.

Set $R' = R \cup \{2\alpha \mid \alpha \in R_1\}$. This R' is the root system of type BC_k. It is *non-reduced*, i.e., it has an element $\alpha \in R'$ with $2\alpha \in R'$. The normalized bilinear form $(,)'$ for R' is given by $(,)' = \frac{1}{2}(,)$. The root basis for R' is the same Δ as above, but here, v_1 has length $1/\sqrt{2}$ and $-v_i + v_{i+1}$ has length 1. The maximal root is $2v_k$, which has length $\sqrt{2}$.

Following is the type BC_3 Dynkin graph: ⊗——•——•, while the following is the extended type BC_3 Dynkin graph: ⊗——•——•——o. The coefficients of the maximal root are 2, 2, 2, and 1 from the left.

Remark. Note the phenomenon called "exceptional deformations" pointed out in Wall [25]. Let $J_{3,0}(a)$ be the subclass of the class of quadrilateral singularity $J_{3,0}$ with a fixed value a of the parameter a in the above normal form of $J_{3,0}$. By Wall, there exist finite exceptional special values a_0, such that the set $PC(J_{3,0}(a))$ for any general value a is a proper subset of $PC(J_{3,0}(a_0))$. Wall's exception seems to have no relation to our exceptions in Theorem 0.3. Note that by our definition $PC(J_{3,0}) = \bigcup_a PC(J_{3,0}(a))$.

For those who are interested in elliptic surfaces, we would like to explain the relation between the above theorem and elliptic K3 surfaces (Kodaira [8]). Let $\Phi : Z \to C(\cong \mathbf{P}^1)$ be an elliptic K3 surface. It has no multiple fibers. By Kodaira's result we have an elliptic K3 surface $\Phi' : Z' \to C'$ with a section $s : C' \to Z'$, $\Phi s =$ the identity on C'

whose combination of singular fibers is same as that of Φ. Therefore, we can assume from the beginning that Φ itself has a section. Then, we can associate each singular fiber with a connected Dynkin graph of type A, D or E naturally.

$$\begin{array}{llll} I_b & \longrightarrow A_{b-1}, & I_b^* & \longrightarrow D_{b+4} \\ II & \longrightarrow \emptyset, & II^* & \longrightarrow E_8 \\ III & \longrightarrow A_1, & III^* & \longrightarrow E_7 \\ IV & \longrightarrow A_2, & IV^* & \longrightarrow E_6 \end{array}$$

Above symbols I_b, ..., IV^* are Kodaira's symbols corresponding to singular fibers of elliptic surfaces.

Let $\hat{G}$ denote the formal sum of all connected Dynkin graphs associated with the singular fibers of Φ. Let PC be the set of all Dynkin graphs $\hat{G}$ obtained from elliptic K3 surfaces $\Phi : Z \to C$. Note that $\hat{G}$ has a component of type D_4 if, and only if, Φ has a fiber of type I_0^*. Now, by Looijenga [9] it is known that $G + D_4$ belongs to PC if, and only if, G belongs to $PC(J_{3,0})$. (See Section 2.) Therefore, one knows by the above theorem that possible combinations of singular fibers in elliptic K3 surfaces with a singular fiber of type I_0^* are subject to the law described above. For one X of the other 5 types of quadrilateral singularities, the set $PC(X)$ describes possible combinations of singular fibers in elliptic K3 surfaces with a I_0^*-fiber satisfying certain additional conditions. (See Section 2.)

The list of all maximal graphs in $PC(J_{3,0})$ with respect to the inclusion relation was first given by F.-J. Bilitewski. Here, I wish to express my sincere thanks to Mr. Bilitewski for showing me his list.

Bilitewski has given the following description for $PC(Z_{1,0})$ and $PC(Q_{2,0})$: First, we consider $PC(Z_{1,0})$. Set

$$\mathcal{M}_1 = \{E_6,\ E_7,\ E_8,\ A_1\} \cup \{D_l \mid l = 4, 5, \ldots\}$$
$$\mathcal{G}_1 = \{(G,\ G_0) \mid G \in PC(J_{3,0}),\ G_0 \in \mathcal{M}_1,\ G_0 \text{ is a component of } G.\}.$$

Consider an element $(G,\ G_0) \in \mathcal{G}_1$. We can write $G = G'' + G_0$. We associate G_0' with G_0 in the following manner, depending on the type of G_0. Then, we set $G' = G'' + G_0'$:

$$\begin{array}{llll} G_0 \longrightarrow G_0' & & & \\ \hline E_8 \longrightarrow E_7, & D_4 \longrightarrow 3A_1, & & \\ E_7 \longrightarrow D_6, & D_5 \longrightarrow A_3 + A_1, & & \\ E_6 \longrightarrow A_5, & D_l \longrightarrow D_{l-2} + A_1 & (l \geq 6), & \\ A_1 \longrightarrow \emptyset. & & & \end{array}$$

Let $\mathcal{G}_1'$ be the set of all G' obtained from elements $(G,\ G_0) \in \mathcal{G}_1$. Then, $\mathcal{G}_1' = PC(Z_{1,0})$.

For $PC(Q_{2,0})$ his description is like the following: Set

$$\mathcal{M}_2 = \{E_6,\ E_7,\ E_8,\ A_2\}$$
$$\mathcal{G}_2 = \{(G,\ G_0) \mid G \in PC(J_{3,0}), G_0 \in \mathcal{M}_2, G_0 \text{ is a component of } G.\}.$$

For $(G,\ G_0) \in \mathcal{G}_2$, we can write $G = G'' + G_0$. Associating G'_0 with G_0 in the following manner, we set $G' = G'' + G'_0$:

$G_0 \longrightarrow G'_0$	
$E_8 \longrightarrow E_6,$	$E_6 \longrightarrow 2A_2,$
$E_7 \longrightarrow A_5,$	$A_2 \longrightarrow \emptyset.$

Let $\mathcal{G}'_2$ be the set of all G' obtained from elements in $\mathcal{G}_2$. Then, $\mathcal{G}'_2 = PC(Q_{2,0})$.

Bilitewski's replacement depends on the theory of singular fibers in elliptic surfaces. It is plain and easy to understand if the set $PC(J_{3,0})$ is known.

To state the theorems for $W_{1,0}$, $S_{1,0}$ and $U_{1,0}$, we must introduce another new concept called "obstruction components". Some of components of type A_k with $k \geq 4$ of a Dynkin graph are distinguished from the others as *obstruction components* and they follow special rules. (See Definition 5.9 (2) and Theorem 5.11.)

Definition 0.4. When a component G_0 of type A_k with $k \geq 4$ of the Dynkin graph G is an *obstruction component*, G_0 follows the below rules:

[The rule under an elementary transformation]

Assume that making the corresponding extended Dynkin graph $\widetilde{G}$ from G, and erasing several vertices and edges issuing from them, we have obtained the Dynkin graph G'.

(1) Let $\widetilde{G}_0$ be the component of $\widetilde{G}$ corresponding to G_0. If the vertex erased from $\widetilde{G}_0$ is unique, then the component G'_0 of G' derived from $\widetilde{G}_0$ is of type A_k. We can make this G'_0 an obstruction component. Also, we can make G'_0 a non-obstruction component, if we so desire.
(2) When two or more vertices are erased from $\widetilde{G}_0$, any component of G' derived from $\widetilde{G}_0$ is *not* an obstruction component.
(3) Obstruction components of G' are only those obtained from the obstruction components of G following the above rules (1) and (2).

[The rule under a tie transformation]

Assume that making the extended Dynkin graph $\widetilde{G}$ from G and choosing subsets A and B of the set of vertices in $\widetilde{G}$ satisfying the condition, we have made the new Dynkin graph G' depending on A and B. Let V_0 be the set of all vertices in the connected component $\widetilde{G}_0$ of the extended Dynkin graph $\widetilde{G}$ corresponding to G_0.

(1) Assume that the sets A and B satisfy the following condition #:

$$\# \quad V_0 \cap B = \emptyset \text{ and } V_0 \cap A \text{ contains only a unique element.}$$

Then, $V_0 - A$ is the set of vertices in a component G'_0 of G'. (G'_0 is also of type A_k.) This G'_0 is *necessarily* an obstruction component of G'.
(2) When the sets A and B do not satisfy the condition #, any component of G' containing a vertex belonging to $V_0 - A$ is *not* an obstruction component.
(3) Obstruction components of G' are only those obtained from obstruction components of G by following rules (1) and (2) above.

We can state the theorem for $W_{1,0}$.

Theorem 0.5. *A Dynkin graph G belongs to $PC(W_{1,0})$ if, and only if, G can be made from one of the following basic Dynkin graphs with distinguished obstruction components by elementary or tie transformations applied 2 times (Four kinds of combinations — i.e., "elementary" twice, "tie" twice, "elementary" after "tie", and "tie" after "elementary" — are all permitted.) and G contains no vertex corresponding to a short root and no obstruction component:*

The basic Dynkin graphs:
A_{11}, $B_9 + G_2$, $E_8 + G_2 + B_1$, $E_7 + B_3 + G_1$.
(A_{11} is the only obstruction component.)

In the above theorem the graph • with only one black vertex is called a type B_1 Dynkin graph, while ◎ with one double vertex is called a type G_1. Their extended Dynkin graphs are •——• and ◎——◎ respectively. The coefficients of the maximal root are 1 and 1 for both graphs.

Our main result for $S_{1,0}$ is as follows.

Theorem 0.6. *A Dynkin graph G belongs to $PC(S_{1,0})$ if, and only if, G contains no vertex corresponding to a short root and no obstruction component, and either following (1) or (2) holds:*

(1) G can be made from one of the following sub-basic Dynkin graphs by one elementary transformation or one tie transformation:

The sub-basic Dynkin graphs:
$B_{10} + A_1$, $B_9 + A_2$, $E_7 + B_4$,
$A_6 + B_5$, $B_6 + A_3 + A_2$.
(No obstruction component.)

(2) G can be made from one of the following basic Dynkin graphs with distinguished obstruction components by elementary or tie transformations applied 2 times (Four kinds of combinations — i.e., "elementary" twice, "tie" twice, "elementary" after "tie", and "tie" after "elementary" — are all permitted.):

The basic Dynkin graphs:
$A_9 + BC_1$, $B_8 + A_1$, $E_8 + BC_1$, $E_7 + BC_2$, $E_6 + B_3$,
(A_9 is the only obstruction component.)

The last remaining case is $U_{1,0}$. We must introduce an additional new concept.

Recall that an extended Dynkin graph is associated with the extended root basis, i.e., the root basis plus (-1) times maximal roots associated with an irreducible component of the basis. Note here that if an irreducible root system contains roots with 2 kinds of different length, the maximal short root ζ is uniquely defined among shorter roots depending on the root basis Δ. When the irreducible root system is the BC_k type with $k \geq 2$, the system contains roots with 3 different lengths and by ζ we denote the maximal root with length 1. ζ is defined among roots with the middle length 1. We call the union $\Delta^{\bullet} = \Delta \cup \{-\zeta\}$ the *dual extended root basis*, and call the associated graph the *dual extended Dynkin graph*.

For example, the dual extended Dynkin graph of type G_2 is similar to the following: ◎——◎——∘.

For the irreducible root system whose roots are of the same length we define that the dual extended Dynkin graph is equal to the extended Dynkin graph.

Definition 0.7. Dual elementary transformation: The following procedure is called a *dual elementary transformation* of a Dynkin graph:

(1) Replace each connected component by the corresponding *dual* extended Dynkin graph.
(2) Choose, in an arbitrary, at least one vertex from each component (of the dual extended Dynkin graph) and then remove these vertices together with the edges issuing from them.

Definition 0.4. (Addition) Under a dual elementary transformation an obstruction component follows the same rule as for an elementary transformation.

Theorem 0.8. *A Dynkin graph G belongs to $PC(U_{1,0})$ if, and only if, G contains no vertex corresponding to a short root and no obstruction component, and either (1), (2) or (3) of the following holds:*

(1) (Exceptions) $G = E_6 + D_4 + A_2$, $E_6 + A_2 + 3A_1$, $D_4 + 3A_2 + A_1$ or $3A_2 + 4A_1$.
(2) G can be made from one of the following basic Dynkin graphs with distinguished obstruction components by elementary or tie transformations applied 2 times (Four kinds of combinations — i.e., "elementary" twice, "tie" twice, "elementary" after "tie", and "tie" after "elementary" — are all permitted):

The basic Dynkin graphs:
$E_8 + A_2(3)$, $E_7 + G_2$, $E_6 + A_2 + A_2(3)$, $A_8 + G_2$.
(A_8 is the only obstruction component.)

(3) G can be made from the following dual basic Dynkin graph by two transformations one of which is a dual elementary transformation and the other is an elementary or a tie transformation (Thus, only four kinds of combinations of transformations — i.e., "dual elementary" after "elementary," "elementary" after "dual elementary," "dual elementary" after "tie," and "tie" after "dual elementary" — are permitted.):

The dual basic Dynkin graphs:
$E_8 + G_2$. *(No obstruction component.)*

Remark. The $A_2(3)$ Dynkin graph in the above theorem is explained herewith.

We can consider a root system $R = \{\alpha, \beta, \alpha+\beta, -\alpha, -\beta, -\alpha-\beta\}$ with 6 roots satisfying the conditions $\alpha^2 = \beta^2 = 2/3$ and $(\alpha, \beta) = -1/3$. In this book, R is denoted to be of type $A_2(3)$ (See Section 5.) since it becomes a root system of type A_2 if the bilinear form $3(\ ,\ )$ is given.

$\Delta = \{\alpha, \beta\}$ is its root basis, and the Dynkin graph is as follows: ◎——◎. The extended type $A_2(3)$ Dynkin graph is a triangle with three double circles ◎ at three angles. The 3 coefficients of the maximal root are all 1.

As for above $PC(W_{1,0})$, $PC(S_{1,0})$ and $PC(U_{1,0})$, Mr. Bilitewski informed me that he had a complete listing of them.

Now, apart from the above theorems, we can consider the set of all general elliptic K3 surfaces. It is not difficult to formulate the corresponding theorem about combinations of singular fibers in them. The basic Dynkin graphs in this case are $2E_8$ and D_{16}. No sub-basic Dynkin graphs appear. Perhaps there are several exceptions such as for $J_{3,0}$ and $U_{1,0}$. Yet, not having given a great deal of consideration to this case we are not certain.

I would like to give a theorem dealing with all elliptic K3 surfaces in a forthcoming article.

Also, there exist similar theorems for 14 exceptional hypersurface singularities with modules number 1 (Arnold [1], [2]). We would like to study them in a forthcoming article.

Mr. Bilitewski informed that he had a complete listing of Dynkin graphs of $PC(X)$ for any one X of 14 exceptional hypersurface singularities.

Our theory seems to have two advantages over Nikulin's arithmetic theory. First, it gives a consistent view-point. Second, we can avoid long tiresome calculations in determining all overlattices of lattices appearing in complicated cases in our problem.

The main ideas of this book are outlined here. (See also Section 1, Summary.)

The starting point is Looijenga's result, which follows from the surjectivity of the period mapping for K3 surfaces. The lattice P of rank $22 - \mu$ is defined for each one X of 6 quadrilateral singularities. By his result our problem is reduced to show the existence of an embedding $S = P \oplus Q(G) \hookrightarrow \Lambda_3$ of lattices satisfying certain conditions (L1) and (L2), where G is a Dynkin graph. $Q(G)$ is the root lattice of type G, and Λ_N stands for the even unimodular lattice with signature $(16+N,\ N)$. If $N \geq 1$, then Λ_N is unique up to isomorphism, and is isomorphic to $Q(E_8) \oplus Q(E_8) \oplus H^{\oplus N}$. Here, $Q(E_8)$ denotes the root lattice of type E_8 and H denotes the hyperbolic plane, i.e., the even unimodular lattice of rank 2 with signature $(1,1)$.

Next, we translate Looijenga's conditions (L1) and (L2) to a simpler equivalent condition. They are satisfied if, and only if, the induced embedding $Q(G) \hookrightarrow \Lambda_3/P$ is *full* and satisfies a certain condition related to obstruction components. Here, we say that the embedding is full, if the root system in $Q(G)$ including short roots and that in the primitive hull of $Q(G)$ coincide. Here is the essential reason we have to introduce the concept of short roots. The condition "no short roots and no obstruction components" implies conditions (L1) and (L2) are satisfied. Short roots play an interesting but mysterious role in our theory.

We can apply the theory of elementary transformations and tie transformations here. Let G' be a Dynkin graph made from a given Dynkin graph G by one elementary transformation or one tie transformation. Assume that a full embedding $Q(G) \hookrightarrow \Lambda_N/P$ is given. Then, we can define another full embedding $Q(G') \hookrightarrow \Lambda_{N+1}/P$. Note that the suffix of Λ has increased by one. A direct summand H is added under the process of one transformation.

Conversely, assume that we have a primitive isotropic vector u in Λ_{N+1}/P in a *nice position* with respect to any given full embedding $Q(G') \hookrightarrow \Lambda_{N+1}/P$. Here, we say that u is in a nice position either if u is orthogonal to $Q(G')$ or for some root basis $\Delta \subset Q(G')$ and for some long root $\theta \in \Delta$ $u \cdot \alpha = 0$ for any $\alpha \in \Delta$ with $\alpha \neq \theta$ and $u \cdot \theta = 1$. (In the case of $U_{1,0}$ we have to assume moreover that if u is not orthogonal to $Q(G')$, then $u \cdot x \in \mathbf{Z}$ for every $x \in \Lambda_N/P$.) Under this assumption we have a Dynkin graph G and another full embedding $Q(G) \hookrightarrow \Lambda_N/P$ such that G' can be made from G by one elementary or tie transformation. Note that the suffix of Λ has decreased by one. A direct summand H vanishes under the process of one transformation in this direction.

For $N = 1$, Λ_1/P is positive definite, and the Dynkin graph of the finite root system of Λ_1/P is called a basic graph. G is a subgraph of a basic graph, if there is a full embedding $Q(G) \hookrightarrow \Lambda_1/P$. Note here that two different embeddings $P \hookrightarrow \Lambda_1$ may define different basic graphs. However, the number of basic graphs is always finite.

The meaning of the number "two" of transformations is now obvious. We can make a full embedding into Λ_3/P after two steps of the procedure starting from Λ_1/P. This number $2\,(= 3-1)$ always appears in our main theorems.

Let $PC^{(2)}(X)$ be the set of Dynkin graphs G with components of type A, D or E only such that it can be made by two transformations starting from one of the basic graphs and such that it has no obstruction component. By our theory the inclusion relation $PC^{(2)}(X) \subset PC(X)$ always holds.

Besides, the opposite inclusion relation holds if there exists a primitive isotropic vector u in Λ_N/P in a *nice position* with respect to any given full embedding $Q(G) \hookrightarrow \Lambda_N/P$ for $N = 3$ and 2.

Here, our problem has been reduced to the following 3 items:

(1) To determine the basic graphs.
(2) To show the existence of a nice isotropic vector for $Q(G) \hookrightarrow \Lambda_2/P$.
(3) The same for $Q(G) \hookrightarrow \Lambda_3/P$.

These three items are very difficult problems, because we have to treat discrete objects. In particular, item (3) is the most difficult, since Λ_3/P has the number of negative signature 2.

To manipulate (1) and (2) we apply the theory of Coxeter-Vinberg graphs related to the hyperbolic geometry. For $X = J_{3,0}$, $Z_{1,0}$, $W_{1,0}$ and $U_{1,0}$ this theory works well. Drawing the Coxeter-Vinberg graph for Λ_2/P, we can obtain the corresponding results. When $X = U_{1,0}$, we notice that the concept of dual basic graphs and the concept of dual elementary transformation are necessary. For $X = Q_{2,0}$ we cannot apply the theory of Coxeter-Vinberg graphs. However, in this case, we can easily reduce it to the $X = J_{3,0}$ case.

The worst case is $X = S_{1,0}$, where for manipulating (1) we apply the naive lattice theory, because the theory of Coxeter-Vinberg graphs is useless. To manipulate (2) we can partially apply the theory of Coxeter-Vinberg graphs. We have to introduce the concept of sub-basic graphs here and we can determine them. To treat the remaining part of (2), ad hoc case-by-case checking is applied. Using the results on $Z_{1,0}$ and $W_{1,0}$, we can list the candidates of the counter-examples of the theory. Then, we show that they are actually not the counter-examples by the arithmetic theory.

To manipulate (3) we can apply the theory of monodromy for elliptic surfaces. First, we consider the case $X = J_{3,0}$, $Z_{1,0}$ or $Q_{2,0}$. In this case this theory works well. If G has a component not of type A, we can construct a certain transcendental 2-cycle using the monodromy. Applying the theory of Coxeter-Vinberg graphs again, we show the existence of a nice isotropic vector. When all components of G are of type A, it is extremely hard to construct any useful cycle by the monodromy. Thus, for preciseness, we apply ad hoc case-by-case checking. For checking we can apply the geometric theory of surfaces effectively. Of the few remaining cases by applying the arithmetic theory we can get the result. For last remaining case we get the exceptions in Theorem 0.3.

To consider (3) for $X = W_{1,0}$, $S_{1,0}$ or $U_{1,0}$ we could not discover any useful method other than case-by-case checking. We regret this point, and hope that someone can discover a more direct approach. However, from the results obtained thus far we can do the checking efficiently.

For (2) and (3) we also apply the theory of the Hasse symbol and the Hilbert norm residue symbol on inner product spaces.

There may be a more direct approach to our main theorems set forth in the Introduction. Verification in this book by case-by-case checking is not very good, though the problem is very difficult. An approach from the representation theory of Lie groups may be useful. We hope that readers can discover the direct approach.

We would like to discuss the possibility of future development of our theory at this point. Readers might notice that our theory depends heavily upon the surjectivity of the period mapping for K3 surfaces. However, we would like note that the concept of elementary and tie transformations is independent of the surjectivity. Of course, it works very well if the period mapping is surjective. Therefore, we cannot apply this concept to surfaces of general type at present. Yet, we can apply it to surfaces of general type if we can give good characterization of points in the image of the period mapping. We have to give the characterization by the lattice theory. It may be a very difficult problem. If it is possible, we can conclude that we can describe singularities occurring on quintic surfaces or on plane curves of higher degree by these transformations. This is a future problem. As for rational surfaces, we can consider the period mapping using rational 2-forms if the surface has an anti-canonical effective divisor. It is known that for describing singularities on such rational surfaces, elementary transformations play an important role. Many concrete cases related to K3 surfaces and rational surfaces remain. We believe we must give clear descriptions of these cases first.

During my stay at Max-Planck-Institute (April 1987 through February 1989) I accomplished approximately half of the work related to this book. I wish to express my thanks to Max-Planck-Institute. Particularly, to Professor F. Hirzebruch and Professor D. Zagier for their warm hospitality. I would also like to thank Professor E. Brieskorn, Mr. F.-J. Bilitewski and the members of the Brieskorn Oberseminar at the University of Bonn for their help. Bilitewski's results on $J_{3,0}$, $Z_{1,0}$ and $Q_{2,0}$ were reported in the Brieskorn Oberseminar on November 20, 1987. In those days I knew it was possible to give a simple description of his list of $PC(J_{3,0})$ from the viewpoint of Dynkin graphs because I had already had analogous results on quartic K3 surfaces and plane sextic curves. Only the concrete description was unknown. And, to obtain it, we had to provide a precise theory. Stimulated by his results, I commenced this work.

I am grateful to Dr. Hiroe Sakata and Mr. George H. Strube Jr. for correcting English in my manuscript. Also, I am grateful to my wife, Yasuko, for preparing the illustrations used in this book.

In this book $H = \mathbf{Z}u + \mathbf{Z}v$ denotes the hyperbolic plane H and its standard basis u, v, i.e., the even unimodular lattice with signature $(1, 1)$ whose bilinear form satisfies $u^2 = v^2 = 0$ and $u \cdot v = v \cdot u = 1$. Λ_N stands for the even unimodular lattice with signature $(16 + N, N)$. We have explained Λ_N above. By $\mathbf{Z}$, $\mathbf{Q}$, $\mathbf{R}$, $\mathbf{C}$ we denote the ring of integers, the rational field, the real field, and the complex field, respectively.

For a prime number p $\mathbf{Z}_p$ denotes the ring of p-adic integers, and $\mathbf{Q}_p$ denotes the field of p-adic numbers. ($\mathbf{Z}_p = \text{proj.lim}_{n\to\infty}\mathbf{Z}/p^n\mathbf{Z}$. $\mathbf{Q}_p$ is the quotient field of $\mathbf{Z}_p$.) For a finite abelian group M M_p stands for the p-Sylow subgroup of M, and $l(M)$ denotes the minimum number of generators of M.

1 Summary

The plan of this book is as follows.

In Chapter 0 Section 0 we stated main results. To state them many words were necessary, because we had to explain many new concepts on graph theory. However, the new concepts were natural and simple. If we have understood them once, their impression remains strongly in our intuition. Section 1 is summary of this book.

Chapter 1 (Section 2 through Section 8) gives the common basis of the theory for 6 quadrilateral singularities.

In Section 2 we review Looijenga's results. The relation between quadrilateral singularities and elliptic K3 surfaces is explained. The lattice P of rank $22 - \mu$ is defined for each one X of 6 quadrilateral singularities. Our problem is reduced to the problem on existence of the embedding $S = P \oplus Q(G) \hookrightarrow \Lambda_3$ of lattices satisfying certain conditions (L1) and (L2), where G is a Dynkin graph, $Q(G)$ is the root lattice of type G.

In Section 3 we begin to study the theory of lattices. Some of terminology used in the theory of lattices are explained. Section 4 is mainly devoted to the calculation to be appled in Section 5.

In Section 5, we first introduce the concept of root modules and develop the general theory of root systems in our situation. Second, by the results of Section 3 and 4 we translate the conditions (L1) and (L2) to the equivalent condition that the induced embedding $Q(G) \hookrightarrow \Lambda_3/P$ is full and has no obstruction component with respect to any associated number $k \geq 4$ of P. Here, the embedding is said to be full if the root system in $Q(G)$ and in the primitive hull of $Q(G)$ are equal.

The meanings of elementary and tie transformations are given in Section 6. At the end, our problem is reduced to showing the existence of a primitive isotropic vector u in Λ_N/P in a *nice position* with respect to any given full embedding $Q(G) \hookrightarrow \Lambda_N/P$ for $N = 3$ and 2. Here, u is said to be in a nice position either if u is orthogonal to $Q(G)$ or for some root basis $\Delta \subset Q(G)$ and for some long root $\theta \in \Delta$ $u \cdot \alpha = 0$ for any $\alpha \in \Delta$ with $\alpha \neq \theta$ and $u \cdot \theta = 1$. (In the case $U_{1,0}$ we have to assume moreover that if u is not orthogonal to $Q(G)$, then $u \cdot x \in \mathbf{Z}$ for every $x \in \Lambda_N/P$. In the other cases this additional condition is automatically satisfied.)

In Section 7 the conditions on isotropic vectors in quasi-lattices written with the Hasse symbol and the Hilbert norm residue symbol are explained. Conditions I2 and I1 are introduced. These conditions have a consistent theoretical meaning independent of the case chosen. The Ik-condition implies that the orthogonal complement of $Q(G)$ in Λ_3/P contains k-dimensional isotropic subspace. (Thus, I2 is stronger than I1.) However, the concrete description depends on which one X of six quadrilateral singularities is chosen, and is a condition including the root lattice $Q(G)$ written with the Hasse symbol and the Hilbert norm residue symbol. For a given G we can check Ik in a few lines. These conditions I2 and I1 are important because if we add one of them to our main theorems, we can obtain partial results giving necessary and sufficient conditions. Indeed, when $X \neq U_{1,0}$ the following holds: A Dynkin graph G belongs to $PC(X)$ and G satisfies I2 if, and only if, G can be obtained from one of the basic Dynkin graphs by two elementary transformations; A Dynkin graph G belongs to $PC(X)$ and G satisfies I1 if, and only if, G is not an exception, and G can be obtained either from one of the basic Dynkin graphs by a combination of two transformations one of which is elementary, or from one of the sub-basic Dynkin graphs by one elementary transformation. Even in the case $X = U_{1,0}$, the corresponding formulation is true if we add the concepts of dual elementary transformation and the dual basic Dynkin graphs.

In Section 8 we explain Vinberg's theory on Coxeter-Vinberg graphs associated with hyperbolic spaces. Coxeter-Vinberg graphs are effective tools to study root systems in quasi-lattices, and are useful in many aspects. This book contains many Coxeter-Vinberg graphs of quasi-lattices with negative signature 1. For example, in the case of $J_{3,0}$, $Z_{1,0}$, $W_{1,0}$ and $U_{1,0}$, they are applied to determine which Dynkin graphs are basic and to show the theorem with the I1-condition.

In Chapter 2 (Section 9 through Section 14) we study 3 singularities $J_{3,0}$, $Z_{1,0}$, and $Q_{2,0}$.

In Section 9 we first draw the Coxeter-Vinberg graph of Λ_2/P for $J_{3,0}$ and $Z_{1,0}$. As the result, we can determine the basic Dynkin graphs and can show the existence of a nice isotropic vector with respect to any full embedding $Q(G) \hookrightarrow \Lambda_2/P$. We obtain the theorems with Ik-conditions ($k = 2$, 1) not only for $J_{3,0}$ and $Z_{1,0}$ but also for $Q_{2,0}$. (Note that for these cases the set of basic Dynkin graphs is the set of the essential basic Dynkin graphs plus the graph B_{12} (for $J_{3,0}$), $B_{10} + BC_1$ (for $Z_{1,0}$) or B_9 (for $Q_{2,0}$) respectively.)

In the remainder of Chapter 2 we study the remaining part of our main theorems, i.e., the cases for $J_{3,0}$, $Z_{1,0}$, and $Q_{2,0}$ not satisfying I1. Considering the corresponding elliptic K3 surface $\Phi : Z \to C$ with a singular fiber F_1 of type I_0^*, we divide the case into three.

((1)) The surface $Z \to C$ has another singular fiber of type I^* apart from F_1.
((2)) $Z \to C$ has a singular fiber of type II^*, III^* or IV^*.
((3)) $Z \to C$ has no singular fiber of type I^*, II^*, III^* or IV^* apart from F_1.

In case ((1)) we can show by the theory of the monodromy that there exists a non-zero transcendental 2-cycle Ξ in Z with $\Xi^2 = 0$ orthogonal to a given section of Φ and to all irreducible components of fibers of $Z \to C$. In particular, the orthogonal complement of $S = P \oplus Q(G)$ in Λ_3 contains an isotropic vector; thus, we have a desired isotropic vector in Λ_3/P. This case ((1)) is treated in Section 10.

Case ((2)) is discussed in Section 11. We can show here that there exists a transcendental 2-cycle Ξ on Z with $\Xi^2 = 4$ such that Ξ is orthogonal to a given section and to all components of fibers of $Z \to C$. Therefore, the orthogonal complement of S contains a vector ξ with $\xi^2 = -4$. (We reverse the sign of the bilinear form when we move to Λ_3.) By drawing the Coxeter-Vinberg graph for K/P where K is the orthogonal complement of $\mathbf{Z}\xi$ in Λ_3, we show the existence of an isotropic vector in a nice position in this case.

To treat the last case ((3)) some arithmetic theory is necessary. In Section 12 we give brief explanation of Nikulin's lattice theory (Nikulin [11]).

In case ((3)) it is difficult to construct a nice transcendental cycle applicable to all examples. Therefore, we list all possible Dynkin graph G here (In this case, all components are of type A.), and we analyze them case-by-case by the theory of K3 surfaces, by the theory of elliptic surfaces, and by the arithmetic theory explained in Section 12. This case is discussed in Section 13.

Finally, we have to show that a Dynkin graph G in the above exception list belongs to $PC(X)$. We show this in Section 14 as an application of Nikulin's theory.

The theme of Chapter 3 (Section 15 through Section 27) is $W_{1,0}$ and $S_{1,0}$.

In Section 15 we develop the theory of obstruction components. A component G_0 of G of type A_k with $k \geq 4$ is defined to be an obstruction component with respect to the embedding $Q(G) \hookrightarrow \Lambda_N/P$ if $Q_1 = Q(G_0)$ ($\subset Q(G)$) satisfies $[\widetilde{Q}_1 : Q_1] \geq k+1$, where $\widetilde{Q}_1$ is the primitive hull. It is shown that obstruction components behave like in Definition 0.4.

In Section 16 first we draw the Coxeter-Vinberg graph of Λ_2/P for $W_{1,0}$. As the corollaries, we determine the basic graphs in this case, and we show the theorems with Ik-conditions.

To show Theorem 0.5 — our main theorem for $W_{1,0}$ without I1-condition —we use the same division of the case into three subcases ((1)), ((2)) and ((3)) as in Chapter 2.

For each case we apply the theory developed in Chapter 2. However, in the case $W_{1,0}$ we have to treat 2 sections of Φ, and the theories in Chapter 2 are not sufficient to study $W_{1,0}$. Thus, in the first step of the proof we write down the list of possible Dynkin graphs, and then we check each item G in the list case-by-case. We show either G can be made from one of the basic graphs by two transformations, or $G \notin PC(W_{1,0})$. To show $G \notin PC(W_{1,0})$ we apply the theory of symmetric bilinear forms, the theory of elliptic surfaces, and the theory of K3 surfaces, etc.

The case ((i)) for $W_{1,0}$ is discussed in Section 16+i.

In Section 20 we determine the basic Dynkin graphs for $S_{1,0}$. Since for $S_{1,0}$ we cannot draw the Coxeter-Vinberg graph of Λ_2/P and since the lattice P has no nice decomposition like in the case $Q_{2,0}$, we determine them by the theory of lattices and by the theory of root systems in them. After tiresome calculations we get the result.

Section 21, 22, 23 and 24 are devoted to showing the existence of an isotropic vector in a nice position with respect to a full embedding $Q(G) \hookrightarrow \Lambda_2/P$ in the case $S_{1,0}$. Because of the absence of the Coxeter-Vinberg graph of Λ_2/P, we spend 4 sections to show it.

In Section 21 we discuss the case where G contains a component of type BC. Let P_C be an even overlattice with index 2 over $P \oplus Q(A_1)$. The isomorphism class of P_C is uniquely determined. Drawing the Coxeter-Vinberg graph of Λ_2/P_C, we get the conclusion.

In Section 22 we deal with the case where G contains a component of type B_k with $k \geq 4$. We consider an even overlattice P_B with index 2 over the lattice $P \oplus Q(D_4)$, which is unique up to isomorphisms. We can draw the Coxeter-Vinberg graph of Λ_2/P_B. By this graph we can determine the sub-basic graphs for $S_{1,0}$ and we get the conclusion.

In Section 23 we deal with the remaining cases related to Λ_2/P. First, note that if G has a full embedding $Q(G) \hookrightarrow \Lambda_2/P$, then G is a subgraph of the Coxeter-Vinberg graph in Section 9 for $Z_{1,0}$ and also a subgraph of the Coxeter-Vinberg graph in Section 16 for $W_{1,0}$. By this fact and by the theory of the Hasse symbol and the Hilbert symbol, we list Dynkin graphs G such that if G has a full embedding $Q(G) \hookrightarrow \Lambda_2/P$, and if there is no nice isotropic vector for any such embedding, then G belongs to the list.

In Section 24 we show for any graph G in the list of Section 23 the existence of a nice isotropic vector, if it has an above full embedding. Our theory of tie transformations and Nikulin's lattice theory are applied to show it. In conclusion, we can show the theorems with Ik-conditions for $S_{1,0}$.

As in the case of $W_{1,0}$, we use the same division of the case into three subcases ((1)), ((2)) and ((3)) in Chapter 2 for the last part of the verification of Theorem 0.6. For $S_{1,0}$ we have similar difficulties to the case of $W_{1,0}$. For example, we have to consider 2 sections of Φ even for $S_{1,0}$. Still, we can apply the same ideas as in $W_{1,0}$ developed in Section 17, 18 and 19, and we manipulate the remaining part of $S_{1,0}$.

The case ((i)) for $S_{1,0}$ is discussed in Section 24+i.

The last chapter, Chapter 4 (Section 28 through Section 34), is devoted to $U_{1,0}$. Here, the corresponding lattice P has a proper even overlattice $\hat{P}$, and for this reason we have to introduce the concept of dual elementary transformations. Moreover, for the corresponding elliptic K3 surfaces $\Phi : Z \to C$ we have to treat 3 sections; thus we cannot apply the theory of transcendental cycles developed in Section 10 and 11.

In Section 28 we first develop the theory of dual elementary transformations. The concept of co-root modules is introduced to formulate the dual transformations. Consequently, as in the above 5 cases, the problem is reduced to showing the existence of a primitive isotropic vector u in Λ_N/P in a nice position with respect to any given full embedding $Q(G) \hookrightarrow \Lambda_N/P$ for $N = 2$ or 3.

In Section 29 we try to draw the Coxeter-Vinberg graph of Λ_2/P, where $P \hookrightarrow \Lambda_2$ is a primitive embedding. However, it turns out that we cannot draw it in a finite step if we follow the original exact definition of it. To overcome this difficulty we introduce dummy roots with length $\sqrt{2}/3$. Then, with this in mind we can draw it. And, in conclusion, we can determine the basic Dynkin graphs and can show the theorems with Ik-conditions.

Finally, to show the remaining part of Theorem 0.8 the theory of elliptic surfaces, the theory of tie transformations and the theory of lattices are applied. In Section 30 we collect the necessary tools from these theorems. Next, we divide the case into three parts. The division is slightly different from the above cases. Let G be a Dynkin graph with type A, D or E components only.

[[1]] G contains a type D component.

[[2]] G contains a type E component but no type D component.

[[3]] all components of G are type A.

The case [[i]] for $U_{1,0}$ is discussed in Section 30+i.

The four exceptions in Theorem 0.8 (1) are discussed in Section 34.

In the appendix we consider the similar theorems for plane sextic curves to our main results. The appendix is a supplement to Urabe [17]. Because type BC Dynkin graphs were not introduced in Urabe [17], we formulated the theorem including type BC graphs in the appendix. Thanks to general theories developed in this book, we can provide more general theorems for plane sextic curves than were contained in Urabe [17].

Chapter 1

2 Quadrilateral singularities and elliptic K3 surfaces

Let $f(x, y, z)$ be the normal form of a fixed one of the 6 types of hypersurface quadrilateral singularities in the Introduction. The polynomial $f(x, y, z)$ contains two parameters a and b. It is easy to see that $f(x, y, z)$ is quasi-homogeneous if $b = 0$, i.e., every term appearing in $f(x, y, z)$ except the one with the coefficient b has the same weight when we give weights to three variables x, y, z respectively as follows. The last two columns show the weight of f when $b = 0$ and the weight of the term with the coefficient b:

	x	y	z	f	b
$J_{3,0}$	6	2	9	18	20
$Z_{1,0}$	4	2	7	14	16
$Q_{2,0}$	4	2	5	12	14
$W_{1,0}$	3	2	6	12	14
$S_{1,0}$	3	2	4	10	12
$U_{1,0}$	3	2	3	9	11

Let $J = (\partial f/\partial x, \partial f/\partial y, \partial f/\partial z)$ be the ideal in the polynomial ring $\mathbf{C}[x, y, z]$ generated by the partial derivatives of f. The dimension $\dim_{\mathbf{C}} \mathbf{C}[x, y, z]/J$ of the quotient ring is called the Milnor number μ of the singularity $f(x, y, z)$. Let $g_1(x, y, z), \ldots, g_\mu(x, y, z)$ be monomials spanning the quotient ring $\mathbf{C}[x, y, z]/J$. We assume that the weight of g_i is less than or equal to the weight of g_{i+1} for $1 \le i \le \mu - 1$. In each case we can assume that g_1 is the constant 1, $g_{\mu-1}$ coincides with the monomial appearing in f with coefficient a, and g_μ coincides with the monomial appearing in f with coefficient b. Set $w = (w_1, w_2, \ldots, w_\mu)$ and

$$F(x, y, z, w) = f_0(x, y, z) + \sum_{i=1}^{\mu} w_i g_i(x, y, z),$$

where $f_0(x, y, z)$ is the polynomial obtained by substituting $a = b = 0$ into $f(x, y, z)$. Let $\mathcal{X} = \{(x, y, z, w) \in \mathbf{C}^3 \times \mathbf{C}^\mu \mid F(x, y, z, w) = 0\}$ be the zero-locus of the polynomial $F(x, y, z, w)$. The restriction to $\mathcal{X}$ of the projection $\mathbf{C}^3 \times \mathbf{C}^\mu \to \mathbf{C}^\mu$ is denoted by $\mathcal{X} \to \mathbf{C}^\mu$. The family $\mathcal{X} \to \mathbf{C}^\mu$ over $\mathbf{C}^\mu$ does not depend on the choice of g_i's, and is called the semi-universal deformation family of the singularity $f(x, y, z)$. It is the main object of our study in this book. The space $\mathbf{C}^\mu$ is called its *parameter space*. The inverse image of a point $p \in \mathbf{C}^\mu$ by the morphism $\mathcal{X} \to \mathbf{C}^\mu$ is called a *fiber*. By definition the fiber over the point $(0, \ldots, 0, a, b)$ in $\mathbf{C}^\mu$ is isomorphic to the singularity defined by $f(x, y, z)$.

The inverse image by $\mathcal{X} \to \mathbf{C}^\mu$ of the hyperplane defined by $w_\mu = 0$ is called the *non-positive weight part* of the semi-universal deformation family. The hyperplane $w_\mu = 0$ is called the non-positive weight part of the parameter space. On the other hand, the inverse image by $\mathcal{X} \to \mathbf{C}^\mu$ of the two-dimensional linear subspace $S_\geq$ defined by $w_1 = w_2 = \cdots = w_{\mu-2} = 0$ is called the *non-negative weight part*, and the subspace

$S_{\geq}$ is called the non-negative weight part of the parameter space. Note that for any point $p = (0, \ldots, 0, \tilde{a}, \tilde{b})$ in $S_{\geq}$, the fiber over p is defined by the polynomial obtained by substituting $a = \tilde{a}$ and $b = \tilde{b}$ into $f(x, y, z)$. Therefore, if $\tilde{a}$ satisfies the condition in the list of the normal forms in the Introduction, (For example, $4\tilde{a}^3 + 27 \neq 0$ for $J_{3,0}$ and $Z_{1,0}$.) then the fiber over p has the quadrilateral singularity at the origin as its unique singularity.

Next, to each of the 6 types of hypersurface quadrilateral singularities we associate a quadruple (p_1, p_2, p_3, p_4) of integers.

$$J_{3,0}\ (2,2,2,3),\quad Z_{1,0}(2,2,2,4),\quad Q_{2,0}(2,2,2,5),$$
$$W_{1,0}(2,2,3,3),\quad S_{1,0}(2,2,3,4),$$
$$U_{1,0}\ (2,3,3,3).$$

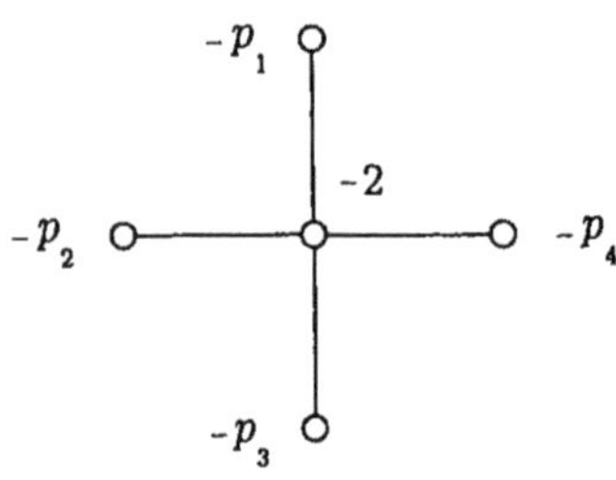

The exceptional curve in the minimal resolution of each singularity has 5 irreducible components. Every component is a smooth rational curve. The dual graph on the left represents how they intersect. Vertices have one-to-one correspondence with components. If two vertices are connected in the graph, the corresponding components intersect transversally at only one point. If two vertices are not connected, the corresponding components are disjoint. The numbers attached to vertices are the self-intersection numbers of the corresponding components.

Now, apart from the above graph, a combination of $\sum_{i=1}^4 p_i - 3 = 22 - \mu$ of smooth rational curves with self-intersection number -2 on a smooth surface is defined by the following dual graph (1). The meaning of vertices and edges is the same as above. Attached numbers -2 are omitted. We here represent particularly the dual graph in the case $Z_{1,0}(2, 2, 2, 4)$:

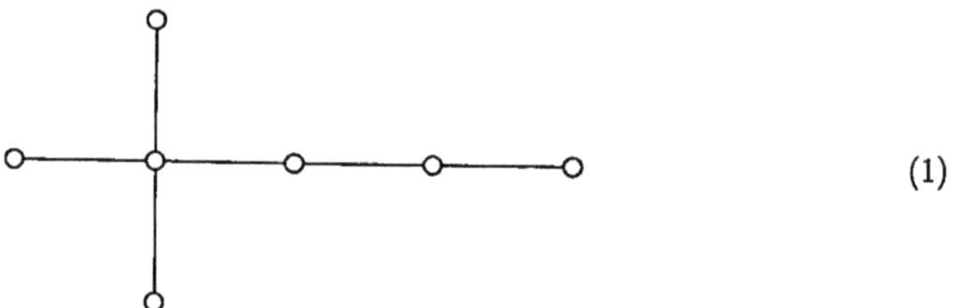

The 4 arms of the dual graph (1) have p_1, p_2, p_3, p_4 of vertices respectively including the common central one. We call this combination *the curve at infinity* $IF = IF(p_1, p_2, p_3, p_4)$.

Let $P = P(p_1, p_2, p_3, p_4)$ be the lattice generated by homology classes of the components of IF. The bilinear form on P is defined to be (-1) times the standard intersection form. By definition P has a basis $B = \{e_0, e_1 \ldots, e_{q-1}\}$ consisting of $q = \sum_{i=1}^4 p_i - 3 = 22 - \mu$ vectors such that each e_i corresponds to a component of IF, and, thus, to a vertex in the graph (1). The bilinear form $\cdot$ on $P = \sum_{e \in B} \mathbf{Z}e$ satisfies the following by definition: For each $e \in B$ $e^2 = e \cdot e = +2$. For two vectors e, $e' \in B$, $e \cdot e' = -1$ if the corresponding vertices to e and e' are connected in the graph (1), and $e \cdot e' = 0$ otherwise. P is an even lattice with signature $(q-1, 1)$.

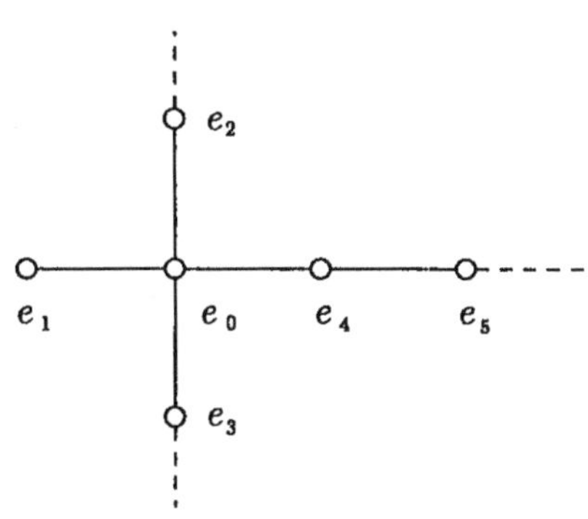

We choose vectors e_0, e_1, e_2, e_3, e_4, $e_5 \in B$ as follows to refer later. Let e_0 be the one corresponding to the central vertex with 4 edges in (1). Four vectors e_1, e_2, e_3, e_4 correspond to the adjacent 4 vertices connected to the central one e_0. Besides we assume that e_i $(1 \leq i \leq 4)$ belongs to the arm with length p_i. Thus, in particular, e_4 belongs to the longest arm with length $p_4 \geq 3$. We assign the vector e_5 to the adjacent vertex to e_4 on the opposite side of e_0.

We here introduce sublattices P_0, H_0, P', and isotropic vectors u_0, v_0 for later application.

Under the above choice, P_0 is the sublattice of P of rank 5 generated by e_0, e_1, e_2, e_3 and e_4. The isotropic vectors $u_0 \in P_0$ and $v_0 \in P$ are defined by

$$\begin{aligned} u_0 &= 2e_0 + e_1 + e_2 + e_3 + e_4 \\ v_0 &= -(u_0 + e_5) = -(2e_0 + e_1 + e_2 + e_3 + e_4 + e_5). \end{aligned}$$

It is easy to check $u_0^2 = v_0^2 = 0$ and $u_0 \cdot v_0 = 1$. The sublattice $H_0 = \mathbf{Z}u_0 + \mathbf{Z}v_0$ is isomorphic to *the hyperbolic plane* H. Let P' denote the orthogonal complement of H_0 in P. One has $P = P' \oplus H_0$ (orthogonal direct sum).

The starting point of this book is Looijenga's result explained below. This is contained in his paper on triangle singularities (Looijenga [9]).

We fix one of 6 types of hypersurface quadrilateral singularities. Let G be a Dynkin graph without a vertex corresponding to a short root. It has components of type A, D or E only. Assume that there exists a K3 surface Z satisfying the following conditions (a) and (b):

(a) Z contains the curve at infinity $IF = IF(p_1, p_2, p_3, p_4)$ corresponding to the quadrilateral singularity as a subvariety.
(b) Let $\mathcal{E}$ be the union of all smooth rational curves on Z disjoint from IF. The dual graph representing mutual intersections among the components of $\mathcal{E}$ coincides with the Dynkin graph G.

An open variety Y and a lattice embedding are defined associated with Z.

An open variety Y is obtained by contracting each connected component of $\mathcal{E}$ to a rational double point and then removing the image of IF.

Let $\Lambda \cong Q(2E_8) \oplus H \oplus H \oplus H$ denote the even unimodular lattice with signature $(19, 3)$. By $Q(G)$ we denote the positive definite root lattice of type G. Choosing an isomorphism $H^2(Z, \mathbf{Z}) \xrightarrow{\sim} \Lambda$ preserving bilinear forms up to sign, we have an embedding of lattices

$$S = P \oplus Q(G) \hookrightarrow \Lambda.$$

S is the image of the sublattice in $H^2(Z, \mathbf{Z})$ generated by the classes of the irreducible components of $\mathcal{E}$ and IF.

Theorem 2.1. (Looijenga) *(1) There exists a fiber isomorphic to Y in the non-positive weight part of the semi-universal deformation family of a quadrilateral singularity of the chosen type.*
(2) The above embedding $S \hookrightarrow \Lambda$ satisfies the following conditions (L1) and (L2). By $\widetilde{S}$ we denote the primitive hull of S in Λ. $\widetilde{S} = \{x \in \Lambda \mid mx \in S$ for some non-zero integer $m.\}$:
(L1) If a vector $\eta \in \widetilde{S}$ with $\eta^2 = +2$ is orthogonal to P, then $\eta \in Q(G)$.
(L2) If a vector $\eta \in \widetilde{S}$ with $\eta^2 = +2$ satisfies $\eta \cdot u_0 = 0$, then either $\eta \in P_0$ or η is orthogonal to P_0.

Theorem 2.2. (Looijenga) *We fix one of 6 types of hypersurface quadrilateral singularities. Let G be a Dynkin graph with components of type A, D or E only. By r we denote the number of vertices in G. The correspondence among Z, Y and the embedding $S \hookrightarrow \Lambda$ defined above induces the equivalence of the following conditions:*
(1) The graph G can be realized as the combination of singularities on a nearby fiber in the semi-universal deformation family of the quadrilateral singularity of the fixed type. In other words, there exists an infinite sequence p_n of points in the parameter space of the semi-universal deformation family of the quadrilateral singularity of the fixed type satisfying the following two conditions; (i) the sequence p_n converges to a point defining a quadrilateral singularity in the non-negative part; (ii) for every n the fiber Y_n over p_n has only rational double points as singularities and the combination of rational double points on Y_n just agrees with G.
(2) There exists a K3 surface Z satisfying the conditions (a) and (b) just before Theorem 2.1.
(2′) There exists a K3 surface Z satisfying the conditions (a) and (b) just before Theorem 2.1 and also the Picard number ρ of Z is equal to $\sum_{i=1}^{4} p_i - 3 + r$.
(3) There exists an embedding of lattices $S = P \oplus Q(G) \hookrightarrow \Lambda$ satisfying the conditions (L1) and (L2) in Theorem 2.1.

In Theorem 2.2 (1) it is to be noticed that the sequence p_n is not necessarily in the non-positive weight part of the deformation family. However, if (1) holds, then we can choose another sequence p'_n lying in the non-positive weight part satisfying the same conditions as p_n.

When we treat only geometric situations, arguments become clearer if we assume that the lattice Λ has the opposite signature (3, 19) and $Q(G)$ is negative definite. Still, in this book we define the sign of the bilinear forms on Λ and $Q(G)$ as above, because we use much algebraic theory of lattices and it is convenient for application. Therefore, note that the isomorphism $H^2(Z, \mathbf{Z}) \xrightarrow{\sim} \Lambda$ *reverses* the sign of bilinear forms.

Next, we explain the relation to the theory of elliptic surfaces. Let Z be a K3 surface satisfying the conditions (a) and (b). Let C_i $(0 \leq i < q)$ be a component of IF corresponding to the vector $e_i \in B$. The divisor $F_1 = 2C_0 + \sum_{i=1}^{4} C_i$ is numerically effective, and satisfies $F_1^2 = 0$. It defines a morphism $\Phi : Z \to \mathbf{P}^1$ to the 1-dimensional projective space $\mathbf{P}^1$. Any general fiber of Φ is a smooth elliptic curve. By definition Φ

has a singular fiber F_1, which is of type I_0^*. Since $F_1 \cdot C_5 = 1$, C_5 defines a section of Φ. (A morphism $s : \mathbf{P}^1 \to Z$ with $\Phi s =$ the identity of $\mathbf{P}^1$ is called a section of Φ. By abuse of terminologies we call also the image $s(\mathbf{P}^1)$ a section of Φ.) Note here that for $J_{3,0}$, $Z_{1,0}$ and $Q_{2,0}$ IF contains only one section C_5. On the other hand, for $W_{1,0}$ and $S_{1,0}$, IF contains another section apart from C_5. For $U_{1,0}$ IF contains 3 sections.

If a smooth rational curve D on Z satisfies $D \cdot F_1 = 0$, then D is a component of a singular fiber. In particular, every connected component of $\mathcal{E}$ is contained in a singular fiber. Conversely, if a singular fiber of Φ contains a component D disjoint from IF, then D is a smooth rational curve (Kodaira [8]). Thus, the claim explained in the Introduction "$G \in PC(J_{3,0})$ if, and only if, $G + D_4 \in PC$" is obvious.

Some readers might notice that the proof of Lemma (4.6) in Looijenga [9] is incomplete (He misses treating the case $\alpha' \in B_0$ and $B' \neq B_0$ in his notation. For example in the case $W_{1,0}(2, 2, 3, 3)$ $\alpha' \in B_0$ and $B' \neq B_0$ hold.). However, we can easily complete the proof and the claim itself is true. On the other hand, the claim of Theorem (4.5) in Looijenga [9] is not complete unless we add a certain condition on an isotropic vector. This mistake is because he has used the false claim (4) on page 367 in his paper. Consider the K3 surface just above. Set $d = \mathcal{O}_Z(kF_1 + C_5) \in \mathrm{Pic}(Z)$ with $k \geq 2$. C_5 is the fixed component of the linear system $|d|$. Therefore, d gives a counter-example to his claim (4).

3
Theory of lattices

In this book we freely use standard terminologies in the theory of lattices, i.e., the theory of integral symmetric bilinear forms (Cassels [4], Milnor-Husemoller [10], Serre [13]).

Let L be a free $\mathbf{Z}$-module of finite rank and M be a submodule. We say that M is *primitive* in L, if the quotient L/M is a free module. An element $x \in L$ is *primitive* in L, if $\mathbf{Z}x$ is primitive in L. On the other hand, if L/M is finite, L is an *over module* of M. We denote the *primitive hull* of M in L by $P(M, L) = \{x \in L \mid mx \in M$ for some non-zero integer $m.\}$ or $\widetilde{M}$ when we need not mention L. $P(M, L)$ is the minimal primitive submodule of L containing M.

Besides, when L has a symmetric bilinear form $(\ ,\ ) : L \times L \to \mathbf{Q}$ with values in rational numbers, the pair $(L, (\ ,\ ))$ is called a *quasi-lattice*. If the values of the bilinear form are all integers, $(L, (\ ,\ ))$ is called a *lattice*. For two quasi-lattices L and L' we denote the orthogonal direct sum $L \oplus L'$ by the symbol $\oplus$.

Let L be a quasi-lattice and M be a submodule. The orthogonal complement $\{x \in L \mid$ For every $y \in M$ $(x, y) = 0.\}$ of M in L is denoted by $C(M, L)$ or $M^\perp$ when we need not mention L. Note that $(M^\perp)^\perp = \widetilde{M}$ when L is non-degenerate.

Next, assume that M is non-degenerate and primitive in L. Then, L is an over-quasi-lattice of $M \oplus M^\perp$. Choose two elements $\overline{x}$, $\overline{y} \in L/M$ in the quotient module, and choose their representatives x, $y \in L$. We can write them in the form $x = x_1 + x_2$ and $y = y_1 + y_2$ with x_1, $y_1 \in M \otimes \mathbf{Q}$, and x_2, $y_2 \in M^\perp \otimes \mathbf{Q}$. Setting $(\overline{x}, \overline{y}) = (x_2, y_2)$, this rational number depends only on $\overline{x}$ and $\overline{y}$, and does not depend on the choice of representatives x, $y \in L$. Therefore, it defines a symmetric bilinear form on L/M with values in rational numbers. In this book we always give the bilinear form in this manner to the quotient module by a primitive non-degenerate submodule.

For simplicity we write $x^2 = (x, x)$. Sometimes we write $(x, y) = x \cdot y$. A vector x with $x \neq 0$, $x^2 = 0$ is called an *isotropic vector.*

Let L be a non-degenerate lattice. The dual module $L^* = \mathrm{Hom}(L, \mathbf{Z})$ is identified with a submodule in $L \otimes \mathbf{Q}$ defined by $\{x \in L \otimes \mathbf{Q} \mid (x, y) \in \mathbf{Z}$ for every $y \in L\}$. Thus, L^* has the canonical structure of a quasi-lattice containing L. We can consider the quotient L^*/L, which we call the *discriminant group* of L. The order of L^*/L equals to the absolute value of the *discriminant* $d(L)$ of L. The *discriminant bilinear form*

$$b_L : L^*/L \times L^*/L \to \mathbf{Q}/\mathbf{Z}$$

is defined by $b_L(x \bmod L, y \bmod L) \equiv (x, y) \bmod \mathbf{Z}$ for $x, y \in L^*$. Note that b_L is also non-degenerate in the sense that if $b_L(\bar{x}, \bar{y}) \equiv 0$ for every $\bar{y} \in L^*/L$, then $\bar{x} = 0$. By b_L we can define the concept of orthogonality for subsets in L^*/L. The p-Sylow subgroup and q-Sylow subgroup of L^*/L are always orthogonal for two prime numbers p, q. Also, to denote the orthogonal direct sum of subgroups in the discriminant group we use the symbol $\oplus$.

A lattice L is an *even* lattice, if x^2 is an even integer for every $x \in L$. Otherwise it is *odd*.

The *hyperbolic plane* $H = \mathbf{Z}u + \mathbf{Z}v$ is an even lattice with discriminant -1 and with signature $(1, 1)$. We have the *standard basis* u, v satisfying $u^2 = v^2 = 0$ and $(u, v) = (v, u) = 1$.

The *discriminant quadratic form*

$$q_L : L^*/L \to \mathbf{Q}/2\mathbf{Z}$$

is defined for a non-degenerate even lattice L. For $x \in L^*$ $q_L(x \bmod L) \equiv x^2 \bmod 2\mathbf{Z}$. Since $q_L(\bar{x} + \bar{y}) - q_L(\bar{x}) - q_L(\bar{y}) \equiv 2b_L(\bar{x}, \bar{y}) \bmod 2\mathbf{Z}$ for $\bar{x}, \bar{y} \in L^*/L$, q_L determines b_L uniquely. Note nevertheless that b_L does not necessarily determine q_L, though $b_L(\bar{x}, \bar{x}) \equiv q_L(\bar{x}) \bmod \mathbf{Z}$. (Note the difference between "mod $\mathbf{Z}$" and "mod $2\mathbf{Z}$". See Nikulin [11].)

Next, let L be a *positive definite* even lattice. By $\pi : L^* \to L^*/L$ we denote the canonical surjective homomorphism. For an element $\bar{x} \in L^*/L$ we define the *expected minimum square length* $m(\bar{x})$ of $\bar{x}$ by

$$m(\bar{x}) = \min\{x^2 \mid x \in L^*, \pi(x) = \bar{x}\}.$$

Lemma 3.1. *(1) $m(\bar{x}) \geq 0$, $m(\bar{x}) = 0 \iff \bar{x} = 0$.*
(2) $m(\bar{x}) \equiv q_L(\bar{x}) \bmod 2\mathbf{Z}$.
(3) Let L and L' be positive definite even lattices. We have the canonical identification $(L^*/L) \oplus (L'^*/L') = (L \oplus L')^*/(L \oplus L')$. Then, for $\bar{x} \in L^*/L$ and $\bar{y} \in L'^*/L'$ we have $m(\bar{x} + \bar{y}) = m(\bar{x}) + m(\bar{y})$.

Let G be a Dynkin graph without a vertex corresponding to a short root. Let $Q = Q(G)$ be the associated root lattice and $\Delta = \{\alpha_1, \alpha_2, \ldots, \alpha_r\} \subset Q$ be a root basis. Δ has one-to-one correspondence with the set of vertices in G. $\alpha_1^2 = \cdots = \alpha_r^2 = +2$ and $\alpha_i \cdot \alpha_j = -1$ or 0 for $i \neq j$ depending on whether the corresponding two vertices are connected in G or not. The dual basis of Δ defined by the condition $\alpha_i \cdot \omega_j = \delta_{ij}$ is denoted by $\omega_1, \omega_2, \ldots, \omega_r$. They are a basis of Q^*. We call ω_i the i-th *fundamental weight.*

Example 3.2. $L = Q(A_r)$ $(r \geq 1)$ (The root lattice of type A_r)

1 2 3 4 $r-1$ r

We assign numbers 1, 2, ..., r to vertices in the Dynkin graph of type A_r from an end in order. By ω_i we denote the fundamental weight associated with the i-th vertex. $L^*/L \cong \mathbf{Z}/(r+1)$. $\pi(\omega_1)$ is a generator of this cyclic group, and $\pi(\omega_i) = i\pi(\omega_1)$, $m(\pi(\omega_i)) = m(i\pi(\omega_1)) = i(r+1-i)/(r+1)$.

Example 3.3. $L = Q(D_r)$ $(r \geq 4)$ (The root lattice of type D_r)
We assign numbers to vertices in the Dynkin graph of type D_r as follows:

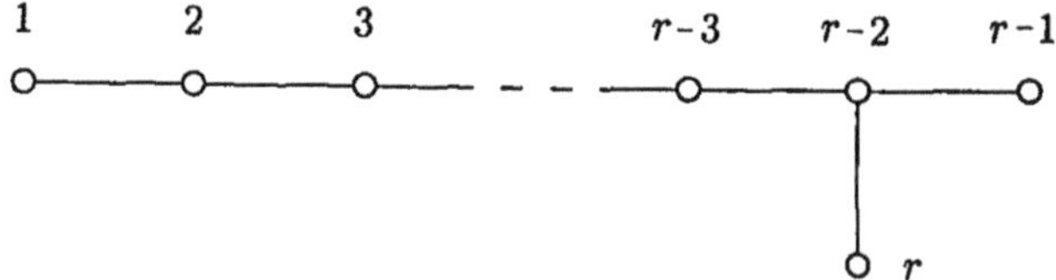

We consider the corresponding fundamental weight ω_i and the fundamental root α_i.

$$\omega_i = \alpha_1 + 2\alpha_2 + \cdots + (i-1)\alpha_{i-1} + i(\alpha_i + \alpha_{i+1} + \cdots + \alpha_{r-2}) + i(\alpha_{r-1} + \alpha_r)/2 \quad (1 \leq i \leq r-2)$$

$$\omega_{r-1} = \{\alpha_1 + 2\alpha_2 + \cdots + (r-2)\alpha_{r-2} + r\alpha_{r-1}/2 + (r-2)\alpha_r/2\}/2$$

$$\omega_r = \{\alpha_1 + 2\alpha_2 + \cdots + (r-2)\alpha_{r-2} + (r-2)\alpha_{r-1}/2 + r\alpha_r/2\}/2$$

Thus, we have $\omega_{r-1} - \omega_r = \alpha_{r-1}/2 - \alpha_r/2$ and $\pi(\omega_1) = \pi(\omega_r) - \pi(\omega_{r-1}) = \pi(\omega_{r-1}) - \pi(\omega_r)$.

If r is even, them $L^*/L \cong \mathbf{Z}/2 + \mathbf{Z}/2$, and $\pi(\omega_{r-1})$ and $\pi(\omega_r)$ are generators of two components.

If r is odd, them $L^*/L \cong \mathbf{Z}/4$, $\pi(\omega_r)$ is a generator, $\pi(\omega_1) = 2\pi(\omega_r)$, and $\pi(\omega_{r-1}) = -\pi(\omega_r)$. We have $m(\pi(\omega_1)) = 1$ and $m(\pi(\omega_{r-1})) = m(\pi(\omega_r)) = r/4$.

Example 3.4. $L = Q(E_6)$ (The root lattice of type E_6)
$L^*/L \cong \mathbf{Z}/3$. If $\bar{x} \neq 0$, $m(\bar{x}) = 4/3$.

Example 3.5. $L = Q(E_7)$ (The root lattice of type E_7)
$L^*/L \cong \mathbf{Z}/2$. If $\bar{x} \neq 0$, $m(\bar{x}) = 3/2$.

Example 3.6. $L = Q(E_8)$ (The root lattice of type E_8)
$L^*/L \cong 0$. L is a unimodular even lattice.

Lemma 3.7. *Let G be a Dynkin graph with only vertices corresponding to long roots. Assume an element $\xi \in Q(G)^*$ in the dual module of the root lattice of type G satisfies $0 < \xi^2 < 1$. Then, $k = \xi^2/(1-\xi^2)$ is a positive integer, and besides G contains a* *component G_0 of type A_k such that ξ is contained in $Q(G_0)^*$. Then, ξ and $Q(G_0)$ together generate $Q(G_0)^*$. In particular, $\xi^2 \geq 1/2$ and $\xi^2 = k/(k+1)$.*

Proof. In Example 3.2 $m(\pi(\omega_i)) = i(r+1-i)/(r+1) \geq 1$ if $2 \leq i \leq r-1$, and $m(\pi(\omega_1)) = m(\pi(\omega_r)) = r/(r+1) \geq 1/2$. Besides, in Example 3.3, 3.4, 3.5 and 3.6, if $\overline{x} \in L^*/L$, $\overline{x} \neq 0$, then $m(\overline{x}) \geq 1$. Our lemma follows from these facts and Lemma 3.1 (3). Q.E.D.

For each of 6 types of quadrilateral singularities the lattice P was defined. It has a decomposition $P = P' \oplus H_0$. P' is an even positive definite lattice. We can define the expected minimum square length $m(\overline{x})$ for every element $\overline{x} \in P'^*/P' \cong P^*/P$.

Definition 3.8. Let $\overline{x}$ be an element in P'^*/P'. By $q = q_{P'}$ we denote the discriminant quadratic form of P'.
(1) We call $\overline{x}$ *an element of the first kind*, if $q(\overline{x}) \equiv t \bmod 2\mathbf{Z}$ for some number t with $0 \leq t < 1$.
(2) If $q(\overline{x}) \equiv 1 \bmod 2\mathbf{Z}$, we call $\overline{x}$ *an element of the second kind*. Besides, if $m(\overline{x}) = 1$ and if $\overline{x}$ has order 2, we call it a *special element of type B*.
(3) If for some positive integer k

$$q(\overline{x}) \equiv 1 + (1/(k+1)) \bmod 2\mathbf{Z},$$

$\overline{x}$ is called an *element of the third kind* and k is called *the associated number* of $\overline{x}$. Besides, if an element $\overline{x}$ of the third kind with the associated number 1 satisfies $m(\overline{x}) = 3/2$ and if it has order 2, then we call it a *special element of type BC*. If an element $\overline{x}$ of the third kind with the associated number 2 satisfies $m(\overline{x}) = 4/3$ and if it has order 3, then we call it a *special element of type G*.
(4) Any element neither of the first kind, of the second kind, nor of the third kind is called *of the fourth kind*.
(5) The associated numbers of the elements of the third kind in P'^*/P' are called the associated numbers of P or P'.

Indeed, special elements are related to short roots in root systems. The above terminologies "type B, BC or G" are used to imply this relation. (See Corollary 5.7.)

The following proposition plays a key role later to translate Looijenga's conditions (L1) and (L2) into a simpler condition:

Proposition 3.9. *Fixing one of the 6 types of hypersurface quadrilateral singularities, we consider the corresponding lattices P' and P_0. Set $P_0' = P' \cap P_0 = \sum_{i=0}^{3} \mathbf{Z}e_i$ ($\cong Q(D_4)$).*
(1) P' has the following property (G′):

(G′) *If $\eta \in P'$, $\eta^2 = +2$ and $\eta \notin P_0'$, then η is orthogonal to P_0'.*

(2) Let $\overline{x} \in P'^*/P'$ be an element of the first kind. Let $x \in P'^*$ be an element such that $x \bmod P' = \overline{x}$. If $x^2 < 2$, then x is orthogonal to P_0'.

(3) For every element $\overline{x} \in P'^{*}/P'$ of the second or third kind, there exists an element $x_0 \in P'^{*}$ satisfying the following three conditions: $x_0 \bmod P' = \overline{x}$, $x_0^2 < 2$ and x_0 is not orthogonal to P'_0.
(4) Every element of the second kind is a special element of type B.
(5) Every element of the third kind with the associated number 2 is a special element of type G.
(6) For the former 5 types except the last one $U_{1,0}$ the following assertion holds:

$$\text{If } q(\overline{x}) \equiv 0 \bmod 2\mathbf{Z} \text{ for } \overline{x} \in P'^{*}/P', \text{ then } \overline{x} = 0.$$

For $U_{1,0}$ this assertion does not hold.

Through long elementary calculations we show this proposition. The verification is given in the next section.

4
Analysis of the lattice P

In this section we give verification of Proposition 3.9. On the way, the discriminant quadratic form of P is calculated. In addition, in the last part we explain the relation between overlattices and the discriminant quadratic form.

Readers admitting Proposition 3.9 can skip most of the tiresome parts of this section except Lemma 4.4 and Proposition 4.5.

First, recall the following facts: $P = P' \oplus H_0$. P_0 is the sublattice of P generated by e_0, e_1, e_2, e_3 and e_4. $P'_0 = P_0 \cap P'$ is the sublattice of P' generated by e_0, e_1, e_2 and e_3. $P'_0 \cong Q(D_4)$.

The case of $J_{3,0}(2, 2, 2, 3)$.

$P' \cong P'_0 \cong Q(D_4)$. Obviously, it has the property (G′). By Example 3.3 ($r = 4$), one knows that any non-zero element in the discriminant group is of the second kind. Referring to Example 3.3 we can check the rest of Proposition 3.9. In particular, P has no associated numbers. Special elements in P'^{*}/P' are of type B.

The case of $Z_{1,0}(2, 2, 2, 4)$

Let e_6 be the member of the basis B corresponding to the end of the longest arm of the dual graph (1). Set $e'_6 = e_6 - u_0$. $e'^2_6 = +2$. $P' = P'_0 \oplus \mathbf{Z}e'_6$. Thus, if the assumption of (G′) is satisfied, then $\eta = \pm e'_6$ and we have the conclusion.

Every element $\overline{x} \in P'^{*}/P'$ can be written uniquely in the form $\overline{x} = \overline{y} + \overline{z}$ where $\overline{y} \in P'^{*}_0/P'_0$ and $\overline{z} \in (\mathbf{Z}e'_6)^{*}/\mathbf{Z}e'_6$. The rest of the proposition follows from Example 3.2 ($r = 1$) and Example 3.3 ($r = 4$).

Consider the case $\overline{y} = \overline{z} = 0$. The element $\overline{x} = 0$ is of the first kind. Here, obviously, if $x \in P'$, $x^2 \equiv 0 \bmod 2\mathbf{Z}$, and $x^2 < 2$, then $x^2 = 0$; thus, $x = 0$. For $\overline{x} = 0$, the assertion (2) holds.

If $\overline{y} = 0$ and $\overline{z} \neq 0$, then $q(\overline{x}) \equiv 1/2 \bmod 2\mathbf{Z}$ and $\overline{x}$ is of the first kind. If $x \in P'^{*}$, $x \bmod P' = \overline{x}$ and $x^2 < 2$, then $x = \pm e'_6/2$; thus, x is orthogonal to P'_0. The assertion (2) holds.

When $\overline{y} \neq 0$ and $\overline{z} = 0$, $q(\overline{x}) \equiv 1 \bmod 2\mathbf{Z}$. The element $\overline{x}$ is of the second kind. By Example 3.3 one knows the assertion (3). By (3) $m(\overline{x}) = 1$. Since $\overline{x}$ has order 2, it is a special element of type B.

When $\overline{y} \neq 0$ and $\overline{z} \neq 0$, $q(\overline{x}) \equiv 3/2 \bmod 2\mathbf{Z}$. The element $\overline{x}$ is of the third kind with the associated number 1. We have a vector $y \in P_0'$ such that $y \bmod P_0' = \overline{y}$ and $y^2 = 1$. Setting $x_0 = y + (e_6'/2)$, one knows this x_0 satisfies the assertion (3). In particular, $m(\overline{x}) = 3/2$. Since $\overline{x}$ has order 2, it is a special element of type BC.

The assertion (6) is obvious by the above.

P has the associated number 1 only, and special elements in P'^*/P' are of type B or of type BC.

The case of $Q_{2,0}(2, 2, 2, 5)$.

We associate vectors e_0, e_4, e_5, e_6, e_7 with the 5 vertices on the longest arm of the dual graph (1) in order from the central one with 4 edges. Set $e_6' = e_6 - u_0$ and $T = \mathbf{Z}e_6' + \mathbf{Z}e_7$. One has $P' = P_0' \oplus T$, and $T \cong Q(A_2)$. (G′) follows easily from this decomposition. Write an element $\overline{x} \in P'^*/P'$ in the form $\overline{x} = \overline{y} + \overline{z}$ where $\overline{y} \in P_0'^*/P_0'$ and $\overline{z} \in T^*/T$. We can apply Example 3.2 ($r = 2$) and Example 3.3 ($r = 4$).

If $\overline{y} = \overline{z} = 0$, then $\overline{x} = 0$ is of the first kind. The assertion (2) holds.

When $\overline{y} = 0$ and $\overline{z} \neq 0$, $q(\overline{x}) \equiv 2/3 \bmod 2\mathbf{Z}$ and $\overline{x}$ is of the first kind. Let $x \in P'^*$ be a vector satisfying $x \bmod P' = \overline{x}$ and $x^2 < 2$. We can write it in the form $x = y + z$ where $y \in P_0'^*$ and $z \in T^*$. One has $z^2 \geq 0$. On the other hand, $y \in P_0'$, since $y \bmod P_0' = \overline{y} = 0$. If $y \neq 0$, then $y^2 \geq 2$, thus, $x^2 = y^2 + z^2 \geq 2$, which is a contradiction. Thus, $y = 0$ and x is orthogonal to P_0'. One knows that the assertion (2) holds in this case.

If $\overline{y} \neq 0$ and $\overline{z} = 0$, then $q(\overline{x}) \equiv 1 \bmod 2\mathbf{Z}$ and $\overline{x}$ is of the second kind. Referring to Example 3.3 one knows that the assertion (3) holds in this case. In particular, $m(\overline{x}) = 1$. Since $\overline{x}$ has order 2, it is a special element of type B. The assertion (4) also holds.

When $\overline{y} \neq 0$ and $\overline{z} \neq 0$, $q(\overline{x}) \equiv 5/3 \bmod 2\mathbf{Z}$ and $\overline{x}$ is of the fourth kind.

By the above, one also knows the assertion (6).

P has no associated numbers. Special elements in P'^*/P' are of type B.

The case of $W_{1,0}(2, 2, 3, 3)$.

We associate the vectors e_0, e_1, ..., e_6 with the vertices in the dual graph like in the left figure below.

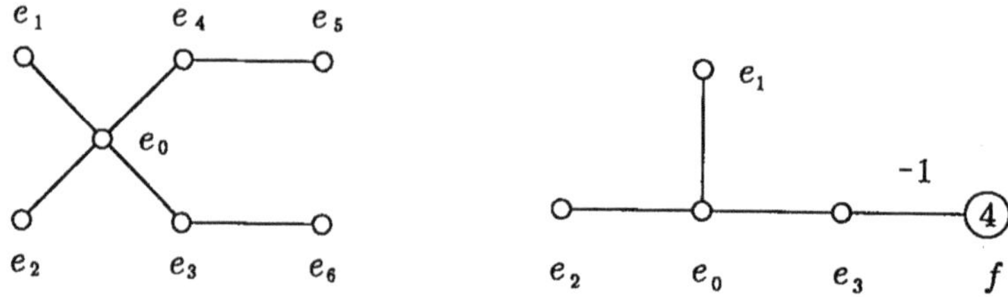

Set $f = e_6 - u_0 + v_0 = e_6 - e_5 - 2u_0$. We have $P' = \sum_{i=0}^{3} \mathbf{Z}e_i + \mathbf{Z}f$. Besides $f^2 = 4$, $f \cdot e_0 = f \cdot e_1 = f \cdot e_2 = 0$, and $f \cdot e_3 = -1$. The dual graph of the lattice P' can be written like in the right figure above.

Set

$$\begin{aligned} w_0 &= (14e_0 + 7e_1 + 7e_2 + 8e_3 + 2f)/6 \\ w_1 &= (14e_0 + 13e_1 + 7e_2 + 8e_3 + 2f)/12 \\ w_2 &= (14e_0 + 7e_1 + 13e_2 + 8e_3 + 2f)/12 \\ w_3 &= (4e_0 + 2e_1 + 2e_2 + 4e_3 + f)/3 \\ z &= (2e_0 + e_1 + e_2 + 2e_3 + 2f)/6. \end{aligned}$$

They are vectors in $P' \otimes \mathbf{Q}$. We can check $w_i \cdot e_j = \delta_{ij}$, $z \cdot e_j = 0$, $w_i \cdot f = 0$ and $z \cdot f = 1$. Thus, $w_0, \ldots, w_3, z$ is a basis of the dual module P'^*. Note that the coefficient of e_i in w_j equals $w_i \cdot w_j$, the coefficient of e_j in z equals to $w_j \cdot z$, and by the same reason $z^2 = 1/3$. One knows the following. For a vector $x \in P'^*$ we denote $\overline{x} = x \bmod P' \in P'^*/P'$:

Proposition 4.1. *For $W_{1,0}(2, 2, 3, 3)$, the discriminant group $P^*/P \cong P'^*/P'$ is a cyclic group of order 12. We can pick $\overline{w}_1$ or $\overline{w}_2$ as its generator. We have $\overline{w}_0 = 2\overline{w}_1$, $\overline{w}_2 = -5\overline{w}_1$, $\overline{w}_3 = -4\overline{w}_1$, and $\overline{z} = 2\overline{w}_1$. For the discriminant quadratic form q,*

$$q(\overline{w}_1) \equiv q(\overline{w}_2) \equiv 13/12 \bmod 2\mathbf{Z}.$$

We continue to check Proposition 3.9. First, we show the property (G′). Assume $\eta \in P'$, $\eta^2 = +2$ and $\eta \notin P'_0$. Set $\eta = \sum_{i=0}^{3} a_i e_i + bf$ with integers a_i and b. $b \neq 0$ by assumption. Corresponding to η, we set $\widetilde{\eta} = \sum_{i=0}^{3} a_i e_i + be_6$. We have $\eta^2 = \widetilde{\eta}^2 + 2b^2$, since $e_i \cdot f = e_i \cdot e_6$ $(0 \le i \le 3)$ and $f^2 = e_6^2 + 2$. Since e_0, e_1, e_2, e_3 and e_6 generate a root lattice of type D_5 and since $\widetilde{\eta} \neq 0$, one has $\widetilde{\eta}^2 \geq 2$. It implies $2 = \eta^2 = \widetilde{\eta}^2 + 2b^2 \geq 4$, which is a contradiction. A vector η satisfying the assumption of (G′) never exists. Thus, (G′) holds.

Next, we show the assertions (2)–(5). Let $\overline{x} \in P'^*/P'$ be an element. We deal with each case separately.
(1) $\overline{x} = 0$. The zero element is of the first kind. The assertion (2) holds in this case.
(2) $\overline{x} = \pm\overline{w}_1$ or $\pm 5\overline{w}_1$. $q(\overline{x}) \equiv 13/12 \bmod 2\mathbf{Z}$. This element $\overline{x}$ is of the third kind with the associated number 11. The vector $x_0 = \pm w_1, \pm w_2$ satisfies $x_0^2 = 13/12$, and it satisfies the assertion (3) in the proposition.
(3) $\overline{x} = \pm 2\overline{w}_1$. $q(\overline{x}) \equiv 1/3 \bmod 2\mathbf{Z}$. This $\overline{x}$ is of the first kind. Note that $2\overline{w}_1 = \overline{z}$. We would like to show the assertion (2) for $\overline{x}$. To show it it suffices to see that if $\eta \in P'$ and $(z + \eta)^2 < 2$, then $\eta = 0$, because z is orthogonal to P'_0. Corresponding to $\eta = \sum_{i=0}^{3} a_i e_i + bf$, set $\widetilde{\eta} = \sum_{i=0}^{3} a_i e_i + be_6$. We have $(z+\eta)^2 = (1/3) + 2b(b+1) + \widetilde{\eta}^2 < 2$. Since b is an integer, $b(b+1) \geq 0$. If $\widetilde{\eta} \neq 0$, then $\widetilde{\eta}^2 \geq 2$ and we have $(1/3) + 2 < 2$, which is a contradiction. Thus, $\widetilde{\eta} = 0$ and $\eta = 0$.
(4) $\overline{x} = \pm 3\overline{w}_1$. $q(\overline{x}) \equiv 7/4 \bmod 2\mathbf{Z}$. This is of the fourth kind.
(5) $\overline{x} = \pm 4\overline{w}_1$. $q(\overline{x}) \equiv 4/3 \bmod 2\mathbf{Z}$. This is of the third kind with the associated number 2. Since $4\overline{w}_1 = -\overline{w}_3$, and since $w_3^2 = 4/3$, the vector $x_0 = \pm w_3$ satisfies the assertion (3). Besides, $\pm 4\overline{w}_1$ has order 3 and also the assertion (5) holds.
(6) $\overline{x} = 6\overline{w}_1$. $q(\overline{x}) \equiv 1 \bmod 2\mathbf{Z}$. This is of the second kind. Set $z_1 = (e_1 - e_2)/2$. Since $z_1 - 6w_1 = -(7e_0 + 6e_1 + 4e_2 + 4e_3 + f)$, $z_1 \in P'^*$ and $\overline{z}_1 = 6\overline{w}_1$. One knows that the assertion (3) holds, since $z_1^2 = 1$ and z_1 is not orthogonal to P_0. Also, the assertion (4) holds, since $6\overline{w}_1$ has order 2.

By the above one sees that also the assertion (6) holds.

The associated numbers of P are 2 and 11, and special elements in P'^*/P' are of type B or of type G.

The case of $S_{1,0}(2, 2, 3, 4)$.

We associate vectors e_0, e_1, ..., e_7 with vertices in the graph as in the following left figure:

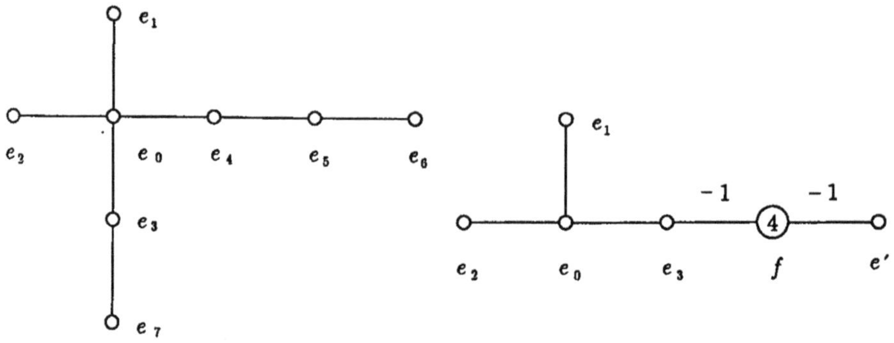

Set $e' = u_0 - e_6$ and $f = e_7 - e_5 - 2u_0$. Vectors e_0, e_1, e_2, e_3, f and e' span P'. We have $f \cdot e_0 = f \cdot e_1 = f \cdot e_2 = 0$, $f \cdot e_3 = -1$, $f^2 = 4$, $e' \cdot e_i = 0$ $(0 \le i \le 3)$, $e' \cdot f = -1$ and $e'^2 = 2$. Thus, the dual graph of the lattice P' is like the right figure above.

Set

$$\begin{aligned} w_0 &= (12e_0 + 6e_1 + 6e_2 + 7e_3 + 2f + e')/5 \\ w_1 &= (12e_0 + 11e_1 + 6e_2 + 7e_3 + 2f + e')/10 \\ w_2 &= (12e_0 + 6e_1 + 11e_2 + 7e_3 + 2f + e')/10 \\ w_3 &= (14e_0 + 7e_1 + 7e_2 + 14e_3 + 4f + 2e')/10 \\ z &= (2e_0 + e_1 + e_2 + 2e_3 + 2f + e')/5 \\ w_4 &= (2e_0 + e_1 + e_2 + 2e_3 + 2f + 6e')/10. \end{aligned}$$

We can check that w_0, w_1, w_2, w_3, z, w_4 is the dual basis of e_0, e_1, e_2, e_3, f, e'. We have the following proposition. We denote $\overline{x} = x \bmod P' \in P'^*/P'$ for $x \in P'^*$:

Proposition 4.2. *For $S_{1,0}(2, 2, 3, 4)$, the discriminant group $P'^*/P' \cong P^*/P$ is the direct sum of three cyclic groups, which have order 5, 2, 2 respectively. The first direct summand of order 5 is generated by $\overline{w}_0 = \overline{z} = 2\overline{w}_1 = 2\overline{w}_2 = -4\overline{w}_3 = 2\overline{w}_4$. The second of order 2 is generated by $\overline{g}_1 = 5\overline{w}_1$ and the third of order 2 by $\overline{g}_2 = 5\overline{w}_2$. For the discriminant quadratic form q*

$$q(a\overline{z} + b_1\overline{g}_1 + b_2\overline{g}_2) \equiv \frac{2}{5}a^2 - \frac{1}{2}(b_1^2 + b_2^2) \bmod 2\mathbf{Z}.$$

Besides, $\overline{w}_0 = \overline{z}$, $\overline{w}_1 = -2\overline{z} + \overline{g}_1$, $\overline{w}_2 = -2\overline{z} + \overline{g}_2$, $\overline{w}_3 = \overline{z} + \overline{g}_1 + \overline{g}_2 = \overline{w}_1 + \overline{w}_2$ and $\overline{w}_4 = -2\overline{z} + \overline{g}_1 + \overline{g}_2$.

We check Proposition 3.9. First, we show (G′). Assume that $\eta \in P'$, $\eta^2 = +2$ and $\eta \notin P'_0$. Corresponding to $\eta = \sum_{i=0}^{3} a_i e_i + bf + ce'$, set $\widetilde{\eta} = \sum_{i=0}^{2} a_i e_i + a_3 e_4 + b e_5 + c e_6$. This $\widetilde{\eta}$ is a vector in the root lattice of type D_6 generated by e_0, e_1, e_2, e_4, e_5 and e_6. Since $\widetilde{\eta} \neq 0$, $\widetilde{\eta}^2$ is a positive even integer. Since $2 = \eta^2 = \widetilde{\eta}^2 + 2b^2 \geq \widetilde{\eta}^2 \geq 2$, we have $b = 0$. Thus, $2 = \left(\sum_{i=0}^{3} a_i e_i\right)^2 + 2c^2$. If $c = 0$, then $\eta = \sum_{i=0}^{3} a_i e_i \in P'_0$, which contradicts the assumption. Therefore, $c \neq 0$; thus, $\sum_{i=0}^{3} a_i e_i = 0$ and $\eta = \pm e'$. So η is orthogonal to P'_0.

Next, we check the assertions (2)–(5). Let $\overline{x} \in P'^*/P'$ be an element.

(1) $\overline{x} = 0$. The zero element is of the first kind and satisfies the assertion (2).

(2) $\overline{x} = \pm\overline{z}$. $q(\overline{x}) \equiv 2/5 \bmod 2\mathbf{Z}$. This $\overline{x}$ is of the first kind. To show the assertion (2) it suffices to see that if $\eta \in P'$ and $(\eta + z)^2 = z^2 = 2/5$, then $\eta = 0$. Set $\widetilde{\eta} = a_0 e_0 + a_1 e_1 + a_2 e_2 + a_3 e_4 + b e_5 + c e_7$, corresponding to $\eta = \sum_{i=0}^{3} a_i e_i + bf + ce'$. If $\eta \neq 0$, then $\widetilde{\eta}^2 > 0$ and $2/5 = (\eta + z)^2 = \widetilde{\eta}^2 + 2b(b+1) + (2/5) > 2/5$, which is a contradiction. Thus, $\eta = 0$.

(3) $\overline{x} = \pm 2\overline{z}$. $q(\overline{x}) \equiv 8/5 \bmod 2\mathbf{Z}$. This is of the fourth kind.

(4) $\overline{x} = \overline{g}_1$ or $\overline{g}_2$. $q(\overline{x}) \equiv -1/2 \equiv 3/2 \bmod 2\mathbf{Z}$. This $\overline{x}$ is of the third kind with the associated number 1. We will see the assertion (3). Indeed, set $x_0 = (e_1 + e_3 + e')/2$. Since $5w_1 - x_0 = 6e_0 + 5e_1 + 3e_2 + 3e_3 + f$, we have $\overline{x}_0 = \overline{g}_1$. This x_0 is not orthogonal to P'_0, since $x_0 \cdot e_1 \neq 0$. Besides, $x_0^2 = 3/2 < 2$. When we treat the element $\overline{g}_2$, we can consider the vector $(e_2 + e_3 + e')/2$ instead. Finally, one sees that $\overline{x}$ is a special element of type BC, since it has order 2.

(5) $\overline{x} = \pm\overline{z} + \overline{g}_1$ or $\pm\overline{z} + \overline{g}_2$. $q(\overline{x}) \equiv 19/10$. This is of the fourth kind.

(6) $\overline{x} = \pm(2\overline{z} + \overline{g}_1)$ or $\pm(2\overline{z} + \overline{g}_2)$. $q(\overline{x}) \equiv 11/10$. This $\overline{x}$ is of the third kind with the associated number 9. We have to show the assertion (3). For $\pm(2\overline{z} + \overline{g}_1)$, set $x_0 = \pm w_1$. For $\pm(2\overline{z} + \overline{g}_2)$, set $x_0 = \pm w_2$. In the both cases, this x_0 satisfies the condition.

(7) $\overline{x} = \overline{g}_1 + \overline{g}_2$. $q(\overline{x}) \equiv 1$. This is of the second kind. The vector $x_0 = (e_1 + e_2)/2$ satisfies the condition in the assertion (3). It is also a special element of type B.

(8) $\overline{x} = \pm\overline{z} + \overline{g}_1 + \overline{g}_2$. $q(\overline{x}) \equiv 7/5$. This is of the fourth kind.

(9) $\overline{x} = \pm 2\overline{z} + \overline{g}_1 + \overline{g}_2$. $q(\overline{x}) \equiv 3/5$. This is of the first kind. Set $x_0 = \pm w_4$. One sees that $\overline{x}_0 = \overline{x}$, $x_0^2 = 3/5$ and x_0 is orthogonal to P'_0. Thus, to show the assertion (2), it suffices to see that if $\eta \in P'$ and $(\eta + w_4)^2 = 3/5$, then η belongs to $\mathbf{Z}e'$. Assume that $\eta^2 + 2\eta \cdot w_4 = 0$. Corresponding to $\eta = \sum_{i+0}^{3} a_i e_i + bf + ce'$, set $\overline{\eta} = \sum_{i=0}^{2} a_i e_i + a_3 + e_4 + b e_5$. $\overline{\eta}^2$ is an even non-negative integer. Set $B = b^2 - bc + c^2 + c$. One sees $\overline{\eta}^2 + 2B = \eta^2 + 2\eta \cdot w_4 = 0$. On the other hand, it is easy to see that $B \geq 0$ for integers b and c. Thus, we have $\overline{\eta}^2 = 0$ and $B = 0$. It implies that $\eta = 0$ or $\eta = -e'$. The assertion (6) is obvious.

The associated numbers of P are 1, 9, and special elements in P'^*/P' are of type B or of type BC.

The case of $U_{1,0}(2, 3, 3, 3)$.

We assign numbers to the basis of P as in the left dual graph on the next page.

Set $f_1 = e_6 - e_5 - 2u_0$ and $f_2 = e_7 - e_5 - 2u_0$. The orthogonal complement P' of $H_0 = \mathbf{Z}u_0 + \mathbf{Z}v_0$ in P is spanned by e_0, e_1, e_2, e_3, f_1 and f_2. One has $f_1 \cdot e_i = 0$ $(i = 0, 1, 3)$, $f_1 \cdot e_2 = -1$, $f_2 \cdot e_i = 0$ $(i = 0, 1, 2)$, $f_2 \cdot e_3 = -1$, $f_1^2 = f_2^2 = 4$ and $f_1 \cdot f_2 = 2$. Thus, the dual graph of P' is like the right figure on the next page.

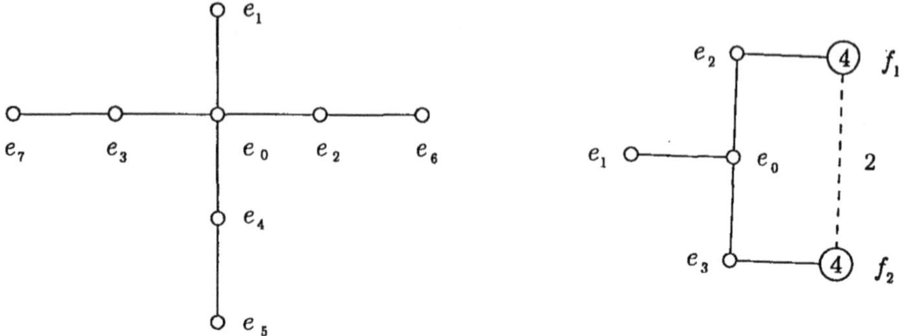

We define a basis of the dual module P'^* by the following:

$$\begin{aligned} w_0 &= (22e_0 + 11e_1 + 12e_2 + 12e_3 + 2f_1 + 2f_2)/9 \\ w_1 &= (11e_0 + 10e_1 + 6e_2 + 6e_3 + f_1 + f_2)/9 \\ w_2 &= (4e_0 + 2e_1 + 4e_2 + 2e_3 + f_1)/3 \\ w_3 &= (4e_0 + 2e_1 + 2e_2 + 4e_3 + f_2)/3 \\ z_1 &= (2e_0 + e_1 + 3e_2 + 4f_1 - 2f_2)/9 \\ z_2 &= (2e_0 + e_1 + 3e_3 - 2f_1 + 4f_2)/9 \end{aligned}$$

We can check that $w_i \cdot e_j = \delta_{ij}$, $w_i \cdot f_j = 0$, $z_i \cdot e_j = 0$ and $z_i \cdot f_j = \delta_{ij}$. On the other hand, $w_0 - (-4z_1 + w_3)$, $w_1 + (2z_1 + w_3)$, $w_2 + (-3z_1 + w_3)$ and $z_2 - (4z_1 + w_3)$ belong to P'. Thus, one knows the following proposition. For a vector $x \in P'^*$ we denote $\overline{x} = x \bmod P' \in P'^*/P'$:

Proposition 4.3. *For $U_{1,0}(2, 3, 3, 3)$, the discriminant group $P'^*/P' \cong P^*/P$ is the direct sum of the cyclic group of order 9 generated by $\overline{z}_1$ and the cyclic group of order 3 generated by $\overline{w}_3$. It can be also represented as the direct sum of the cyclic group of order 9 generated by $\overline{z}_2$ and the cyclic group of order 3 generated by $\overline{w}_2$. For the discriminant quadratic form q, we have*

$$q(a\overline{z}_1 + b\overline{w}_3) \equiv \frac{4}{9}a^2 + \frac{4}{3}b^2 \bmod 2\mathbf{Z}.$$

Besides, $\overline{w}_0 = -4\overline{z}_1 + \overline{w}_3$, $\overline{w}_1 = -2\overline{z}_1 - \overline{w}_3$, $\overline{w}_2 = 3\overline{z}_1 - \overline{w}_3$ and $\overline{z}_2 = 4\overline{z}_1 + \overline{w}_3$.

We check Proposition 3.9. First, we show (G′). Assume that $\eta \in P'$, $\eta^2 = +2$ and $\eta \notin P'_0$. Set $\widetilde{\eta} = \sum_{i=0}^3 a_i e_i + b_1 e_6 + b_2 e_7$, corresponding to $\eta = \sum_{i=0}^3 a_i e_i + b_1 f_1 + b_2 f_2$. This $\widetilde{\eta}$ belongs to the root lattice generated by e_0, e_1, e_2, e_3, e_6 and e_7. Since $\widetilde{\eta} \neq 0$, $\widetilde{\eta}^2$ is a positive even integer. Since $2 = \eta^2 = \widetilde{\eta}^2 + 2(b_1 + b_2)^2 \geq \widetilde{\eta}^2 \geq 2$, we have $b_1 + b_2 = 0$.

Setting $b = b_1 = -b_2$ and $f_3 = f_1 - f_2 = e_6 - e_7$, one can write $\widetilde{\eta} = \eta = \sum_{i=0}^3 a_i e_i + b f_3$. Note that $f_3^2 = 4$, $f_3 \cdot e_0 = f_3 \cdot e_1 = 0$, $f_3 \cdot e_2 = -1$ and $f_3 \cdot e_3 = +1$.

Let V be the 5-dimensional Euclid space with a basis ϵ_i (i = 1, 2, 3, 4, 5). We assume the bilinear form on V is defined by the following conditions: $\epsilon_1^2 = \epsilon_2^2 = \epsilon_3^2 = \epsilon_4^2 = +1$, $\epsilon_5^2 = 3$, and $\epsilon_i \cdot \epsilon_j = 0$ if $i \neq j$.

We can check that under the following identification the bilinear forms are preserved:

$$e_0 = \epsilon_2 - \epsilon_3, \qquad e_1 = -\epsilon_1 - \epsilon_2, \qquad e_2 = \epsilon_3 - \epsilon_4, \qquad e_3 = \epsilon_1 - \epsilon_2,$$
$$f_3 = \frac{1}{2}\epsilon_1 - \frac{1}{2}\epsilon_2 - \frac{1}{2}\epsilon_3 + \frac{1}{2}\epsilon_4 + \epsilon_5.$$

Under this base change the equality $\eta^2 = 2$ is equivalent to

$$(-a_1 + a_3 + \frac{b}{2})^2 + (a_0 - a_1 - a_3 - \frac{b}{2})^2 + (-a_0 + a_2 - \frac{b}{2})^2 + (-a_2 + \frac{b}{2})^2 + 3b^2 = 2.$$

Thus, $b = 0$ and $\eta \in P'_0$, which contradicts the assumption. There exists no η satisfying the conditions, and (G′) holds.

Next, we show the assertions (2)–(5) for each element $\overline{x} \in P'^*/P'$.

(1) $\overline{x} = 0$. This $\overline{x}$ is of the first kind. The assertion (2) holds.

(2) $\overline{x} = \pm\overline{z}_1$. $q(\overline{x}) \equiv 4/9 \bmod 2\mathbf{Z}$. This $\overline{x}$ is of the first kind.

To show the assertion (2) it suffices to see that if $(\eta + z_1)^2 = z_1^2 = 4/9$ for $\eta \in P'$, then $\eta = 0$, because z_1 is orthogonal to P'_0. Assume $(\eta + z_1)^2 = 4/9$. Corresponding to $\eta \in P'$, we define $\tilde{\eta}$ as above. We have $\tilde{\eta}^2 + 2b_1 + 2(b_1 + b_2)^2 = 0$.

Note here that $\tilde{\eta}$ belongs to the root lattice Q_6 of type E_6 with a root basis $\Delta = \{e_0, e_1, e_2, e_3, e_6, e_7\}$. Let ω be the fundamental weight of Q_6 corresponding to e_6. Referring to Example 3.4, one knows $(\tilde{\eta} + \omega)^2 \geq 4/3 = \omega^2$, which implies $\tilde{\eta}^2 + 2b_1 \geq 0$. One concludes $b_1 + b_2 = 0$ and $\tilde{\eta}^2 + 2b_1 = 0$.

Setting $b = b_1 = -b_2$, one can write $\tilde{\eta} = \eta = \sum_{i=0}^{3} a_i e_i + b f_3$. By the above base change one knows easily that the equality $\eta^2 + 2b = 0$ is equivalent to

$$(-a_1 + a_3 + \frac{b}{2})^2 + (a_0 - a_1 - a_3 - \frac{b}{2})^2 + (-a_0 + a_2 - \frac{b}{2})^2 + (-a_2 + \frac{b}{2})^2 + 3(b + \frac{1}{3})^2 = \frac{1}{3}.$$

Thus, $b = a_0 = \cdots = a_3 = 0$ and $\eta = 0$.

(3) $\overline{x} = \pm 2\overline{z}_1$. $q(\overline{x}) \equiv 16/9$. This is of the fourth kind.

(4) $\overline{x} = \pm 3\overline{z}_1$. $q(\overline{x}) \equiv 0$. This is of the first kind. But $m(\overline{x}) \geq 2$ by Lemma 3.1 (1), (2). Since the assumption of the assertion (2) never satisfied, (2) holds.

(5) $\overline{x} = \pm 4\overline{z}_1$. $q(\overline{x}) \equiv 10/9$. This is of the third kind with the associated number 8. Set $x_0 = w_3 - z_2$. One has $\overline{x}_0 = -4\overline{z}_1$. This x_0 is not orthogonal to P'_0, since $x_0 \cdot e_3 \neq 0$. On the other hand, $x_0^2 = w_3^2 - 2w_3 \cdot z_2 + z_2^2 = (4/3) - (2/3) + (4/9) = 10/9 < 2$. One sees that the assertion (3) holds.

(6) $\overline{x} = \pm\overline{w}_3$. $q(\overline{x}) \equiv 4/3$. This is of the third kind with the associate number 2. Since w_3 is not orthogonal to P'_0 and since $w_2^2 = 4/3$, one knows the assertion (3). Besides, $\overline{x}$ has order 3 and is a special element of type G.

(7) $\overline{x} = \pm\overline{z}_1 \pm \overline{w}_3$. $q(\overline{x}) \equiv 16/9$. This is of the fourth kind.

(8) $\overline{x} = \pm(2\overline{z}_1 + \overline{w}_3)$. $q(\overline{x}) \equiv 10/9$. This is of the third kind with the associated number 8. Now, $\overline{w}_1 = -(2\overline{z}_1 + \overline{w}_3)$, $w_1^2 = 10/9 < 2$, and w_1 is not orthogonal to P'_0, since $w_1 \cdot e_1 \neq 0$. The assertion (3) holds.

(9) $\overline{x} = \pm(-2\overline{z}_1 + \overline{w}_3)$. $q(\overline{x}) \equiv 10/9$. This is of the third kind with the associated number 8. Setting $x_0 = z_1 - w_2$, one can show the assertion (3) by the same argument as in the above case (5).

(10) $\overline{x} = \pm(3\overline{z}_1 + \overline{w}_3)$. $q(\overline{x}) \equiv 4/3$. This is of the third kind with the associated number 2. Set $x_0 = z_1 - w_1$. One can show that $\overline{x}_0 = 3\overline{z}_1 + \overline{w}_3$, $x_0^2 = 4/3 < 2$ and $x_0 \cdot e_1 = -1 \neq 0$. One knows the assertion (3). The element $\overline{x}$ has order 3, and one has the assertion (5), too.
(11) $\overline{x} = \pm(-3\overline{z}_1 + \overline{w}_3)$. $q(\overline{x}) \equiv 4/3$. This is of the third kind with the associated number 2. Now, $\overline{w}_2 = (3\overline{z}_1 - \overline{w}_3)$, $w_2^2 = 4/3$ and w_2 is not orthogonal to P'_0. One knows the assertion (3). Since $\overline{x}$ has order 3, one knows also the assertion (5).
(12) $\overline{x} = \pm(4\overline{z}_1 + \overline{w}_3)$. $q(\overline{x}) \equiv 4/9$. This is of the first kind. Note here that $\overline{z}_2 = 4\overline{z}_1 + \overline{w}_3$ and $z_2^2 = 4/9$. Thus, one can show the assertion (2) by the same argument as in the above case (2) $\overline{x} = \pm\overline{z}_1$.
(13) $\overline{x} = \pm(-4\overline{z}_1 + \overline{w}_3)$. $q(\overline{x}) \equiv 4/9$. This is of the first kind. We would like to show the assertion (2). First, note that $\overline{z_1 + z_2} = -4\overline{z}_1 + \overline{w}_3$, $(z_1 + z_2)^2 = 4/9$ and $z_1 + z_2$ is orthogonal to P'_0. Thus, to show (2) it suffices to see that if $(\eta + z_1 + z_2)^2 = 4/9$ for $\eta \in P'$, then $\eta = 0$. Set $\tilde{\eta} = \sum_{i=0}^{3} a_i e_i + b_1 e_6 + b_2 e_7$, corresponding to $\eta = \sum_{i=0}^{3} a_i e_i + b_1 f_1 + b_2 f_2$. Assume that $(\eta + z_1 + z_2)^2 = 4/9$ and $\eta \neq 0$. Since $\tilde{\eta}^2 > 0$, we have $4/9 = \tilde{\eta}^2 + 2(b_1 + b_2)(b_1 + b_2 + 1) + (4/9) > 4/9$, a contradiction.

As for the assertion (6) in the proposition, it is obvious, since $q(\overline{x}) \equiv 0 \bmod 2\mathbf{Z}$ for $\overline{x} = \pm 3\overline{z}_1 \neq 0$.

For $U_{1,0}(2, 3, 3, 3)$, the associated numbers of P are 2 and 8, and special elements in P'^*/P' are of type G.

We have established Proposition 3.9. This proposition is the basis of the following arguments.

In the remainder of this section, we try to make the meaning of Proposition 3.9 (6) clearer. Consider the last case $U_{1,0}$ in particular. Let $u_1 = 3e_0 + 2(e_2 + e_3 + e_4) + e_5 + e_6 + e_7$. This u_1 is an isotropic vector in P with $u_1 \cdot e_i = 0$ for $i \neq 1$ and $u_1 \cdot e_1 = -3$. Set

$$
\begin{aligned}
y_1 &= -e_1 + e_4 + e_5 - u_1/3 \\
y_2 &= -e_1 + e_2 + e_6 - u_1/3 \\
y_3 &= -e_1 + e_3 + e_7 - u_1/3.
\end{aligned}
$$

We can check $y_i \in P^*$ and $y_i^2 = 2$ $(i = 1, 2, 3)$. Since $(u_1/3) - 3z_1 = 3e_0 + e_1 + e_2 + 2e_3 + 2e_4 + e_5 - e_6 + e_7$, $\overline{y}_1 = \overline{y}_2 = \overline{y}_3 = 3\overline{z}_1 \in P'^*/P' \cong P^*/P$.

Recall some general theory on discriminant quadratic forms here.

Let L be a non-degenerate even lattice with the discriminant quadratic form q_L. An element $\overline{x} \in L^*/L$ is called *isotropic*, if $q_L(\overline{x}) \equiv 0 \bmod 2\mathbf{Z}$ and if $\overline{x} \neq 0$. If every non-zero element in a subgroup $I \subset L^*/L$ is isotropic, I is called an *isotropic subgroup*. For any subset $S \subset L^*/L$ the orthogonal complement with respect to the discriminant bilinear form b_L is denoted by $S^\perp$. $S^\perp = \{\overline{x} \in L^*/L \mid b_L(\overline{x}, \overline{y}) \equiv 0 \bmod \mathbf{Z}$ for every $\overline{y} \in S.\}$ is a subgroup. If I is an isotropic subgroup, then $I \subset I^\perp$, and the order of I is equal to the index of $I^\perp$ in L^*/L.

Lemma 4.4. (Nikulin [11]) *Let K be an even overlattice of a non-degenerate lattice L. We have natural inclusion relation $L \subset K \subset K^* \subset L^*$. Thus, the quotient $I = K/L$ is regarded as a subgroup of the discriminant group L^*/L.*
(1) I is an isotropic subgroup.

(2) $I^{\perp} = K^*/L$.

(3) Let $\sigma : I^{\perp} \to K^*/K$ *denote the canonical surjective homomorphism. Then,* $q_K \sigma = q_L \mid I^{\perp}$, *where* $q_L \mid I^{\perp}$ *denotes the restriction of* q_L *to the subgroup* $I^{\perp}$.

Conversely, for any isotropic subgroup $I \subset L^*/L$ *in the discriminant group of a non-degenerate even lattice* L, *the inverse image* K *of* I *by the natural surjective homomorphism* $L^* \to L^*/L$ *is an even overlattice of* L.

Proposition 4.5. *(1) For* $J_{3,0}$, $Z_{1,0}$, $Q_{2,0}$, $W_{1,0}$, *and* $S_{1,0}$, *the lattice* P *has no even overlattice except* P *itself.*

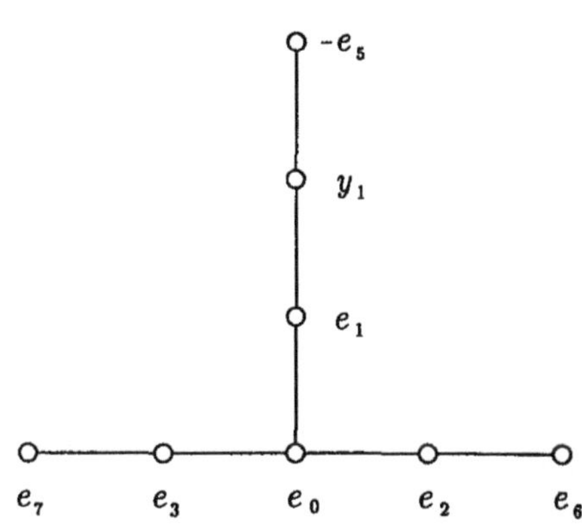

(2) For $U_{1,0}$, P *has a unique overlattice* $\hat{P}$ *except* P *itself.* $\hat{P}$ *has index 3 over* P, *and* $\hat{P} = P + \mathbf{Z}y_1 = P + \mathbf{Z}y_2 = P + \mathbf{Z}y_3$. *Choosing one of* y_1, y_2, y_3 *corresponds to choosing one of the 3 arms with length 3 in the dual graph of the basis of* P. *If we choose* y_1, *the bilinear form on* $\hat{P}$ *is described by the left dual graph. In particular, the following holds:*

⟨a⟩ $\hat{P} \cong Q(E_6) \oplus H$.

⟨b⟩ $\hat{P}^*/\hat{P} \cong \mathbf{Z}/3$.

⟨c⟩ *Any non-zero element* $\overline{x}$ *in the discriminant group of* $\hat{P}$ *satisfies* $q_{\hat{P}}(\overline{x}) \equiv 4/3 \bmod 2\mathbf{Z}$.

Proof. Only the last part of (2) is not obvious. Set $u = 3e_0 + 2(e_1 + e_2 + e_3) + e_6 + e_7 + y_1$ and $v = e_5 - u$. We can check $u^2 = v^2 = 0$ and $u \cdot v = 1$. Thus, $H = \mathbf{Z}u + \mathbf{Z}v \subset \hat{P}$ is a hyperbolic plane. The orthogonal complement of H is spanned by e_0, e_1, e_2, e_3, e_6 and e_7 and is isomorphic to the root lattice of type E_6. One has ⟨a⟩. Assertion ⟨b⟩ and ⟨c⟩ follows from ⟨a⟩. Q.E.D.

5
Root modules

We develop general theory of root systems in this section, and translate Looijenga's conditions (L1) and (L2) into a simpler one by the theory introduced here and by the results in Section 3, 4.

We always work fixing arbitrary one of 6 types of quadrilateral singularities. By P we denote the corresponding lattice to the fixed type. The sublattices P', P_0 and P_0' and the isotropic vectors $u_0, v_0 \in P$ are also defined.

We say that an embedding $P \hookrightarrow \Lambda$ into another even lattice Λ is *good*, if it satisfies the following condition (G) (Looijenga [9]):

(G) Let $\widetilde{P} = P(P, \Lambda)$ be the primitive hull of P in Λ. If $\eta \in \widetilde{P}$, $\eta^2 = 2$, $\eta \cdot u_0 = 0$ and $\eta \notin P_0$, then η is orthogonal to P_0.

On the other hand, if the image of P in Λ is primitive in Λ, then the embedding is called *primitive*. By Proposition 4.5 every embedding of P into an even lattice is primitive except for $U_{1,0}$.

Proposition 5.1. *An embedding of our lattice P into an even lattice is good if, and only if, it is primitive.*

Proof. For the former 5 cases, except $U_{1,0}$, it suffices to show that a primitive embedding is good. Now, by Proposition 3.9 (1), P has the property (G′). This (G′) implies the desired claim.

Next, we consider the case $U_{1,0}$. Similarly by Proposition 3.9 (1), a primitive embedding is good. Any non-primitive embedding into an even lattice Λ can be factored in the form $P \hookrightarrow \hat{P} \hookrightarrow \Lambda$ (See Proposition 4.5.). The vector $y_1 \in \hat{P} = \widetilde{P}$ satisfies $y_1^2 = 2$, $y_1 \cdot u_0 = 0$ and $y_1 \notin P_0$, but y_1 is not orthogonal to P_0, since $y_1 \cdot e_1 \neq 0$. Thus, the embedding is not good. Q.E.D.

Recall that Λ_N denotes the even unimodular lattice with signature $(16 + N, N)$.

Proposition 5.2. *(1) If $N \geq 1$, then the lattice P has a primitive embedding into Λ_N.*
(2) If $N \geq 2$, for any two primitive embeddings ι, $\iota' : P \hookrightarrow \Lambda_N$, we have an integral orthogonal transformation $\phi : \Lambda_N \to \Lambda_N$ with $\iota' = \phi\iota$.
(3) For $U_{1,0}$, the same statements as (1) and (2) hold also for $\hat{P}$.

Proof. (1) By $l(M)$ we denote the minimum number of generators of a finite abelian group M. Obviously, $l(P^*/P) \leq \text{rank } P = q$. For our 6 cases $q \leq 8$. The signature of P is $(q - 1, 1)$. Comparing it with the signature of Λ_N, one has:

difference of positive signatures $(16 + N) - (q - 1) \geq 9 + N > 0$
difference of negative signatures $N - 1 \geq 0$.

Comparing the rank of Λ_N and that of P, one has:

$$\begin{aligned} \text{rank } \Lambda_N - \text{rank } P &= 16 + 2N - q \geq 2 + q \\ &\geq 2 + l(P^*/P) > l(P^*/P). \end{aligned}$$

Applying Nikulin [11] Theorem 1. 12. 2, one knows the existence of an embedding (See Lemma 12.4 (1) in Section 12.).
(2) If $N \geq 2$, one has a stronger inequality $N - 1 > 0$ about the negative signature. Thus, by Nikulin [11] Theorem 1. 14. 4 one has the uniqueness (See Lemma 12.4 (2) in Section 12.).
(3) The reasoning is similar. Q.E.D.

Remark. When $N = 1$, we cannot claim the uniqueness.

In Urabe [15] we have introduced the concept of root modules as quasi-lattices satisfying certain conditions. But the conditions in [15] are too strong for our 6 cases under consideration. Therefore, we would like to define the concept of root modules again in this book as a more general concept. In Urabe [15] we have had irreducible root systems of type A, B, D or E only. According to the definition here, we have root systems of all types A, B, C, D, E, F and G, and, moreover, we have non-reduced root systems of type BC (Bourbaki [3]).

Besides these generalized root modules, we will introduce the concept of obstruction components later.

Now, let L be a quasi-lattice, FL be a submodule of L such that the index $\#(L/FL)$ is finite. We define the set $R = R(L, FL)$ for this pair (L, FL) as follows:

$$R = \{\alpha \in FL \mid \alpha^2 = 2\} \cup \{\beta \in L \mid \beta^2 = 1 \text{ or } 2/3\} \cup \{\gamma \in L \mid \gamma^2 = 1/2, 2\gamma \in FL\}.$$

We call the set R the *root system* of (L, FL), and call a vector in R a *root*. For any root $\alpha \in R$, $\sqrt{\alpha^2}$ is the *length* of α. A root with length $\sqrt{2}$ is called a *long root*, while a root with length shorter than $\sqrt{2}$ is called a *short root*. Setting $\alpha^\vee = 2\alpha/\alpha^2$ for a root $\alpha \in R$, we call $\alpha^\vee$ the *co-root* of α. We have $\alpha^\vee \in \mathbf{Z}\alpha$. Consider the following condition (R1):

(R1) $2(x, \alpha)/\alpha^2 = (x, \alpha^\vee)$ is an integer for every $x \in L$ and $\alpha \in R$.

Under (R1), for every $\alpha \in R$ we can define an isomorphism $s_\alpha : L \to L$ preserving the bilinear forms, by setting for $x \in L$

$$s_\alpha(x) = x - 2(x, \alpha)\alpha/\alpha^2 = x - (x, \alpha^\vee)\alpha = x - (x, \alpha)\alpha^\vee.$$

We call s_α the *reflection* with respect to α. Indeed, on $L \otimes \mathbf{R}$ s_α defines the reflection whose mirror is the hyperplane orthogonal to α. In particular, s_α^2 is the identity and $s_\alpha = s_{-\alpha}$.

(R2) $s_\alpha(FL) = FL$ for every $\alpha \in R$.

Assume that the pair (L, FL) satisfies the conditions (R1), and (R2). Then, we call this pair (L, FL) a *root module*. When $L = FL$, particularly we say that this root module is *regular*, and we do not mention FL and abbreviate it. (Sometimes we abbreviate FL for simplicity even if $L \neq FL$.) Any lattice is a regular root module.

The root system of the root module satisfies the most important axioms $\mathrm{SR_{II}}$ and $\mathrm{SR_{III}}$ of the 4 axioms for root systems in Bourbaki [3] (Chap. VI, n° 1. 1, Def. 1 and Résumé in the last part); thus, it suits the name.

Let (L, FL) be a root module. For a root $\beta \in R$ with length $1/\sqrt{2}$, $\alpha = 2\beta$ is a long root, $\beta^\vee = 2\alpha^\vee$, and $s_\alpha = s_\beta$ for the reflections. Setting

$$\hat{R} = \hat{R}(L, FL) = \{\alpha \in R \mid \alpha^2 \neq 1/2\},$$

we call $\hat{R}$ the *reduced root system* of the root module (L, FL). This satisfies the axioms $(\mathrm{SR_{II}})$, $(\mathrm{SR_{III}})$ and the reduced axiom

$$(\mathrm{SR_{IV}}) \qquad \text{If } \alpha \in \hat{R}, \text{ then } 2\alpha \notin \hat{R}.$$

of the axioms for root systems.

The subgroup generated by all reflections s_α with $\alpha \in R$ in the group of all integral orthogonal transformations on L is called the *Weyl group* of (L, FL) and is denoted by $W(L, FL)$ or $W(R)$. It is obviously equal to the subgroup $W(\hat{R})$ generated by reflections corresponding to the reduced root system $\hat{R}$. $W(L, FL) = W(R) = W(\hat{R})$.

The submodule of L generated by R (respectively $\hat{R}$) is denoted by $Q(R)$ (resp. $Q(\hat{R})$) and is called the *root quasi-lattice* of R (resp. $\hat{R}$). $Q(R) \supset Q(\hat{R})$. These are not necessarily lattices, When it is a lattice, we call it a *root lattice*. Sometimes we write $Q(L, FL) = Q(R(L, FL))$ for simplicity, which is the submodule of L generated by roots, and we call it the root quasi-lattice of the root module (L, FL).

On the other hand, the submodule generated by all co-roots $\alpha^\vee$ ($\alpha \in R$) is denoted by $Q(R^\vee)$ and is called the *co-root lattice*. Indeed $Q(R^\vee)$ is always an *even* lattice. For $\alpha, \beta \in R$,

$$(\alpha^\vee, \beta^\vee) \in \mathbf{Z}$$
$$(\alpha^\vee, \alpha^\vee) = 4/\alpha^2 \in 2\mathbf{Z}.$$

(We can consider also the submodule $Q(\hat{R}^\vee)$ generated by all co-roots corresponding to roots in the reduced root system $\hat{R}$. But $Q(\hat{R}^\vee) = Q(R^\vee)$.)

The reflection s_α ($\alpha \in R$) induces an isomorphism $s_\alpha : Q(R^\vee) \to Q(R^\vee)$ of co-root lattices. Indeed, for $\beta \in R$,

$$s_\alpha(\beta^\vee) = \beta^\vee - (\alpha, \beta^\vee)\alpha^\vee \in Q(R^\vee).$$

The Weyl group $W(R)$ acts on $Q(R^\vee)$.

Next, let M be a submodule of L. Then, setting $FM = FL \cap M$, (M, FM) is a root module. A submodule is always regarded as a root module in this manner. In other words, we define that a homomorphism $\phi : (M, FM) \to (L, FL)$ between root modules is a homomorphism $\phi : M \to L$ of modules preserving the bilinear forms satisfying $\phi^{-1}(FL) = FM$.

Lemma 5.3. *Let $\hat{R} = \hat{R}(L, FL)$ be the reduced root system of a root module (L, FL). For every $\alpha \in \hat{R}$,*

$$\mathbf{R}\alpha^\vee \cap Q(R^\vee) = \mathbf{Z}\alpha^\vee.$$

Proof. Easy. (See Urabe [15] Lemma 2.2.)

Remark. The above equality does not hold for a short root α with length $1/\sqrt{2}$.

Note that if L is positive definite, then the root system $R(L, FL)$ is a finite set.

In the following we freely use standard concepts and terminology in the theory of root systems (Bourbaki [3]). Any finite root system is uniquely decomposed into a direct sum of irreducible ones.

Proposition 5.4. *(1) Finite irreducible root systems containing a long root are classified into the following types. The lower index represents the rank of the root system:*

$$A_k \ (k \geq 1), \quad B_k \ (k \geq 2), \quad C_k \ (k \geq 4), \quad D_k \ (k \geq 4),$$
$$E_6, \quad E_7, \quad E_8, \quad F_3, \quad F_4, \quad G_2, \quad BC_k \ (k \geq 1).$$

(In Bourbaki [3] F_3 is not used as the name of the type. It is called C_3 instead of F_3. "$F_3 = C_3$". Yet, we use F_3 in this book, because the Dynkin graph of type F_3 is a subgraph of the Dynkin graph of type F_4 which appears in our main theorem and because of convenience to state the following (3).)

Let R denote a root system of type BC_k. Every long root α in R is divisible, i.e., $\alpha/2 \in R$. The reduced root system $\hat{R}$ consisting all long roots and all short roots with length 1 in R has the following type:

$$A_1 \quad (k=1), \quad B_2 \quad (k=2), \quad F_3 \quad (k=3), \quad C_k \quad (k \geq 4).$$

(2) Any finite root system of a root module has at most one component containing a short root with length $1/\sqrt{2}$.
(3) Consider a finite root system of a regular root module. Any irreducible component of it is never of type C. If an irreducible component of it is of type BC_k, then $1 \leq k \leq 3$. Besides, it has at most one component containing a short root with length 1.

Proof. (1) The main parts follow from Bourbaki [3]. Note that by the axiom (R1) $\hat{R}$ cannot be of type B_k with $k \geq 3$.
(2) If $\gamma_1^2 = \gamma_2^2 = 1/2$ and $\gamma_1 \cdot \gamma_2 = 0$, then $(\gamma_1 + \gamma_2)^2 = 1$. It follows from this fact.
(3) Assume $k \geq 4$. Consider a free module $F = \sum_{i=1}^{k} \mathbf{Z}\epsilon_i$ of rank k. Set $L = \sum_{i=1}^{k} \mathbf{Z}(\epsilon_i/2)$. L is an overmodule of F with index 2^k. We define a bilinear form by $\epsilon_i^2 = 2$ $(1 \leq i \leq k)$, $\epsilon_i \cdot \epsilon_j = 0$ $(i \neq j)$. Then, (L, F) is a root module whose root system is of type BC_k. Set $L' = \{\sum a_i(\epsilon_i/2) \in L \mid \sum a_i \text{ is an even integer.}\}$. The pair (L', F) is also a root module, whose root system is of type C_k. Set $\beta_1 = (\epsilon_1 + \epsilon_2)/2$ and $\beta_2 = (\epsilon_3 + \epsilon_4)/2$. Then, we have $\beta_1^2 = \beta_2^2 = 1$, $\beta_1 \cdot \beta_2 = 0$ and $\beta_1, \beta_2 \in L'$. On the other hand, $(\beta_1 + \beta_2)^2 = 2$ and $\beta_1 + \beta_2 \notin F$. By this fact one sees the former half of the assertion. By the similar argument to (2) one has the latter half. Q.E.D.

We introduce here three agreements in order to make the descriptions in this book clearer. Consider the symbols for root systems in Proposition 5.4 (1). In general, we can use these symbols (in particular, those of type A, D or E) for root systems containing no long root. But in this book we would like to follow the agreements below:

(Agreement 1) When we use the symbols for irreducible root systems in Proposition 5.4 (1), they also imply that the root system contains a long root.

(Agreement 2) Consider the case where an irreducible root system R contains no long root. Then, we have an irreducible root system R' with a long root and a positive real number t such that $R' = \{t\alpha \mid \alpha \in R\}$. If R' is type X, then we denote that R is of type $X(t^2)$.

(Agreement 3) (Exceptions) The type of a reduced root system $\{\alpha, -\alpha\}$ of rank 1 is B_1 if $\alpha^2 = 1$, and G_1 if $\alpha^2 = 2/3$. The type of a reduced irreducible root system of rank 2 consisting of only short roots with length 1 is F_2.

Therefore, $B_1 = A_1(2)$, $G_1 = A_1(3)$, and $F_2 = A_2(2)$.

Let R be a finite root system. We can choose a root basis $\Delta = \{\alpha_1, \alpha_2, \ldots, \alpha_k\} \subset R$ ("une base de racines" in Bourbaki [3]. Sometimes it is also called a fundamental system of roots.) when we fix a Weyl chamber. Each *fundamental root* α_i is indivisible, i.e., $\alpha_i/2 \notin R$. Δ is a basis of $Q(R)$.

We would like to explain the concept of Dynkin graphs here. Recall that for a finite subset S of a Euclid space consisting of vectors with length $\sqrt{2}$, 1, $\sqrt{2/3}$ or $1/\sqrt{2}$, we can draw a graph Γ depending on S by the rules (1)–(5) explained at the beginning part of the Introduction. When $S = \Delta$, the corresponding graph G is the Dynkin graph of the root system R. Since vectors in Δ are linearly independent, the fifth rule (5) is never applied, and any edge in G is never bold. It depends only on the isomorphism class of R plus the Euclidean bilinear form on the ambient space of R. It does not depend on the choice of Δ, since for another root basis Δ' we have an element $w \in W(R)$ of the Weyl group such that $\Delta' = w(\Delta)$. The isomorphism class of a finite root system plus the Euclidean bilinear form is uniquely determined by the Dynkin graph.

Our Dynkin graphs are slightly different from standard Dynkin graphs in Bourbaki [3] used commonly. We explain the difference:

Standard Dynkin graphs have only one kind of vertices, but three different kinds of edges — a single segment, a double segment, and a triple segment —. Besides, a double segment and a triple segment carry an arrow on them.

Also in standard graphs vertices have one-to-one correspondence with members in Δ. The difference of the edge depends on the angle $\arccos(\alpha, \beta)/\sqrt{\alpha^2}\sqrt{\beta^2}$ between the corresponding fundamental roots $\alpha, \beta \in \Delta$. (single — $2\pi/3$, double — $3\pi/4$, triple — $5\pi/6$.) For fundamental roots the angle determines the ratio of the length uniquely. (single — $\alpha^2 = \beta^2$, double — $\alpha^2/\beta^2 = 2$ or $1/2$, triple — $\alpha^2/\beta^2 = 3$ or $1/3$.) The arrow shows that the root at the bottom is longer than the root at the top.

Thus, in the standard graphs the absolute length of each root is ignored. We can know only the mutual ratio of the length of fundamental roots belonging to the same connected component of the graph. Besides, customarily we do not associate a Dynkin graph with any non-reduced root system.

Our Dynkin graph of type BC_k is the following:

BC_1 : ⊗

BC_k $(k \geq 2)$: ⊗——•——•······•——• (k vertices)

Our Dynkin graphs carry more information on length of roots. Corresponding to a root system of rank 1, we have four different graphs o, •, ◎, ⊗ depending on the length of the root. Yet the corresponding standard graphs are the same o.

As explained in the beginning part of the Introduction, we can recover the corresponding standard graph from our graph by the procedure explained there.

When a finite root system R has a_k of components of type A_k, b_k of components of type B_k, ..., we identify the formal sum $G = \sum a_k A_k + \sum b_k B_k + \cdots$ with the Dynkin graph of R, and we say that R is of type G. We use also abbreviations like $R = R(G)$, $Q(R) = Q(G)$, $W(R) = W(G)$, etc.

Next, we explain the concept of extended Dynkin graphs. ("graphs de Dynkin complété" in Bourbaki [3].)

Let R be an *irreducible* finite root system and $\Delta \subset R$ be a root basis. The maximal root $\eta \in R$ is uniquely defined depending on Δ. We can write it in the form $\eta = \sum_{\alpha \in \Delta} n_\alpha \alpha$. The coefficient n_α is necessarily a positive integer, and is called the *coefficient of the maximal root* corresponding to α. The union $\Delta^+ = \Delta \cup \{-\eta\}$ is called the *extended root basis*. We define the coefficient $n_{-\eta}$ of the maximal root corresponding to $-\eta$ to be $n_{-\eta} = 1$. We have $\sum_{\alpha \in \Delta^+} n_\alpha \alpha = 0$.

The graph obtained by applying the rules (1)–(5) in the beginning of the Introduction to $S = \Delta^+$ is the extended Dynkin graph of R. The fifth rule (5) is applied only for the case where R has rank 1. Therefore, if the rank is greater than 1, no bold edge appears in the extended Dynkin graph. The extended Dynkin graphs and the coefficients of the maximal root in the case of rank 1 are like the following:

A_1 : 1 — 1		B_1 : 1 — 1
G_1 : 1 — 1		BC_1 : 2 — 1

The edges in the above four graphs are bold edges. (Sometimes we use a single segment $\overset{\infty}{\text{———}}$ accompanied with the mark ∞ instead of a bold edge.)

Note, in particular, that the maximal root for type BC is a long root.

For a reducible finite root system R, the union of the extended root bases of the irreducible components is called the extended root basis of R. Applying the rules (1)–(5) in the beginning of the Introduction to the extended root basis, the extended Dynkin graph of R is obtained. Therefore, it is the disjoint union of the extended Dynkin graphs of the irreducible components of R. The number of vertices minus the number of connected components is called the *rank* of the extended Dynkin graph.

The following figures are the extended Dynkin graphs for main types. Numbers are the coefficients of the maximal root:

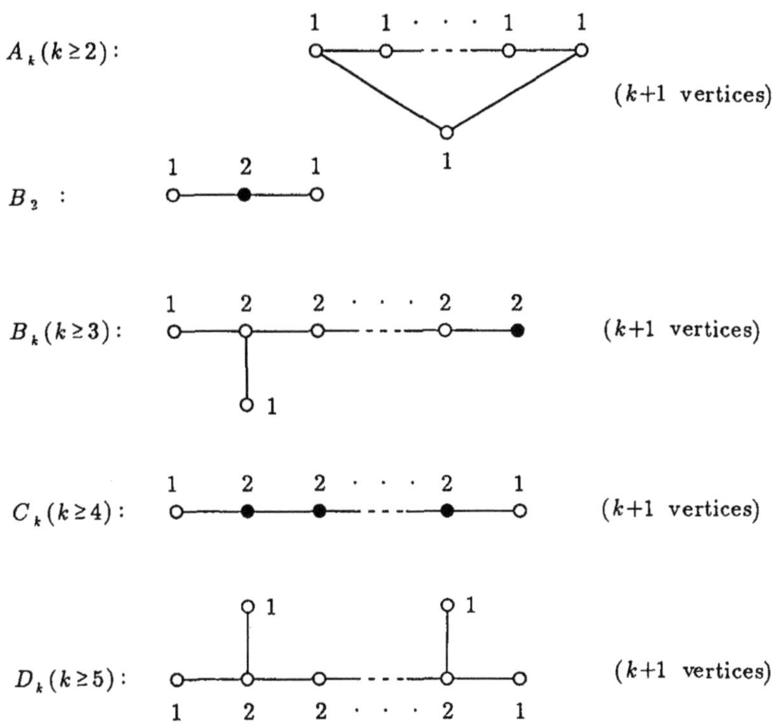

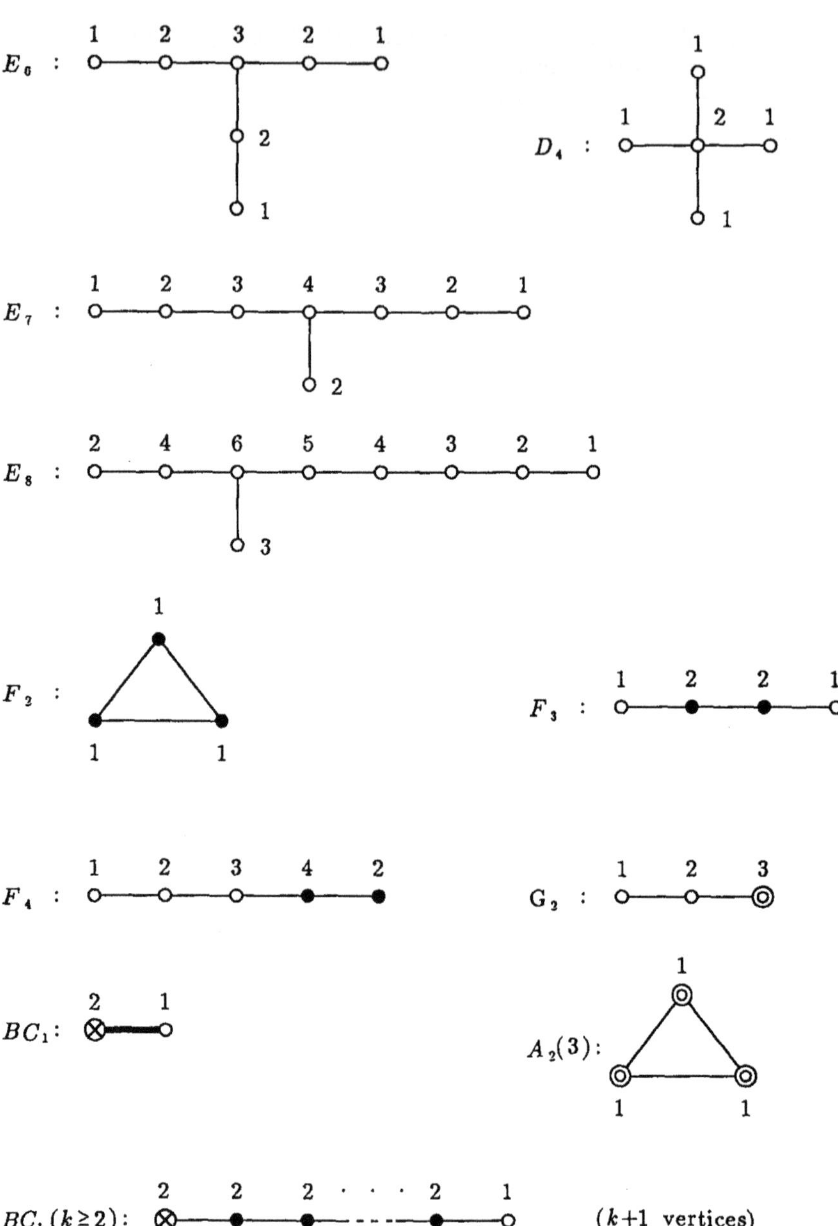

When we pick an arbitrary vertex with the attached number 1 in a connected extended Dynkin graph, the graph obtained by erasing the picked vertex and the edges issuing from it is the corresponding Dynkin graph.

Lemma 5.5. *Let Λ be an even unimodular lattice, L be a non-degenerate primitive sublattice, and $M = C(L, \Lambda)$ be the orthogonal complement of L in Λ.*
(1) Let $\Lambda \to \mathrm{Hom}(M, \mathbf{Z}) = M^*$ be the homomorphism associating a vector $x \in \Lambda$ with a homomorphism $M \to \mathbf{Z}$ defined by $y \in M \mapsto (x, y)$. This induces an isomorphism $\Lambda/L \cong M^*$ of quasi-lattices.
(2) The composition $M \hookrightarrow \Lambda \to \Lambda/L$ of the natural homomorphisms is injective and it induces an embedding $M \hookrightarrow \Lambda/L$ of quasi-lattices such that the composition $M \hookrightarrow \Lambda/L \cong M^*$ coincides with the natural inclusion $M \hookrightarrow M^*$.
(3) We can define an isomorphism $r : L^*/L \to M^*/M$ whose graph $\{(\overline{x}, r(\overline{x})) \mid \overline{x} \in L^*/L\}$ in $L^*/L \oplus M^*/M \cong (L \oplus M)^*/(L \oplus M)$ coincides with the subgroup $\Lambda/(L \oplus M) \subset (L \oplus M)^*/(L \oplus M)$. This r satisfies $q_M r = -q_L$ for discriminant quadratic forms.
(4) Conversely, if there is a group-isomorphism $r : L^*/L \to M^*/M$ between the discriminant groups of non-degenerate even lattices L and M satisfying $q_M r = -q_L$, then we have an embedding $L \oplus M \hookrightarrow \Lambda$ of lattices into an even unimodular lattice Λ such that L and M are the orthogonal complement of each other in Λ and such that the graph of r coincides with $\Lambda/(L \oplus M)$.

Proof. Easy.

Proposition 5.6. *Let P be the lattice corresponding to the fixed type of our quadrilateral singularities. Let Λ be an even unimodular lattice. Assume that we have a primitive embedding $P \hookrightarrow \Lambda$. We identify the orthogonal complement F of P in Λ with the image of it under the natural surjective homomorphism $\Lambda \to \Lambda/P$.*
(1) $F = \{x \in \Lambda/P \mid (x, y) \in \mathbf{Z}$ for every $y \in \Lambda/P\} \cong \mathrm{Hom}(\Lambda/P, \mathbf{Z})$.
(2) For the 5 cases $J_{3,0}$, $Z_{1,0}$, $Q_{2,0}$, $W_{1,0}$, $S_{1,0}$ except $U_{1,0}$, also the following holds: $F = \{x \in \Lambda/P \mid x^2$ is an even integer.$\}$.
(3) The pair $(\Lambda/P, F)$ is a root module.
(4) For the 5 cases except $U_{1,0}$, Λ/P is a regular root module.

Proof. We identify Λ/P with F^* by the canonical isomorphism. The canonical surjective homomorphism $\Lambda/P \to (\Lambda/P)/F \cong F^*/F$ is denoted by π. The composition $F^*/F \xrightarrow{\sim} P^*/P \xrightarrow{\sim} P'^*/P'$ of the canonical isomorphisms is denoted by $\widetilde{r}$. The discriminant quadratic forms of P' and F are denoted by $q_{P'}$ and q_F respectively.
(1) It follows from Lemma 5.5.
(2) The claim is equivalent to $q_F^{-1}(0 \bmod 2\mathbf{Z}) = \{0\}$. Since $q_F = -q_{P'}\widetilde{r}$, this is equivalent to $q_{P'}^{-1}(0 \bmod 2\mathbf{Z}) = \{0\}$. The last claim is equivalent to Proposition 3.9 (6).
(3) We check the axioms for root modules. First, let $\alpha \in F$ be a long root. By (1) $2(\alpha, x)/\alpha^2 = (\alpha, x) \in \mathbf{Z}$ for every $x \in \Lambda/P$.

Second, assume $\beta^2 = 1$ for $\beta \in \Lambda/P$. Set $\overline{\beta} = \widetilde{r}(\pi(\beta)) \in P'^*/P'$. $q_{P'}(\overline{\beta}) \equiv -q_F(\pi(\beta)) \equiv 1 \bmod 2\mathbf{Z}$. $\overline{\beta}$ is an element of the second kind. By Proposition 3.9 (4) $2\overline{\beta} = 0$. Since $\widetilde{r}$ is an isomorphism, $2\pi(\beta) = 0$; thus, $2\beta \in F$. By (1) $2(\beta, x)/\beta^2 = (2\beta, x) \in \mathbf{Z}$ for every $x \in \Lambda/P$.

Third, assume $\gamma^2 = 2/3$ for $\gamma \in \Lambda/P$. Set $\overline{\gamma} = \widetilde{r}(\pi(\gamma)) \in P'^*/P'$. $q_{P'}(\overline{\gamma}) \equiv -q_F(\pi(\gamma)) \equiv -\gamma^2 \equiv 1 + (1/3) \bmod 2\mathbf{Z}$. $\overline{\gamma}$ is an element of the third kind with the associated number 2. By Proposition 3.9 (5), $3\overline{\gamma} = 0$. Thus, $3\gamma \in F$. By (1) $2(\gamma, x)/\gamma^2 = (3\gamma, x) \in \mathbf{Z}$ for every $x \in \Lambda/P$.

Finally, assume $\delta^2 = 1/2$ and $2\delta \in F$ for $\delta \in \Lambda/P$. By (1) $2(\delta, x)/\delta^2 = 2(2\delta, x) \in \mathbf{Z}$ for every $x \in \Lambda/P$.

Now, the axiom (R2) is also satisfied, since F is invariant under all integral orthogonal transformations on Λ/P by (1).

(4) Any element $\alpha \in \Lambda/P$ with $\alpha^2 = 2$ belongs to F by (2) in the 5 cases except $U_{1,0}$. Thus, (4) follows from (3). Q.E.D.

In the following we consider the regular root module Λ/P for simplicity, when we treat the 5 cases $J_{3,0}$, $Z_{1,0}$, $Q_{2,0}$, $W_{1,0}$ and $S_{1,0}$. Only when we deal with $U_{1,0}$, for example, in Chapter 4, we consider the root module $(\Lambda/P, F)$. Sometimes F is abbreviated.

Corollary 5.7. *If the root module Λ/P contains a short root with length 1 (respectively $\sqrt{2/3}$), then P^*/P contains a special element of type B (resp. type G).*

Remark. *Λ/P can contain a short root with length $1/\sqrt{2}$ only for the cases $Z_{1,0}$ and $S_{1,0}$. It is easy to see that P^*/P contains a special element of type BC for these two cases.*

The following proposition is obvious by Proposition 4.5 (2).

Proposition 5.8. *For $U_{1,0}$ the quotient $\Lambda/\hat{P}$ is a regular root module for any embedding $\hat{P} \hookrightarrow \Lambda$ into an even unimodular lattice Λ.*

Definition 5.9. (1) Let (L, FL) be a root module, and M be a submodule of L. We consider the primitive hull $\widetilde{M} = P(M, L)$. We say that M is *full* in L, if $R(M, M \cap FL) = R(\widetilde{M}, \widetilde{M} \cap FL)$ for root systems. An embedding of root modules whose image is full is called a *full embedding*.

(2) Let k be a fixed integer. Let G be a Dynkin graph with connected components G_i $(1 \le i \le m)$. The corresponding root quasi-lattice has the orthogonal decomposition $Q(G) = \oplus_{i=1}^{m} Q(G_i)$. We consider an embedding $Q(G) \hookrightarrow L$ into a root module (L, FL). (By definition the root system of $(Q(G), Q(G) \cap FL)$ is of type G.) We say that a component G_i of G is an *obstruction component* with respect to k for this embedding, if G_i is of type A_k and if the index of the primitive hull satisfies $[P(Q(G_i), L) : Q(G_i)] \ge k+1$.

Lemma 5.10. *We consider the situation in Definition 5.9 (2). If a component G_i is of type A_k, the following three conditions are equivalent. Set $Q_i = Q(G_i)$ and $\widetilde{Q}_i = P(Q_i, L)$:*

(1) G_i is an obstruction component, i.e., $[\widetilde{Q}_i : Q_i] \ge k+1$.

(2) $[\widetilde{Q}_i : Q_i] = k+1$.

(3) $\widetilde{Q}_i = Q_i^*$.

Proof. By the axiom (R1) of root modules $(\alpha, x) \in \mathbf{Z}$ for any $x \in \widetilde{Q}_i$ and for any $\alpha \in Q_i \cap FL$ with $\alpha^2 = 2$. Since Q_i is spanned by such α's, $Q_i^* \supset \widetilde{Q}_i \supset Q_i$. Since $[Q_i^* : Q_i] = k+1$, the above (1), (2) and (3) are equivalent. Q.E.D.

Theorem 5.11. *Let P be the lattice corresponding to one of 6 types of hypersurface quadrilateral singularities. By $\Lambda = \Lambda_N$ we denote an even unimodular lattice with*

signature $(16+N, N)$. Let G be a Dynkin graph without a vertex corresponding to a short root, and $Q(G)$ be the root lattice of type G. We consider an embedding $P \oplus Q(G) \hookrightarrow \Lambda$ such that P is primitive in Λ. The induced embedding $Q(G) \hookrightarrow \Lambda/P$ is defined as the composition $Q(G) \hookrightarrow \Lambda \to \Lambda/P$ of natural homomorphisms. Then, the following (A) and (B) are equivalent:
(A) The embedding $P \oplus Q(G) \hookrightarrow \Lambda$ satisfies Looijenga's conditions (L1) and (L2) in Theorem 2.1.
(B) The embedding $Q(G) \hookrightarrow \Lambda/P$ is full and for every associated number k of P with $k \geq 4$, G has no obstruction component with respect to k.

Proof. Set $S = P \oplus Q(G)$ and $F = C(P, \Lambda)$. We denote $\widetilde{S} = P(S, \Lambda)$ $(\subset \Lambda)$ and $\widetilde{Q}(G) = P(Q(G), \Lambda/P)$ $(= \widetilde{S}/P \subset \Lambda/P)$. By $\pi : \Lambda \to \Lambda/P$ we denote the canonical surjective homomorphism.

(1) We will show that $\widetilde{Q}(G)$ contains no short root under the condition (L2).

Let $\eta \in \widetilde{Q}(G)$ be a short root. Choose a vector $\alpha \in \widetilde{S}$ with $\pi(\alpha) = \eta$. We can write it in the form $\alpha = x + y$ with $x \in P^*$, $y \in F^*$, since $P \oplus F \subset \Lambda \subset P^* \oplus F^*$. By the definition of the bilinear form on Λ/P, $y^2 = \eta^2 = 1, 2/3$ or $1/2$.

On the other hand, the vector $\alpha_1 = \alpha - (\alpha \cdot v_0)u_0 - (\alpha \cdot u_0)v_0 \in \widetilde{S}$ also satisfies $\pi(\alpha_1) = \eta$. Thus, we can assume further that α satisfies $\alpha \cdot u_0 = \alpha \cdot v_0 = 0$. Under this assumption one has $x \in P'^*$. Since $\alpha^2 \equiv 0 \bmod 2\mathbf{Z}$, $x^2 = \alpha^2 - y^2 \equiv -y^2 \equiv -\eta^2 \equiv 1$, $4/3$ or $3/2 \bmod 2\mathbf{Z}$. The element $x \bmod P' \in P'^*/P'$ is of the second or third kind. By Proposition 3.9 (3) we have a vector $z \in P'$ such that $(x-z)^2 = 2 - \eta^2$ and $x - z$ is not orthogonal to P'_0. By exchanging α for $\alpha - z$, we can assume that $x^2 = 2 - \eta^2$ and x is not orthogonal to P'_0. So one has a vector $\alpha \in \Lambda$ such that $\pi(\alpha) = \eta$, $\alpha^2 = 2$, $\alpha \cdot u_0 = \alpha \cdot v_0 = 0$, $\alpha \in \widetilde{S}$ and α is not orthogonal to P'_0. Then, by the condition (L2), $\alpha \in P_0$. This implies $y = 0$, which contradicts the fact $y^2 = \eta^2 \neq 0$. Therefore, we have no short root η.

(2) Assume that a component G_0 of type A_k in G is an obstruction component with respect to an associated number $k \geq 4$ of P. We will deduce a contradiction from the condition (L2).

Let $\Delta_0 \subset Q(G_0)$ be a root basis of $Q(G_0)$. Let ω be the fundamental weight corresponding to the vertex at one of the two ends of the Dynkin graph of Δ_0. One has $\omega^2 = k/(k+1)$. On the other hand, by Lemma 5.10 and by assumption one has $Q(G_0)^* = P(Q(G_0), \Lambda/P)$; thus, $\omega \in P(Q(G_0), \Lambda/P)$.

It follows that we have a vector α in $\widetilde{S}$ such that $\pi(\alpha) = \omega$ and $\alpha \cdot u_0 = \alpha \cdot v_0 = 0$. We can write it in the form $\alpha = x + \omega$ $(x \in P'^*)$. Since $\alpha^2 \equiv 0 \bmod 2\mathbf{Z}$,

$$x^2 \equiv -\omega^2 \equiv 1 + (1/(k+1)) \bmod 2\mathbf{Z}.$$

The element $x \bmod P' \in P'^*/P'$ is of the third kind with the associated number k. By Proposition 3.9 (3) we have a vector $z \in P'$ such that $(x-z)^2 = 1 + (1/(k+1)) = 2 - \omega^2$ and $x - z$ is not orthogonal to P'_0. Exchanging α for $\alpha - z$, one has a vector $\alpha \in \widetilde{S}$ such that $\pi(\alpha) = \omega$, $\alpha^2 = 2$, $\alpha \cdot u_0 = \alpha \cdot v_0 = 0$ and α is not orthogonal to P'_0. By the condition (L2), $\alpha \in P'_0$, and $0 = \pi(\alpha) = \omega \neq 0$, which is a contradiction.

(3) We will show that the condition (L2) is satisfied, if $\widetilde{Q}(G)$ does not contain a short root, and if G has no obstruction component with respect to any associated number $k \geq 4$ of P.

Let α be a vector in $\widetilde{S}$ such that $\alpha \cdot u_0 = 0$, $\alpha^2 = 2$ and α is not orthogonal to P_0. Replacing α by $\alpha - (\alpha \cdot v_0)u_0$ one can assume further that it satisfies $\alpha \cdot v_0 = 0$. We can write it in the form $\alpha = x + y$ with $x \in P'^*$, $y \in Q(G)^*$. Here, x is not orthogonal to P'_0. $2 = \alpha^2 = x^2 + y^2$, $y^2 = \pi(\alpha)^2$. Since both of P' and $Q(G)$ are positive definite, we have $0 \leq y^2 \leq 2$.

We divide the case into four subcases.

⟨i⟩ $1 < y^2 \leq 2$.

Here, $0 \leq x^2 < 1$ and $x \bmod P' \in P'^*/P'$ is of the first kind. By Proposition 3.9 (2) x is orthogonal to P'_0, which contradicts the choice of α.

⟨ii⟩ $y^2 = 1$.

The vector $\pi(\alpha)$ belongs to $\widetilde{Q}(G)$. On the other hand, since $\pi(\alpha)^2 = y^2 = 1$, $\pi(\alpha)$ is a short root, which contradicts the assumption.

⟨iii⟩ $0 < y^2 < 1$.

Since G has no vertex corresponding to a short root, $Q(G) \subset \pi(F)$ and we have $Q(G)^* \supset \widetilde{Q}(G) \ni \pi(\alpha)$. By Lemma 3.7 $k = y^2/(1-y^2)$ is a positive integer and G has a component G_0 of type A_k such that $P(Q(G_0), \Lambda/P) = \mathbf{Z}\pi(\alpha) + Q(G_0) = Q(G_0)^*$. By Lemma 5.10 G_0 is an obstruction component with respect to k. On the other hand, $x^2 = 2 - y^2 = 2 - (k/(1+k)) = 1 + (1/(1+k))$. The element $x \bmod P' \in P'^*/P'$ is of the third kind with the associated number k, and k is an associated number of P. By assumption $k \leq 3$.

If $k = 3$, then the second fundamental weight $\omega_2 \in Q(G_0)^* = P(Q(G_0), \Lambda/P)$ satisfies $\omega_2^2 = 1$ and it is a short root in $\widetilde{Q}(G)$. It contradicts the assumption.

If $k = 2$, then $\pi(\alpha)^2 = y^2 = 2/3$ and $\pi(\alpha)$ is a short root with length $\sqrt{2/3}$ lying in $\widetilde{Q}(G)$, which contradicts the assumption.

If $k = 1$, then $\pi(\alpha)^2 = 1/2$ and $\pi(\alpha) \in \widetilde{Q}(G) \subset Q(G)^*$. By Lemma 3.7 G has a component G_0 of type A_1 and $\pi(\alpha) \in Q(G_0)^*$. Thus, $2\pi(\alpha) \in Q(G_0)$ is a long root and $\widetilde{Q}(G)$ contains a short root $\pi(\alpha)$, which contradicts the assumption as well.

⟨iv⟩ $y^2 = 0$.

We have $y = 0$ and $\alpha \in P'^*$. By the property (G′) in Proposition 3.9 (1), one has $\alpha \in P'_0$. Thus, the conclusion of the condition (L2) holds.

(4) Under the condition (L1), if $\eta \in \widetilde{Q}(G)$ for a long root $\eta \in \Lambda/P$, then $\eta \in Q(G)$.

For $U_{1,0}$ by the definition of a long root we have $\eta \in \pi(F)$ $(\subset \Lambda/P)$. For the other five cases we have $\eta \in \pi(F)$ by Proposition 5.6 (2). Thus, there is a vector $\widetilde{\eta} \in F$ $(\subset \Lambda)$ with $\pi(\widetilde{\eta}) = \eta$. This $\widetilde{\eta}$ is orthogonal to P and is contained in the primitive hull of $Q(G)$ in Λ. By the condition (L1) one has $\widetilde{\eta} \in Q(G)$; thus, $\eta = \pi(\widetilde{\eta}) \in \pi(Q(G)) = Q(G)$.

(5) If the condition $\eta \in \widetilde{Q}(G)$ for a long root $\eta \in \Lambda/P$ implies $\eta \in Q(G)$, then the condition (L1) holds.

Let α be a vector in $\widetilde{S}$ with $\alpha^2 = 2$ such that α is orthogonal to P. Then, $\pi(\alpha) \in \pi(F)$, $\pi(\alpha)^2 = 2$ and $\pi(\alpha) \in \widetilde{Q}(G)$. Thus, $\pi(\alpha)$ is a long root in Λ/P. By the assumption one has $\pi(\alpha) \in Q(G)$. On the other hand, since $\alpha \in F$ and since $\pi|F$ is injective, one has $\alpha \in Q(G)$. Thus, (L1) holds.

Theorem 5.11 follows from above (1)–(5). Q.E.D.

6
Elementary transformations and tie transformations

Throughout this book for $N \geq 1$ by Λ_N we denote the even unimodular lattice with signature $(16+N, N)$. It is unique up to isomorphisms, and isomorphic to $Q(2E_8) \oplus H^{\oplus N}$ and also to $\Gamma_{16} \oplus H^{\oplus N}$. Here, Γ_{16} is the even overlattice with index 2 over the root lattice $Q(D_{16})$ of type D_{16}.

For a fixed primitive embedding $P \hookrightarrow \Lambda_N$ by F we denote the orthogonal complement of P in Λ_N. F is identified with the image of it under the canonical surjective homomorphism $\Lambda_N \to \Lambda_N/P$.

We have defined the concept of elementary transformations for finite root systems and Dynkin graphs in Urabe [15]. The concept of root modules in [15] is more restricted than that in this book. Moreover, even the concept of Dynkin graphs is slightly different from that in this book.

In spite of such difference, the same definition of elementary transformations as before is effective even in our present situation.

An example of a non-reduced root system is given here. As readers might not be familiar with this concept, an explanation will be provided. Consider BC_3 in particular. We use the notations shown in the Remark, that follows Theorem 0.3 in the Introduction.

Let $L = \sum_{i=1}^{3} \mathbf{Z}v_i$ be a quasi-lattice with the bilinear form $(\ ,\ )'$ defined by $(v_i, v_i)' = 1/2$, and $(v_i, v_j)' = 0$ for $1 \leq i, j \leq 3$, $i \neq j$. Set $R'_{1/2} = \{\pm v_i \mid i = 1, 2, 3\}$, $R'_1 = \{\pm v_i \pm v_j \mid i, j = 1, 2, 3, i \neq j\}$, and $R'_2 = \{\pm 2v_i \mid i = 1, 2, 3\}$. Vectors in R'_m have length $\sqrt{m}$. The union $R' = R'_{1/2} \cup R'_1 \cup R'_2$ is the root system of type BC_3. Note that $R'_{1/2} \cup R'_1$ is a root system of type $B_3(2)$, while $R'_1 \cup R'_2$ is a root system of type so-called C_3 $(= F_3)$. The system of type BC_3 is the union of them. $\Delta = \{v_1, -v_1+v_2, -v_2+v_3\}$ is a root basis for R', and $2v_1 + 2(-v_1+v_2) + 2(-v_2+v_3) = 2v_3$ is the maximal root associated with Δ. This is a long root. Corresponding to the extended root basis $\Delta^+ = \{v_1, -v_1+v_2, -v_2+v_3, -2v_3\}$, we have the following extended Dynkin graph and the coefficients of the maximal root:

```
2       2       2       1
⊗-------•-------•-------o
```

Let us consider the root system R'' generated by a proper subset $\Delta' \subset \Delta^+$. The component of R'' containing v_1 is non-reduced and of type BC. However, note that any component of R'' not containing v_1 is reduced and is of type A_1, B_1, B_2, F_2 or F_3. In particular, if $v_1 \notin \Delta'$, then R'' is a reduced root system.

Proposition 6.1. *Let (L, FL) be a positive definite root module and M be a submodule of L.*
(1) If M is primitive in L, then any root basis of the root system of $(M, M \cap FL)$ can be extended to a root basis of the root system of (L, FL).

(2) If the torsion group of the quotient L/M is cyclic, then the root system of $(M, M \cap FL)$ is obtained from that of (L, FL) by one elementary transformation.

Proof. In the proof we abbreviate FL and $M \cap FL$ for simplicity.
(1) The Weyl group $W(M)$ of M acts transitively on the set of all root bases of M, and the action of $W(M)$ can be extended naturally to L. Thus, it suffices to show that there is a root basis Δ_L for L such that Δ_L contains a root basis Δ_M for M.

By assumption we have a linear homomorphism $\xi : L \to \mathbf{R}$ from L to the real field $\mathbf{R}$ such that the kernel $\xi^{-1}(0)$ coincides with M. Regarding ξ as an element in $L^* \otimes \mathbf{R}$, we consider the action of the Weyl groups $W(M)$ and $W(L)$. Let $C \subset L^* \otimes \mathbf{R}$ be a Weyl chamber for $W(L)$ such that the closure $\overline{C}$ contains ξ. Let Δ_L be the root basis for L corresponding to C. Set $\Delta' = \{\alpha \in \Delta_L \mid \xi(\alpha) = 0\} = \{\alpha \in \Delta_L \mid$ The hyperplane orthogonal to α passes through $\xi\}$. Let W' be the subgroup of $W(L)$ generated by reflections corresponding to roots in Δ'. If $\alpha \in \Delta'$, then $\xi(\alpha) = 0$; thus, $\alpha \in M$. One knows $W' \subset W(M)$. Let $I(\xi) = \{g \in W(L) \mid g(\xi) = \xi\}$ be the isotropic subgroup of $W(L)$ with respect to ξ. If $\alpha \in R(M)$, then $s_\alpha(\xi) = \xi - 2\xi(\alpha)\alpha/\alpha^2 = \xi$. One has $W(M) \subset I(\xi)$. Now, by Bourbaki [3] Chap. 5 §3 n° 3 Prop. 2, $W' = W(M) = I(\xi)$. This implies that the Weyl chamber for W' is the Weyl chamber for $W(M)$, and Δ' is a root basis for M.
(2) First, note that Lemma 5.3 does not hold for a short root with length $1/\sqrt{2}$.

By assumption we have a linear homomorphism $\xi : L \to \mathbf{R}$ with $\xi^{-1}(\mathbf{Z}) = M$.

Let Δ_M be a root basis for M. If Δ_M contains a short root β with length $1/\sqrt{2}$, then replacing β with 2β we can make the set $\hat{\Delta}_M$. This $\hat{\Delta}_M$ is a root basis of the reduced root system $\hat{R}(M)$. By Lemma 5.3 and by results in Urabe [15] (Prop. 2.5, Cor. 2.6, Prop. 2.9 (4), Lemma 2.10 in [15]) there is a root basis $\hat{\Delta}_L$ of the reduced root system $\hat{R}(L)$ for L such that $\hat{\Delta}_M \subset \hat{\Delta}_L^+$ for the corresponding extended root basis $\hat{\Delta}_L^+$, that is, $\hat{\Delta}_M$ is obtained from $\hat{\Delta}_L$ by an elementary transformation.

First, we consider the case where Δ_M contains no short root with length $1/\sqrt{2}$. Then, $\Delta_M = \hat{\Delta}_M$. If $\hat{\Delta}_L$ does not contain a divisible root, then $\hat{\Delta}_L$ is a root basis for $R(L) = \hat{R}(L)$ and we have the desired claim $\Delta_M \subset \Delta_L^+$. Thus, in the following we consider the case where $\hat{\Delta}_L$ contains a divisible root. By Proposition 5.4 (2) we have a unique component $\hat{\Delta}_1$ of $\hat{\Delta}_L$ containing a divisible root. We can write $\hat{\Delta}_1^+$ in the form $\hat{\Delta}_1^+ = \{2\beta_1, \gamma_1, \ldots, \gamma_{k-1}, 2\beta_2\}$ with $\beta_1^2 = \beta_2^2 = 1/2$, $\gamma_1^2 = \cdots = \gamma_{k-1}^2 = 1$.

Assume here that $2\beta_1 \in \Delta_M$ and $2\beta_2 \in \Delta_M$. We will deduce a contradiction. Now, the graph of $\hat{\Delta}_1^+$ is the extended Dynkin graph of type A_1, B_2, F_3 or C_k ($k \geq 4$). The vertices corresponding to $2\beta_1$ and $2\beta_2$ are at the both ends. The subgraph G_0 consisting vertices corresponding to $\Delta_M \cap \hat{\Delta}_1^+$ does not contain at least one vertex corresponding to a short root. Thus, $2\beta_1$ and $2\beta_2$ belong to different connected components of G_0. This implies that any root of M is orthogonal to either $2\beta_1$ or $2\beta_2$.

Now, on the other hand, since any two roots in Δ_M are linearly independent, the rank k of $\hat{\Delta}_1$ satisfies $k \geq 2$. Under our assumption $\beta_1, \beta_2 \in L$ and $2\beta_1, 2\beta_2 \in M$. Besides, since Δ_M contains no short root with length $1/\sqrt{2}$, $\beta_1 \notin M$ and $\beta_2 \notin M$. The torsion of L/M is cyclic and L/M has only one element of order 2. Thus, $\beta_1 - \beta_2 \in M$. On the other hand, since $\beta_1 + \gamma_1 + \cdots + \gamma_{k-1} + \beta_2 = 0$, one knows $\gamma_1 + \cdots + \gamma_{k-1} + 2\beta_1 \in M$ and $\gamma = \gamma_1 + \cdots + \gamma_{k-1} \in M$. This γ is a short root with length 1, and satisfies $2\beta_1 \cdot \gamma \neq 0$ and $2\beta_2 \cdot \gamma \neq 0$. One has a contradiction.

Therefore, either $2\beta_1 \notin \Delta_M$ or $2\beta_2 \notin \Delta_M$ holds. If $2\beta_1 \notin \Delta_M$, set $\Delta_1 = \{\beta_1, \gamma_1, \ldots, \gamma_{k-1}\}$. If $2\beta_2 \notin \Delta_M$, set $\Delta_1 = \{\gamma_1, \ldots, \gamma_{k-1}, \beta_2\}$. Since $\hat{\Delta}_L - \hat{\Delta}_1$ contains no divisible root, $\Delta_L = (\hat{\Delta}_L - \hat{\Delta}_1) \cup \Delta_1$ is a root basis for L and it satisfies $\Delta_M \subset \Delta_L^+$.

Next, we consider the case where Δ_M contains a short root β with length $1/\sqrt{2}$. In this case, $\hat{\Delta}_M$, $\hat{\Delta}_L$ and $\hat{\Delta}_L^+$ have a unique component containing a divisible root. Let $\hat{\Delta}_1^+$ be the component of $\hat{\Delta}_L^+$ containing a divisible root. We can write it in the form $\hat{\Delta}_1^+ = \{2\beta, \gamma_1, \ldots, \gamma_{k-1}, 2\beta'\}$, with $\beta^2 = \beta'^2 = 1/2$, $\gamma_1^2 = \cdots = \gamma_{k-1}^2 = 1$. Set $\Delta_1 = \{\beta, \gamma_1, \ldots, \gamma_{k-1}\}$ and $\Delta_L = (\hat{\Delta}_L - \hat{\Delta}_1) \cup \Delta_1$. This Δ_L is a root basis of $R(L)$ and satisfies $\Delta_M \subset \Delta_L^+$. Q.E.D.

Lemma 6.2. *(1) For any primitive isotropic vector u in Λ_{N+1}/P belonging to F, there exists another isotropic vector v in F with $u \cdot v = 1$.*
(2) Set $H = \mathbf{Z}u + \mathbf{Z}v$ and $J = C(H, \Lambda_{N+1}/P)$. One has the decomposition $\Lambda_{N+1}/P \cong J \oplus H$, and there is a primitive embedding $P \hookrightarrow \Lambda_N$ with $\Lambda_N/P \cong J$.
(3) In the 5 cases $J_{3,0}$, $Z_{1,0}$, $Q_{2,0}$, $W_{1,0}$ and $S_{1,0}$ except in the case of $U_{1,0}$, any isotropic vector in Λ_{N+1}/P belongs to F.

Proof. (1) By primitiveness we have a homomorphism $f : \Lambda_{N+1}/P \to \mathbf{Z}$ with $f(u) = 1$. By Proposition 5.6 (1) we have a vector $v' \in F$ with $f(x) = x \cdot v'$ for $x \in \Lambda_{N+1}/P$. In particular, $u \cdot v' = 1$. Since F is an even lattice, $v'^2 = 2m$ for some integer m. Setting $v = v' - mu$, we have $v^2 = 0$ and $u \cdot v = 1$.
(2) Since $H \subset F$, $x \cdot u$ and $x \cdot v$ are integers for all $x \in \Lambda_{N+1}/P$ by Proposition 5.6 (1). Thus, we can define an isomorphism $\phi : \Lambda_{N+1}/P \to J \oplus H$ by

$$\phi(x) = (x - (x \cdot v)u - (x \cdot u)v,\ (x \cdot v)u + (x \cdot u)v) \in J \oplus H.$$

Now, let $\tilde{u}$ and $\tilde{v} \in F \subset \Lambda_{N+1}$ be the lifts of u and $v \in \Lambda_{N+1}/P$ respectively. By Lemma 5.5 (2) we have $\tilde{u}^2 = \tilde{v}^2 = 0$ and $\tilde{u} \cdot \tilde{v} = 1$. Setting $\tilde{H} = \mathbf{Z}\tilde{u} + \mathbf{Z}\tilde{v}$ and $K = C(\tilde{H}, \Lambda_{N+1})$, one knows that K is an even unimodular lattice with signature $(16 + N, N)$. The existence of u implies $N \geq 1$. Thus, we have an isomorphism $K \cong \Lambda_N$. The composition $P \hookrightarrow K \cong \Lambda_N$ is a primitive embedding with $\Lambda_N/P \cong J$.
(3) It follows from Proposition 5.6 (2). Q.E.D.

By Proposition 6.1 and Lemma 6.2 the theory of elementary transformations in Urabe [15] is effective even in our general situation.

Theorem 6.3. *(The fundamental theorem of elementary transformations.)*
Assume $N \geq 1$.
(1) Assume that for a primitive embedding $P \hookrightarrow \Lambda_{N+1}$ and for a positive definite full root submodule $L \subset \Lambda_{N+1}/P$, the orthogonal complement of L contains a primitive isotropic vector u belonging to F. Then, there exist a primitive embedding $P \hookrightarrow \Lambda_N$ with $(\mathbf{Z}u)^\perp/\mathbf{Z}u \cong \Lambda_N/P$ and the image $M_0 \subset \Lambda_N/P$ of the root quasi-lattice $Q(L)$ of L under the composition of homomorphisms $Q(L) \subset (\mathbf{Z}u)^\perp \to (\mathbf{Z}u)^\perp/\mathbf{Z}u \cong \Lambda_N/P$ has the following property:

For every positive definite full root submodule M with $M_0 \subset M \subset \Lambda_N/P$, the root system of L is obtained from that of M by one elementary transformation. In particular, the Dynkin graph of L is obtained from that of M by one elementary transformation.

(2) Conversely, let $P \hookrightarrow \Lambda_N$ be a primitive embedding and $M \subset \Lambda_N/P$ be a positive definite full submodule. Let G' be a Dynkin graph obtained from the Dynkin graph of M by one elementary transformation. Then, there is a full embedding $Q(G') \hookrightarrow (\Lambda_N/P) \oplus H \cong \Lambda_{N+1}/P$ of root quasi-lattices such that the image is contained in the orthogonal complement $(\Lambda_N/P) \oplus \mathbf{Z}u$ of the isotropic vector $u \in H$. Moreover, if a submodule $J \subset \Lambda_N/P$ is orthogonal to M, then the image of $Q(G')$ is orthogonal to $J \oplus 0 \subset (\Lambda_N/P) \oplus H$.

Let us proceed to the theory of tie transformations. The key parts in the theory of tie transformations in Urabe [17] are the theory of elementary transformations and Fact 1.5 in Urabe [17] section 1. It is easy to check that Fact 1.5 is essentially true even under the general definition of root modules in this book. Of course, we have to replace the expression

$\gamma = \sum_{\alpha \in \Delta^+} m_\alpha \alpha$ satisfies $\gamma^2 = 1$ (when Q is of type B) or $\gamma^2 = 2$ (when Q is of type A, D or E)

in the statement of Fact 1.5 by

$\gamma = \sum_{\alpha \in \Delta^+} m_\alpha \alpha$ has equal length to the shortest root in Q

It is easy to check that we can pick above γ among roots in the proof of Proposition 1.4 in [17]. (Recall the definition of roots in Section 5.)

Theorem 6.4. *(The fundamental theorem of tie transformations.)* *Assume $N \geq 1$. (1) Assume that a primitive embedding $P \hookrightarrow \Lambda_{N+1}$ is given. Let $L \subset \Lambda_{N+1}/P$ be a positive definite full submodule satisfying the following condition $\langle * \rangle$:*

$\langle * \rangle$ *For some root basis $\Delta \subset R(L)$, for some long root $\theta \in \Delta$ and for some isotropic vector u belonging to F, $u \cdot \theta = 1$ and $u \cdot \alpha = 0$ for every $\alpha \in \Delta$ with $\alpha \neq \theta$.*

Then, there are a primitive embedding $P \hookrightarrow \Lambda_N$ such that $(\mathbf{Z}u)^\perp/\mathbf{Z}u \cong \Lambda_N/P$ and the image $M_0 \subset \Lambda_N/P$ of the root quasi-lattice Q' generated by $\Delta - \{\theta\}$ under the composition of homomorphisms $Q' \subset (\mathbf{Z}u)^\perp \to (\mathbf{Z}u)^\perp/\mathbf{Z}u \cong \Lambda_N/P$ has the following property:

For every positive definite full root submodule M with $M_0 \subset M \subset \Lambda_N/P$, the Dynkin graph of L is obtained from the Dynkin graph of M by one tie transformation.

(2) Conversely, for a primitive embedding $P \hookrightarrow \Lambda_N$ and for a positive definite full submodule $M \subset \Lambda_N/P$, if a Dynkin graph G' can be obtained from the Dynkin graph of M by one tie transformation, then there is a full embedding $Q(G') \hookrightarrow (\Lambda_N/P) \oplus H \cong \Lambda_{N+1}/P$ such that the above condition $\langle * \rangle$ is satisfied for $L = Q(G')$ and $u \in H$. Moreover, if a submodule $J \subset \Lambda_N/P$ is orthogonal to M, then $J \oplus 0 \subset (\Lambda_N/P) \oplus H$ is orthogonal to $Q(G')$.

Definition 6.5. An isotropic vector $u \in \Lambda_{N+1}/P$ is said to be *in a nice position* with respect to a positive definite root submodule $L \subset \Lambda_{N+1}/P$, if either u is orthogonal to all roots in L, or the above condition $\langle * \rangle$ is satisfied.

Lemma 6.6. *We consider a primitive embedding $P \hookrightarrow \Lambda_{N+1}$, a positive definite root submodule $L \subset \Lambda_{N+1}/P$, and an isotropic vector $u \in \Lambda_{N+1}/P$*
(1) If u is orthogonal to all roots in L, then there exists a primitive isotropic vector in Λ_{N+1}/P orthogonal to all roots in L.
(2) If u satisfies $\langle * \rangle$, then u is primitive in Λ_{N+1}/P.
(3) If L has an isotropic vector in a nice position with respect to L, so does any full submodule M of L.

Proof. (1) The generator of $\mathbf{Q}u \cap \Lambda_{N+1}/P$ satisfies the condition.
(2) Assume that we can write $u = mw$ for some $m \in \mathbf{Z}$ and $w \in \Lambda_{N+1}/P$. By the axiom (R1) of root modules $(\theta, w) \in \mathbf{Z}$. Since $1 = m(\theta, w)$ we have $m = \pm 1$.
(3) By u we denote an isotropic vector in a nice position with respect to L. If u is orthogonal to all roots in L, then obviously u is orthogonal to all roots in M. Assume that u satisfies $\langle * \rangle$. By Proposition 6.1 (1) a root basis Δ_M for M is contained in a root basis Δ_L for L. We have an element $w \in W(L)$ of the Weyl group of L such that $\Delta_L = w(\Delta)$. The isotropic vector $w(u)$ is in a nice position with respect to M. Q.E.D.

Now, if we have an isotropic vector in a nice position, we can assume moreover that it is primitive by Lemma 6.6. Furthermore, if it belongs to F, we can reduce the problem from Λ_{N+1}/P to Λ_N/P by an elementary or a tie transformation. (Theorem 6.3, Theorem 6.4.) By Lemma 6.2 (3) in the 5 cases $J_{3,0}$, $Z_{1,0}$, $Q_{2,0}$, $W_{1,0}$ and $S_{1,0}$ except the case $U_{1,0}$ it always belongs to F. Therefore, existence of isotropic vectors in a nice position comes into question. The remainder of this book is devoted to showing the existence of such vectors.

As for the case of $U_{1,0}$, it turns out that if a primitive isotropic vector orthogonal to all roots in L does not belong to F, then we can apply the dual elementary transformation. See Chapter 4.

7
The Hasse symbol and the Hilbert symbol

When the orthogonal complement of $L \subset \Lambda_{N+1}/P$ contains an isotropic vector, we can apply an elementary transformation, and we can reduce the problem from Λ_{N+1}/P to Λ_N/P. Therefore, it is important to write this condition on an isotropic vector explicitly. In this section we see that we can write it by the Hasse symbol and the Hilbert norm residue symbol (Cassels [4], Serre [12]).

Let V be a finite dimensional vector space over $\mathbf{Q}$ equipped with a symmetric bilinear form $V \times V \to \mathbf{Q}$ (an *inner product space* in short). We can define the *discriminant* $\overline{d}(V)$ and the Hasse symbol $\epsilon_p(V)$. The discriminant $\overline{d}(V)$ is an element in $\mathbf{Q}/\mathbf{Q}^{*2}$ $(\mathbf{Q}^* = \mathbf{Q} - \{0\})$. $\overline{d}(V) = 0$ if, and only if, V is degenerate. If V is non-degenerate, the Hasse symbol $\epsilon_p(V) = \pm 1$ is defined for every prime number p and $p = \infty$. When V

has k of negative eigen values of the bilinear form, $\epsilon_\infty(V) = (-1)^{k(k-1)/2}$. It satisfies the product formula $\prod_{all\ p\ incl.\ \infty} \epsilon_p(V) = 1$.

The following lemma is well-known (Serre [12]).

Lemma 7.1. *A non-degenerate innner product space V contains an isotropic vector if, and only if, V is indefinite and satisfies one of the following 4 conditions:*
(1) dim $V = 2$ and $-\overline{d}(V) \equiv 1 \bmod \mathbf{Q}^{*2}$.
(2) dim $V = 3$ and for every prime number p $\epsilon_p(V) = (-1, -\overline{d}(V))_p$.
(3) dim $V = 4$ and for every prime number p $\overline{d}(V) \not\equiv 1 \bmod \mathbf{Q}_p^{2*}$ or $\epsilon_p(V) = (-1, -1)_p$.
(4) dim $V \geq 5$.

Here, $(\ ,\ )_p$ is the Hilbert symbol. The subscript p is a prime number or $p = \infty$. For a, b, $c \in \mathbf{Q}^*$, it satisfies $(a, b)_p = \pm 1$, $(a, b)_p = (b, a)_p$, $(a, bc)_p = (a, b)_p(a, c)_p$, and $(a, b)_p = 1$ if $a + b = 0$ or 1. Thus, in particular, $(a, b^2)_p = 1$ and $(-1, -\overline{d}(V))_p$ is well-defined. For non-zero integers $a = p^\alpha u$ and $b = p^\beta v$ where u and v are integers not divisible by a prime number p,

$$(a, b)_2 = (-1)^{(u-1)(v-1)/4 + \alpha(v^2-1)/8 + \beta(u^2-1)/8},$$

if $p = 2$, and

$$(a, b)_p = (-1)^{\alpha\beta(p-1)/2} \left(\frac{u}{p}\right)^\beta \left(\frac{v}{p}\right)^\alpha,$$

if $p \neq 2$. Here, $\left(\frac{-}{p}\right)$ is Legendre's quadratic residue symbol. (For any integer x with $x \not\equiv 0 \pmod p$, $\left(\frac{x}{p}\right) = \pm 1$ is defined. $\left(\frac{x}{p}\right) = +1 \iff x \equiv y^2 \pmod p$ for some integer y.) Note that if a and b are integers not divisible by p and if $p \neq 2$, then $(a, b)_p = +1$.

For $p = \infty$

$$(a, b)_\infty = \begin{cases} -1, & \text{if } a < 0 \text{ and } b < 0; \\ +1, & \text{otherwise.} \end{cases}$$

Also the Hilbert symbol satisfies the product formula $\prod_{all\ p\ incl.\ \infty} (a, b)_p = 1$.

For two inner product spaces V and V', $\overline{d}(V \oplus V') \equiv \overline{d}(V)\overline{d}(V') \bmod \mathbf{Q}^{*2}$ and $\epsilon_p(V \oplus V') = \epsilon_p(V)\epsilon_p(V')(\overline{d}(V), \overline{d}(V'))_p$.

Let L be a non-degenerate quasi-lattice. We write $\epsilon_p(L)$ instead of $\epsilon_p(L \otimes \mathbf{Q})$. Obviously, the discriminant $d(L) \in \mathbf{Q}$ satisfies $\overline{d}(L \otimes \mathbf{Q}) \equiv d(L) \bmod \mathbf{Q}^{*2}$.

For two non-degenerate quasi-lattices L and M we consider the following condition I(M, L). We assume that L has signature (l_+, l_-) and M has signature (m_+, m_-):

I(M, L): $l_+ \geq m_+ + 1$, $l_- \geq m_- + 1$ and one of the following 4 conditions holds:

(1) $(l_+ + l_-) - (m_+ + m_-) = 2$ and $-d(L)d(M) \in \mathbf{Q}^{*2}$.

(2) $(l_+ + l_-) - (m_+ + m_-) = 3$ and for every prime number p $\epsilon_p(L)\epsilon_p(M) = (-1, -1)_p(d(L), -d(M))_p$.

(3) $(l_+ + l_-) - (m_+ + m_-) = 4$ and for every prime number p $\epsilon_p(L)\epsilon_p(M) = (-1, -1)_p$ or $d(L)d(M) \notin \mathbf{Q}_p^{*2}$.

(4) $(l_+ + l_-) - (m_+ + m_-) \geq 5$

Lemma 7.2. *Assume that there exists an embedding $M \hookrightarrow L$ of non-degenerate quasi-lattices. The following three conditions are equivalent:*
(1) $I(M, L)$ holds.
(2) For every embedding $M \hookrightarrow L$ the orthogonal complement of M in L contains an isotropic vector.
(3) For some embedding $M \hookrightarrow L$ the orthogonal complement of M in L contains an isotropic vector.

Proof. Let T be the orthogonal complement of M in L. Note that T contains an isotropic vector if and only if $T \otimes \mathbf{Q}$ contains an isotropic vector. The discriminant $d(T)$ and the Hasse symbol $\epsilon_p(T)$ satisfies $d(T) \equiv d(L)d(M) \bmod \mathbf{Q}^{*2}$ and $\epsilon_p(T) = \epsilon_p(L)\epsilon_p(M)(-d(L), d(M))_p$. Thus, by Lemma 7.1 we have (1) $\Rightarrow$ (2) and (3) $\Rightarrow$ (1). (2) $\Rightarrow$ (3) is obvious. Q.E.D.

The following table shows the signature, the discriminant $d(P)$, and the Hasse symbol $\epsilon_p(P)$ (Here, p is a prime number or $p = \infty$.) for the lattice P corresponding to 6 types of our quadrilateral singularities:

	signature	$d(P)$	$\epsilon_p(P)$
$J_{3,0}$	$(5,1)$	-4	1
$Z_{1,0}$	$(6,1)$	-8	1
$Q_{2,0}$	$(7,1)$	-12	1
$W_{1,0}$	$(6,1)$	-12	$(-1,3)_p$
$S_{1,0}$	$(7,1)$	-20	$(-2,5)_p$
$U_{1,0}$	$(7,1)$	-27	$(-1,3)_p$

As for Λ_N, $d(\Lambda_N) = (-1)^N$ and $\epsilon_p(\Lambda_N) = (-1,-1)_p^{N(N-1)/2}$.

Note that $\mathrm{I}(M, \Lambda_N/\widetilde{P})$ is equivalent to $\mathrm{I}(M \oplus P, \Lambda_N)$, and it does not depend on the choice of the embedding $P \hookrightarrow \Lambda_N$. The condition $\mathrm{I}(M \oplus P, \Lambda_3) = \mathrm{I}(M, \Lambda_3/\widetilde{P})$ is called simply the *I1-condition* for M.

In each case the I1-conditions can be rewritten more explicitly by the above data. Let Q be a positive definite lattice. We have three explicit conditions depending on the case such that the I1-condition for Q holds if, and only if, one of the three holds. By Lemma 7.2 the orthogonal complement of Q in Λ_3/P contains an isotropic vector if, and only if, one of them holds, when there is an embedding $Q \hookrightarrow \Lambda_3/P$.

The I1-condition in the case of $J_{3,0}$, $Z_{1,0}$, $Q_{2,0}$.

According as the value of m, the following gives the I1-condition in the case of $J_{3,0}$ $(m = 1)$, $Z_{1,0}$ $(m = 2)$, and $Q_{2,0}$ $(m = 3)$. We assume that Q is positive definite:
(1) rank $Q = 14 - m$, and for every prime number p $\epsilon_p(Q) = (m, -d(Q))_p$.
(2) rank $Q = 13 - m$, and for every prime number p $\epsilon_p(Q) = 1$ or $md(Q) \notin \mathbf{Q}_p^{*2}$
(3) rank $Q \leq 12 - m$.

The I1-condition in the case of $W_{1,0}$, $U_{1,0}$.

According as the value of m, the following gives the I1-condition in the case of $W_{1,0}$ $(m = 0)$, and $U_{1,0}$ $(m = 1)$:
(1) rank $Q = 12 - m$, and for every prime number p $\epsilon_p(Q) = (3, d(Q))_p$.
(2) rank $Q = 11 - m$, and for every prime number p $\epsilon_p(Q) = (-1, 3)_p$ or $3d(Q) \notin \mathbf{Q}_p^{*2}$.
(3) rank $Q \leq 10 - m$.

The I1-condition in the case of $S_{1,0}$.

(1) rank $Q = 11$, and for every prime number p $\epsilon_p(Q) = (5, 2d(Q))_p$.
(2) rank $Q = 10$, and for every prime number p $\epsilon_p(Q) = (-2, 5)_p$ or $5d(Q) \notin \mathbf{Q}_p^{*2}$.
(3) rank $Q \leq 9$.

Lemma 7.3. *Assume that there exists an embedding $M \hookrightarrow L$ of non-degenerate quasi-lattices. The following conditions are equivalent. H denotes the hyperbolic plane:*
(1) Both $I(M, L)$ and $I(M \oplus H, L)$ hold.
(2) For every embedding $M \hookrightarrow L$ the orthogonal complement of M in L contains an isotropic submodule of dimension 2.
(3) For some embedding $M \hookrightarrow L$ the orthogonal complement of M in L contains an isotropic submodule of dimension 2.
A submodule $I \subset L$ is said to be isotropic if $x \cdot y = 0$ for any $x, y \in I$.

Proof. (1)⇒(2). Let T be the orthogonal complement of M in L for a fixed embedding. By I(M, L) T contains an isotropic vector $u \in T$. We can pick a vector $v' \in L \otimes \mathbf{Q}$ with $u \cdot v' = 1$. Set $v = v' - (v'^2/2)u \in L \otimes \mathbf{Q}$ and $U = \mathbf{Q}u + \mathbf{Q}v \subset L \otimes \mathbf{Q}$. Let Ξ be the orthogonal complement of U in $L \otimes \mathbf{Q}$. Since $U \cong H \otimes \mathbf{Q}$, I($M$, Ξ), I(M, L/H) and I($M \oplus H$, L) are equivalent. Since there is an embedding $M \hookrightarrow (\mathbf{Q}u)^{\perp} \to (\mathbf{Q}u)^{\perp}/\mathbf{Q}u \cong \Xi$, one can conclude that Ξ contains an isotropic vector orthogonal to M. It is equivalent to the existence of an isotropic vector $u_2 \in (\mathbf{Q}u)^{\perp} \cap T$ with $u_2 \notin \mathbf{Q}u$. $(\mathbf{Q}u + \mathbf{Q}u_2) \cap L$ is the desired submodule.
(2)⇒(3) Trivial.
(3)⇒(1) Obviously, I(M, L) holds. Let I be a 2-dimensional primitive isotropic submodule in L orthogonal to M. Let $u \in I$ be a primitive vector. Since u is isotropic, $(\mathbf{Z}u)^{\perp} \supset \mathbf{Z}u$. Since $I \cap (\mathbf{Z}u)^{\perp} \neq \mathbf{Z}u$, I($M$, $(\mathbf{Z}u)^{\perp}/\mathbf{Z}u$) holds. Choosing a vector $v \in L \otimes \mathbf{Q}$ as in the above, one sees that I(M, $(\mathbf{Z}u)^{\perp}/\mathbf{Z}u$) is equivalent to I($M \oplus H$, L).
Q.E.D.

Lemma 7.4. *Let L and M be a non-degenerate quasi-lattices with signature (l_+, l_-) and (m_+, m_-) respectively. Both $I(M, L)$ and $I(M \oplus H, L)$ hold if and only if $l_+ \geq m_+ + 2$, $l_- \geq m_- + 2$ and one of the following 4 conditions holds:*
(1) $(l_+ + l_-) - (m_+ + m_-) = 4$, $d(L)d(M)$ is a square number, and for every prime number p $\epsilon_p(L)\epsilon_p(M) = (-1, -1)_p$.
(2) $(l_+ + l_-) - (m_+ + m_-) = 5$, and for every prime number p $\epsilon_p(L)\epsilon_p(M) = (-1, -1)_p(-d(L), d(M))_p$.
(3) $(l_+ + l_-) - (m_+ + m_-) = 6$, and for every prime number p $\epsilon_p(L)\epsilon_p(M) = (-1, d(L))_p$ or $-d(L)d(M) \notin \mathbf{Q}_p^{*2}$.
(4) $(l_+ + l_-) - (m_+ + m_-) \geq 7$.

Again, let Q be a positive definite lattice. We say that the *I2-condition* for Q is satisfied, if both I($P \oplus Q$, Λ_3) ($=$ I(Q, $\Lambda_3/\widetilde{P}$)) and I($P \oplus Q \oplus H$, Λ_3) ($=$ I($P \oplus Q$, Λ_2) $=$ I(Q, $\Lambda_2/\widetilde{P}$)) hold simultaneously.

This I2-condition is important because if it is satisfied, we can reduce our problem from Λ_3/P to Λ_1/P by elementary transformations (and by dual elementary transformations in the case of $U_{1,0}$). The following shows the explicit forms of the I2-condition. We use them later.

The orthogonal complement of a positive definite lattice Q in Λ_3/P contains a 2-dimensional isotropic submodule if, and only if, one of the following 4 conditions holds, when we have an embedding $Q \hookrightarrow \Lambda_3/P$:

The I2-condition in the case of $J_{3,0}$, $Z_{1,0}$, $Q_{2,0}$.

According as the value of m, the following gives the I2-condition in the case $J_{3,0}$ $(m=1)$, $Z_{1,0}$ $(m=2)$, and $Q_{2,0}$ $(m=3)$. We assume that Q is positive definite:

(1) rank $Q = 13 - m$, $md(Q)$ is a square number, and for every prime number p $\epsilon_p(Q) = 1$.
(2) rank $Q = 12 - m$, and for every prime number p $\epsilon_p(Q) = (-m,\ d(Q))_p$.
(3) rank $Q = 11 - m$, and for every prime number p $\epsilon_p(Q) = (-m,\ -1)_p$ or $-md(Q) \notin \mathbf{Q}_p^{*2}$.
(4) rank $Q \leq 10 - m$.

The I2-condition in the case of $W_{1,0}$, $U_{1,0}$.

According as the value of m, the following gives the I2-condition in the case of $W_{1,0}$ $(m=0)$, and $U_{1,0}$ $(m=1)$:

(1) rank $Q = 11 - m$, $3d(Q)$ is a square number, and for every prime number p $\epsilon_p(Q) = (-1,\ 3)_p$.
(2) rank $Q = 10 - m$, and for every prime number p $\epsilon_p(Q) = (-3,\ d(Q))_p(-1,\ 3)_p$.
(3) rank $Q = 9 - m$, and for every prime number p $\epsilon_p(Q) = (-1,\ -1)_p$ or $-3d(Q) \notin \mathbf{Q}_p^{*2}$.
(4) rank $Q \leq 8 - m$.

The I2-condition in the case of $S_{1,0}$.

(1) rank $Q = 10$, $5d(Q)$ is a square number, and for every prime number p $\epsilon_p(Q) = (-2,\ 5)_p$.
(2) rank $Q = 9$, and for every prime number p $\epsilon_p(Q) = (5,\ -2)_p(-5,\ d(Q))_p$.
(3) rank $Q = 8$, and for every prime number p $\epsilon_p(Q) = (-1,\ -1)_p(2,\ 5)_p$ or $-5d(Q) \notin \mathbf{Q}_p^{*2}$.
(4) rank $Q \leq 7$.

Also the following is used later:

The $\mathrm{I}(Q,\ \Lambda_2/\widetilde{P})$-condition in the case of $S_{1,0}$.

The condition $\mathrm{I}(Q,\ \Lambda_2/\widetilde{P})$ $(= \mathrm{I}(Q \oplus P,\ \Lambda_2))$ is satisfied if, and only if, one of the following 4 conditions is satisfied:

(1) rank $Q = 10$, and $5d(Q)$ is a square number.
(2) rank $Q = 9$, and for every prime number p $\epsilon_p(Q) = (5,\ -2)_p(-5,\ d(Q))_p$.
(3) rank $Q = 8$, and for every prime number p $\epsilon_p(Q) = (-1,\ -1)_p(2,\ 5)_p$ or $-5d(Q) \notin \mathbf{Q}_p^{*2}$.
(4) rank $Q \leq 7$.

Lemma 7.5. *(1) $d(Q(A_k)) = k+1$, $d(Q(B_k)) = 1$ $(k = 1, 2, 3, \cdots)$, $d(Q(D_l)) = 4$ $(l = 4, 5, 6, \cdots)$, $d(Q(E_m)) = 9 - m$ $(m = 6, 7, 8)$.*
(2) $\epsilon_p(Q(A_k)) = (-1, k+1)_p$, $\epsilon_p(Q(B_k)) = \epsilon_p(Q(D_l)) = \epsilon_p(Q(E_m)) = 1$.
(3) Let G be a Dynkin graph with components of type A, B, D or E only. If a power p^s of a prime number p divides $d(Q(G))$, then $s(p-1) \leq \text{rank } Q(G)$.
(4) Let G be the same as in (3). If an odd prime number p does not divide $d(Q(G))$, then $\epsilon_p(Q(G)) = 1$.

In the above there appeared conditions claiming that an integer belongs to $\mathbf{Q}_p^{*2}$. We can apply the following lemmas to them (Serre [12]).

Lemma 7.6. *Let p be an odd prime number and $N = p^k u$ be an integer, where u is an integer not divisible by p. $N \in \mathbf{Q}_p^{*2}$ if, and only if, k is even and $\left(\dfrac{u}{p}\right) = +1$.*

Lemma 7.7. *Let $N = 2^k u$ (u is an odd integer.) be an integer. We associate N with an element $e(N) = (x, y, z) \in \mathbf{Z}/2 + \mathbf{Z}/2 + \mathbf{Z}/2$ by the following conditions:*
$x \neq 0 \iff k$ is odd.
$y \neq 0 \iff u \equiv 3 \pmod 4$
$z \neq 0 \iff u \equiv \pm 3 \pmod 8$.
(1) For two integers N and N', $e(NN') = e(N) + e(N')$.
(2) $N \in \mathbf{Q}_2^{*2} \iff k$ is even and $u \equiv 1 \pmod 8 \iff e(N) = (0, 0, 0)$.

For a Dynkin graph G, we write $e(G) = e(d(Q(G)))$. We have $e(G + G') = e(G) + e(G')$. The following table shows the values of $e(G)$:

G	$e(G)$				
A_{10}	(0, 1, 1)	A_9	(1, 0, 1)	A_8	(0, 0, 0)
A_7	(1, 0, 0)	A_6	(0, 1, 0)	A_5	(1, 1, 1)
A_4	(0, 0, 1)	A_3	(0, 0, 0)	A_2	(0, 1, 1)
A_1	(1, 0, 0)	$\emptyset$	(0, 0, 0)		
E_8	(0, 0, 0)	E_7	(1, 0, 0)	E_6	(0, 1, 1)

8
The Coxeter-Vinberg graph

In this section we explain a very effective method to treat root modules such that the number of negative eigen values of the bilinear form is 1. The main tool in this method is the Coxeter-Vinberg graph (Vinberg [18], [19], [20], [21], [22], Conway-Sloane [5]). It is closely related to the geometry on the hyperbolic space.

For example, we can apply it to choose basic Dynkin graphs, and we can apply it to show the existence of isotropic vectors in a nice position, when we consider the reduction from Λ_2/P to Λ_1/P. It has other applications, and this book contains many Coxeter-Vinberg graphs for various root modules.

Now, let (L, FL) be a root module with $l = \text{rank } L$. We assume the bilinear form on L is non-degenerate and it has signature $(l-1, 1)$. The *negative cone* $\Sigma_L \subset L \otimes \mathbf{R}$ of L is defined by

$$\Sigma_L = \{x \in L \otimes \mathbf{R} \mid x^2 < 0\}.$$

The cone Σ_L has two connected components by the assumption on the signature. Choosing one of two and fixing it, we denote it by Σ_+. The other component is $\Sigma_- = -\Sigma_+$. The quotient $\Sigma_+/\mathbf{R}_+$ by the multiplicative group $\mathbf{R}_+$ of positive real numbers can be regarded as a Lobačevskiĭ space of dimension $l-1$. The Weyl group $W = W(L, FL)$ acts properly discontinuously on $\Sigma_\pm$ and $\Sigma_+/\mathbf{R}_+$. Let $R = R(L, FL)$ be the root system. By $H_\alpha = \{x \in L \otimes \mathbf{R} \mid (x, \alpha) = 0\}$ we denote the hyperplane in $L \otimes \mathbf{R}$ orthogonal to a root $\alpha \in R$. A connected component of $\Sigma_+ - \bigcup_{\alpha \in R} H_\alpha$ is called a *fundamental polyhedron* of W or a *Weyl chamber* of W. The Weyl group W acts transitively on the set of all fundamental polyhedrons. Choose and fix one fundamental polyhedron C. By $\overline{C}$ we denote the closure of C. Corresponding to the walls of C, a set $\Delta \subset R$ of indivisible roots can be chosen as follows. (Note that $H_\alpha = H_{-\alpha}$.)

$$\Delta = \{\alpha \in R \mid \alpha \text{ is indivisible, } H_\alpha \cap \overline{C} \text{ contains an open set of } H_\alpha, \alpha \text{ is directed outwards from } C.\}$$

We call a vector in Δ a *fundamental root*. This set Δ is defined depending on C. However, note that Δ and another set of fundamental roots depending on another polyhedron C' are conjugate with respect to the Weyl group W.

Let $S \subset R$ be a subset. Following the rules below, we can draw a graph Γ from S. The graph associated with $S = \Delta$ is called the *Coxeter-Vinberg graph* of the root module (L, FL). Indeed, it is defined by (L, FL) and does not depend on the choice of the fundamental polyhedron. Note that the rules are very similar to those for Dynkin graphs and extended Dynkin graphs, but slightly different.

(1) Vertices in the graph have one-to-one correspondence with vectors in S.

(2) Any vertex in Γ has one of the 4 different expressions depending on the length of the corresponding vector in S:

length	$\sqrt{2}$	1	$\sqrt{2/3}$	$1/\sqrt{2}$
expression	$\circ$	$\bullet$	$\circledcirc$	$\otimes$

(3) If two vectors $\alpha, \beta \in S$ are orthogonal, then we do not connect the corresponding two vertices in Γ. $\alpha * \quad\quad * \beta$. ($*$ denotes one of $\circ$, $\bullet$, $\circledcirc$, and $\otimes$.)

(4) If α and $\beta \in S$ are not orthogonal and if the quasi-lattice $\mathbf{Z}\alpha + \mathbf{Z}\beta$ generated by them is positive-definite, then the corresponding two vertices in Γ are connected by a single segment. α *——* β.

(5) If the quasi-lattice $\mathbf{Z}\alpha + \mathbf{Z}\beta$ generated by two roots $\alpha, \beta \in S$ is degenerate, then the corresponding two vertices are connected by a bold segment. α *━━━* β.

(6) If the quasi-lattice $\mathbf{Z}\alpha + \mathbf{Z}\beta$ generated by two roots $\alpha, \beta \in S$ is non-degenerate and indefinite, then we connect the corresponding two vertices by a dotted segment. α *·······* β. Moreover, if necessary, the intersection number $\alpha \cdot \beta$ is added to the dotted segment.

As a practical method to construct the set of fundamental roots, we have an algorithm due to Vinberg [19].

At the first step of Vinberg's algorithm, we choose and fix a vector $v_0 \in \Sigma_+$ called the *controlling vector* in the fixed component of the negative cone. Let L_0 be the set of vectors in L orthogonal to v_0. L_0 is a positive definite root module. Let $e_1, e_2, \ldots, e_k$ be a root basis for the root system of L_0.

At the second step we choose $e_l \in L$ for an integer $l > k$ inductively. Assume that $e_1, \ldots, e_{l-1}$ have been chosen. Set

$$R_l = \{\alpha \in R \mid (\alpha, e_i) \leq 0 \ \ (1 \leq i < l), \quad (\alpha, v_0) \neq 0\}.$$

If R_l is empty, set $\Delta_V = \{e_1, \ldots, e_{l-1}\}$. If $R_l \neq \emptyset$, we define e_l to be an indivisible element α in R_l attaining the minimal value for $\mu(\alpha) = (\alpha, v_0)^2/\alpha^2$, and satisfying $(\alpha, v_0) < 0$.

Note that the set $\{\mu(\alpha) \mid \alpha \in R_l\}$ is discrete and bounded by $\mu(e_{l-1})$ from below. Moreover, $R \cap \mu^{-1}(t)$ is a finite set for any t. Thus, repeating checking whether $R_l \cap \mu^{-1}(t)$ is empty or not for a value t in $\{\mu(\alpha) \mid \alpha \in R\}$ with $t \geq \mu(e_{l-1})$ from below in order, e_l can be determined in finite steps.

Finally, when $R_l \neq \emptyset$ for all $l > k$, set $\Delta_V = \{e_i \mid i \geq 1\}$ (an infinite set).

Then, Δ_V coincides with the set Δ of fundamental roots associated with some fundamental polyhedron C (Vinberg [19] Proposition 4).

Lemma 8.1. *Let $M \subset L$ be a positive definite full root submodule. A root basis Δ_M ($\subset R(M) \subset M$) for M is conjugate to some subset of Δ with respect to the Weyl group $W(L)$ for L. In particular, the Dynkin graph of M is a subgraph of the Coxeter-Vinberg graph of L.*

Remark. If a graph Γ' is a *subgraph* of a graph Γ, in this book we always assume that any two vertices in Γ' connected in Γ are connected also in Γ'. We use the word "subgraph" in this strong sense.

Proof. Note that the real field $\mathbf{R}$ is a vector space of infinite dimension over the rational field $\mathbf{Q}$. Because of this reason, we can choose a vector $v_0 \in \Sigma_+$ in the negative cone such that $\widetilde{M} = \{x \in L \mid (x, v_0) = 0\}$. Using v_0 as the controlling vector, we carry out Vinberg's algorithm. The root basis of $\widetilde{M}$ becomes a subset of the constructed set Δ_V. By fullness, $\widetilde{M}$ and M have the common root basis, and Δ_V is conjugate over W to the given set Δ of fundamental roots. Thus, we get the lemma. Q. E. D.

Let $u \in L$ be a primitive isotropic vector. Let us consider the action of W in the neighborhood of $\mathbf{R}u$. Set $I = \mathbf{Z}u$, $I^\perp = \{x \in L \mid (x, u) = 0\}$, $J = I^\perp/I$ and $FJ = (FL \cap I^\perp) + I/I$. The pair (J, FJ) is a positive definite root module.

Now, set $U = \{x \in L \otimes \mathbf{R} \mid (x, u) = 1\}$. U is an affine space over the vector space $I^\perp \otimes \mathbf{R}$. Thus, the quotient affine space $V = U/\mathbf{R}u$ is defined. V is an affine space over the vector space $J \otimes \mathbf{R}$. A bilinear form $V \times I^\perp \to \mathbf{R}$ is induced.

Note here that a root $\alpha \in I^\perp \cap R$ induces an affine reflection $\tilde{s}_\alpha : V \to V$. If $\alpha - mu \in I^\perp \cap R(L, FL)$, $\tilde{s}_{\alpha - mu}(z) = z - (z, \alpha)\bar{\alpha}^\vee + m\bar{\alpha}^\vee$ for $z \in V$. (For $\xi \in I^\perp$, $\bar{\xi} \in J$ denotes the canonical image.) A connected component of $V - \bigcup_{\alpha \in I^\perp \cap R(L, FL)} \widetilde{H}_\alpha$

is called a small Weyl chamber, where $\widetilde{H}_\alpha$ is the reflecting hyperplane of $\widetilde{s}_\alpha$. Let D be a small Weyl chamber. Set

$$\Delta_u = \{\alpha \in I^\perp \cap R(L, FL) \mid \alpha \text{ is indivisible. } \widetilde{H}_\alpha \cap \overline{D} \text{ contains an open set of } \widetilde{H}_\alpha, \\ \overline{\alpha} \in J \text{ is directed outwards from } D.\}.$$

Δ_u is called the root basis of $I^\perp$.

Lemma 8.2. *If $u \in FL$, then the graph associated with $S = \Delta_u$ is the extended Dynkin graph associated with the Dynkin graph of (J, FJ).*

Proof. Let $K \subset I^\perp$ be a submodule with $I^\perp = K + I$ and $K \cap I = 0$. The pair $(K, K \cap FL)$ can be identified with (J, FJ) by the assumption $u \in FL$. Let $R_K = R \cap K$ be the root system of K. For $\overline{\alpha} \in R_K$ and $m \in \mathbf{Z}$ we can define an affine reflection $s_{\overline{\alpha},m}$ on V by $s_{\overline{\alpha},m}(z) = z - (z, \overline{\alpha})\overline{\alpha}^\vee + m\overline{\alpha}^\vee$. The group generated by all $s_{\overline{\alpha},m}$'s is the affine Weyl group $W_a(R_K)$ of R_K and is the semi-direct product of $W(R_K)$ over $Q(R_K^\vee)$.

Now, any root $\alpha \in R \cap I^\perp$ can be written in the form $\alpha = \overline{\alpha} - mu$ with $\overline{\alpha} \in R_K = R(J, FJ)$, $m \in \mathbf{Z}$. Then, $\widetilde{s}_\alpha = s_{\overline{\alpha},m}$. Thus, the group generated by $\widetilde{s}_\alpha$'s $(\alpha \in R \cap I^\perp)$ coincides with the affine Weyl group for $R_K = R(J, FJ)$. By the theory of affine Weyl groups (Bourbaki [3]), one gets the conclusion. Q.E.D.

In above Vinberg's algorithm a vector v_0 with $v_0^2 < 0$ is used as the controlling vector. We have another similar algorithm using an isotropic vector as the controlling vector, which is also due to Vinberg. ([21] section 1.4.)

Let $u \in L$ be a primitive isotropic vector. We use the above notation. We assume that the root system of (J, FJ) spans $J \otimes \mathbf{Q}$ over $\mathbf{Q}$. (This condition is called the *compactness property*. See Vinberg [21] section 1.3.)

Let $e_1, e_2, \ldots, e_k$ be the members in the root basis Δ_u of $I^\perp$. We choose $e_l \in L$ for $l > k$ inductively. Assume that we have chosen $e_1, \ldots, e_{l-1}$. If the set of roots

$$R_l = \{\alpha \in R \mid (\alpha, e_i) \leq 0 \ \ (1 \leq i < l), \quad (\alpha, u) \neq 0\}$$

is empty, set $\Delta_V = \{e_1, e_2, \ldots, e_{l-1}\}$. If $R_l \neq \emptyset$, we define e_l to be an element in R_l attaining the minimal value for $(\alpha, u)^2/\alpha^2$ and satisfying $(\alpha, u) < 0$.

When $R_l \neq \emptyset$ for all $l > k$, we set $\Delta_V = \{e_i \mid i \geq 1\}$ (an infinite set).

Even under this algorithm Δ_V coincides with the set Δ of fundamental roots associated with some fundamental polyhedron C (Vinberg [19], [21]).

Corollary 8.3. *The root basis Δ_u of $I^\perp$ is conjugate over the Weyl group W to a subset of the set Δ of fundamental roots of L. In particular, if $u \in FL$, then the extended Dynkin graph associated with the Dynkin graph of (J, FJ) is a subgraph of the Coxeter-Vinberg graph of L.*

Proposition 8.4. (Vinberg [21] section 2.4 and section 3.2.)
The following conditions are equivalent:
(1) The Weyl group $W(L, FL)$ has finite index in the group of all integral orthogonal transformations on L.

(2) The set Δ of fundamental roots of (L, FL) is finite.
(3) The polyhedron $C/\mathbf{R}_+$ in the Lobačevskiĭ space $\Sigma_+/\mathbf{R}_+$ associated with the fundamental polyhedron C has finite volume.
(4) There are a finite number of vectors $v_1, \ldots, v_l$ in the closure of Σ_+ such that the fundamental polyhedron C coincides with the interior of the minimum convex body containing the set $\bigcup_{i=1}^{l} \mathbf{R}_+ v_i$.

For application of Vinberg's algorithm, further some practical method is necessary to determine whether or not the obtained set $\{e_1, \ldots, e_l\}$ equals to Δ_V. For this purpose we have the following (Vinberg [19] Proposition 1).

A graph each of whose connected component is either an extended Dynkin graph or a dual extended graph is called a *generalized extended Dynkin graph.* The number of vertices of a generalized extended Dynkin graph minus the number of connected components of it is called the *rank* of it.

Proposition 8.5. *Consider one of the above 2 kinds of Vinberg's algorithm. Let S be a finite subset of Δ_V spanning $L \otimes \mathbf{Q}$ over $\mathbf{Q}$. Let G be the graph associated with S. If G satisfies the following condition $\langle a \rangle$ and $\langle b \rangle$, then $S = \Delta_V$. In particular, then, (L, FL) satisfies the equivalent conditions in Proposition 8.4.*

Conversely, if (L, FL) satisfies a condition in Proposition 8.4, then the Coxeter-Vinberg graph G of (L, FL) satisfies the following $\langle a \rangle$ and $\langle b \rangle$ for $S = \Delta_V$:

- $\langle a \rangle$ *For any subgraph Γ of G isomorphic to a generalized extended graph we can find a subgraph Γ' of G containing Γ such that Γ' is a generalized extended Dynkin graph whose rank equals to rank $L - 2$.*
- $\langle b \rangle$ *Let $S(\Xi)$ denote the subset of S corresponding to the vertices in a subgraph Ξ of G. Let Γ be an arbitrary subgraph of G isomorphic to one of the following Lannér graphs. Let $x \in L \otimes \mathbf{R}$ be a vector. If $(x, \alpha) = 0$ for every $\alpha \in S(\Gamma)$ and if $(x, \beta) \leq 0$ for every $\beta \in S - S(\Gamma)$, then $x = 0$.*

Lannér graphs

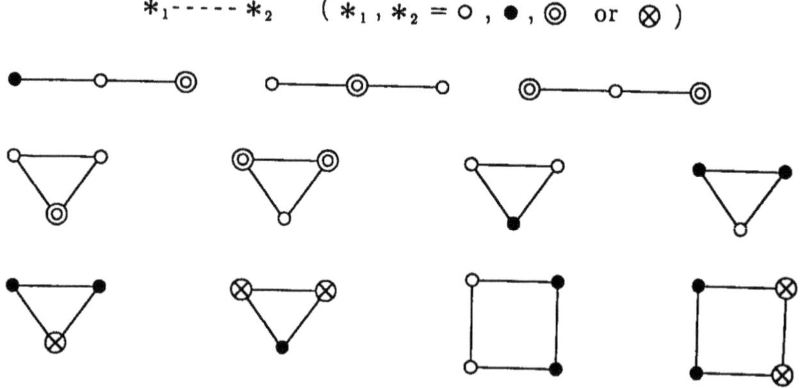

Remark. The first is a graph with only two vertices and with a dotted edge. In spite that other graphs can be found in the list of Lannér graphs in Vinverg [19] (Table 3), only the above 11 can appear in our theory. This is because the angle between two roots α, $\beta \in \Delta$ is either $\pi/2$, $2\pi/3$, $3\pi/4$ or $5\pi/6$ and the angle determines the ratio of the length of roots uniquely, if the quasi-lattice $\mathbf{Z}\alpha + \mathbf{Z}\beta$ is positive definite.

The following lemma makes the checking of the condition $\langle b \rangle$ easier.

Lemma 8.6. (Vinberg [19] Proposition 2) *We use the notations in Proposition 8.5. Let Γ be a Lannér subgraph of G. Let Ξ be the subgraph of G consisting of vertices not connected with any vertex in Γ and not belonging to Γ. If the following condition on a vector $y \in L \otimes \mathbf{R}$ is satisfied, then the condition $\langle b \rangle$ in Proposition 8.5 is also satisfied: The condition: If $(y, \alpha) = 0$ for every $\alpha \in S(\Gamma) \cup S(\Xi)$ and if $(y, \beta) \leq 0$ for every $\beta \in S - (S(\Gamma) \cup S(\Xi))$, then $y = 0$.*

Lemma 8.7. *Assume that the equivalent conditions in Proposition 8.4 are satisfied. Then, the number of vertices in any maximal Dynkin subgraph G of the Coxeter-Vinberg graph is equal to* $\operatorname{rank} L - 1$ *or* $\operatorname{rank} L - 2$. *If it is* $\operatorname{rank} L - 2$, *then G is contained in a generalized extended Dynkin subgraph whose rank equals* $\operatorname{rank} L - 2$.

Proof. Let $C \subset \Sigma_+ \subset L \otimes \mathbf{R}$ be a fixed fundamental polyhedron. Let $H_1, \ldots, H_m$ be the set of hyperplanes in $L \otimes \mathbf{R}$ such that $H_i \cap C = \emptyset$ and $H_i \cap \overline{C}$ contains an open set of H_i. (The bar $\overline{}$ denotes the closure.) $F_i = H_i \cap \overline{C}$ is called a *wall* of C. Walls correspond vertices in the Coxeter-Vinberg graph. Let $T \subset \{1, 2, \ldots, m\}$ be a non-empty subset. Set $F_T = \bigcap_{i \in T} F_i$. If $F_T \neq \emptyset$ and if $F_T \neq \{0\}$, then F_T is called a *facet* of C. Every facet F determines a subgraph $\Gamma(F)$ of the Coxeter-Vinberg graph whose vertices correspond to walls containing F. If $F \not\subset \partial\Sigma_+ = \overline{\Sigma}_+ - \Sigma_+$, then $\Gamma(F)$ is a Dynkin subgraph, else F is a one-dimensional ray on $\partial\Sigma_+$, and $\Gamma(F)$ is a generalized extended Dynkin subgraph whose rank equals $\operatorname{rank} L - 2$.

By the above 2 kinds of Vinberg's algorithms, one knows that the converse also holds: every such subgraph occurs once in this way. As this correspondence reverses the inclusion relation, the maximal Dynkin subgraphs correspond to minimal facets with $F \not\subset \partial\Sigma_+$. Since $C/\mathbf{R}_+$ is a polyhedron with $\overline{C}/\mathbf{R}_+ \subset \overline{\Sigma}_+/\mathbf{R}_+$, every minimal facet has dimension 1. So the minimal facet with $F \not\subset \partial\Sigma_+$ has dimension 1 or 2. In the last case the facet is spanned by two rays on $\partial\Sigma_+$ and each of them corresponds to a generalized extended Dynkin subgraph whose rank is $\operatorname{rank} L - 2$. Q.E.D.

In the following chapters we apply propositions and lemmas explained in this section.

Chapter 2

9 Theorems with the Ik-conditions for $J_{3,0}$, $Z_{1,0}$ and $Q_{2,0}$

In Chapter 2 $J_{3,0}$, $Z_{1,0}$ and $Q_{2,0}$ are treated.

In this section we draw the Coxeter-Vinberg graph for the root module Λ_2/P in the case of $J_{3,0}$ and $Z_{1,0}$, and several results are deduced from the graph. Also, the case of $Q_{2,0}$ is discussed.

The case of $J_{3,0}$.

In this case, $P \cong Q(D_4) \oplus H$. Recall that F stands for the orthogonal complement of P in Λ_2. We have $F \cong C(Q(D_4), \Lambda_1)$ and $\Lambda_1 \cong \Gamma_{16} \oplus H$. Γ_{16} is the even overlattice over $Q(D_{16})$ with index 2. An embedding $Q(D_4) \hookrightarrow \Gamma_{16}$ is unique up to orthogonal transformations (Dynkin [7]. See Lemma 20.3.) and $C(Q(D_4), \Gamma_{16}) \cong Q(D_{12})$. Thus, by Proposition 5.2 (2) we have $F \cong Q(D_{12}) \oplus H$ and $\Lambda_2/P \cong F^* \cong Q(D_{12})^* \oplus H$.

Now, let K be the odd unimodular lattice with signature (13, 1). We can write it in the form $K = \sum_{i=0}^{13} \mathbf{Z} v_i$, where $v_0^2 = -1$, $v_i^2 = +1$ $(1 \le i \le 13)$, and $(v_i, v_j) = 0$ $(i \ne j)$. We define the vectors w, $f_1, \dots, f_{12}$, g, h as follows:

$$
\begin{aligned}
&w = v_0 + v_1 + \cdots + v_{13},\\
&g = v_0 + v_{13}, \quad h = -(v_0 + v_{12}),\\
&f_i = -v_i + v_{i+1} \quad (1 \le i \le 10),\\
&f_{11} = v_0 - v_{11} + v_{12} + v_{13}, \quad f_{12} = -(v_0 + v_{11} + v_{12} + v_{13}).
\end{aligned}
$$

Set $M = \{x \in K \mid (x, w) \equiv 0 \pmod 2\} = \{x = \sum x_i v_i \mid \sum x_i \equiv 0 \pmod 2\}$. The vectors $f_1, \dots, f_{12}$ g, h are a basis for M. The vectors $f_1, \dots, f_{12}$ are a root basis of type D_{12}, and g and h generate a hyperbolic plane H orthogonal to f_i's $(1 \le i \le 12)$. Thus, we have $M \cong Q(D_{12}) \oplus H$ and

$$\Lambda_2/P \cong M^* \cong K + \mathbf{Z}(w/2).$$

For Vinberg's algorithm to draw the Coxeter-Vinberg graph of Λ_2/P the above basis of the quasi-lattice $K + \mathbf{Z}(w/2)$ is useful. We choose v_0 as the controlling vector. As the root basis for the orthogonal complement of v_0, we pick

$$
\begin{aligned}
e_i &= -v_i + v_{i+1} \quad (1 \le i \le 12),\\
e_{13} &= -v_{13}.
\end{aligned}
$$

Successively we get

$$
\begin{aligned}
e_{14} &= v_0 + v_1 + v_2 + v_3\\
e_{15} &= -v_{13}.
\end{aligned}
$$

Drawing the graph for these 15 vectors, we get:

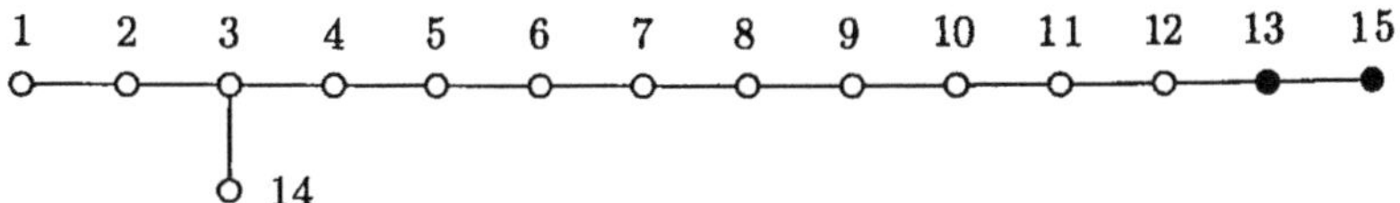

This contains no Lannér subgraph. By Proposition 8.5, the above is the Coxeter-Vinberg graph for Λ_2/P.

Corollary 9.1. *Let P be the lattice defined for the case $J_{3,0}$. For every positive definite full root submodule $L \subset \Lambda_2/P$, there exists a primitive isotropic vector $u \in \Lambda_2/P$ in a nice position with respect to L such that the root system of the positive definite root module $(\mathbf{Z}u)^\perp/\mathbf{Z}u$ is of type $E_8 + F_4$.*

Proof. By the action of the Weyl group $W(\Lambda_2/P)$, we can assume that the root basis Δ_L for L is a subset of above $\{e_1, \ldots, e_{15}\}$. On the other hand, setting

$$u = -(e_{10} + 2e_{11} + 3e_{12} + 4e_{13} + 2e_{15}),$$

one has $(u, e_9) = 1$ and $(u, e_i) = 0$ for $i \neq 9$, $1 \leq i \leq 15$. Thus, u is in a nice position with respect to L. Moreover, by Lemma 6.6 u is primitive. One can read off the above graph that the root system of $(\mathbf{Z}u)^\perp/\mathbf{Z}u$ is of type $E_8 + F_4$, since vertices except 9 form the extended Dynkin graph of type $E_8 + F_4$. Q.E.D.

Set $u' = -(e_2 + 2\sum_{i=3}^{13} e_i + e_{14})$. This u' is a primitive isotropic vector and the root system of $(\mathbf{Z}u')^\perp/\mathbf{Z}u'$ is of type B_{12}. The above Coxeter-Vinberg graph contains only two types — $E_8 + F_4$ and B_{12} — of (generalized) extended Dynkin graphs of rank 12. (No dual extended Dynkin graph appears, since Λ_2/P is a regular root module.) Combining this with Lemma 6.2, one gets the following:

Corollary 9.2. *Let P be the lattice associated with $J_{3,0}$. The root system of the quotient quasi-lattice Λ_1/P is of type G for some primitive embedding $P \hookrightarrow \Lambda_1$, if, and only if, $G = E_8 + F_4$, or B_{12}.*

The case of $Z_{1,0}$.

In this case, $P = P_0' \oplus T \oplus H_0$, $P_0' \cong Q(D_4)$ and $T \cong Q(A_1)$. Consider the orthogonal complement F of P in Λ_2 for a fixed primitive embedding $P \hookrightarrow \Lambda_2$. Since every embedding $P \hookrightarrow \Lambda_2$ is equivalent by Proposition 5.2, we can choose a convenient one for our purpose. Since $\Lambda_2 = \Gamma_{16} \oplus H \oplus H$, we can take the direct sum of the embeddings for each component $P_0' \hookrightarrow \Gamma_{16}$, $T \hookrightarrow H$, $H_0 \hookrightarrow H$. Since $C(P_0', \Gamma_{16}) \cong Q(D_{12})$ and $C(T, H) \cong \mathbf{Z}v_0$ ($v_0^2 = -2$), we have $F \cong Q(D_{12}) \oplus \mathbf{Z}v_0$ with $v_0^2 = -2$.

Let K be the root lattice of type B_{12}. We can write it in the form $K = \sum_{i=1}^{12} \mathbf{Z}v_i$ where $v_i^2 = 1$ $(1 \leq i \leq 12)$, $(v_i, v_j) = 0$ $(i \neq j)$. Set $w = v_1 + v_2 + \cdots + v_{12} \in K$.

The root lattice of type D_{12} can be identified with the sublattice $\{x \in K \mid (x, w) \equiv 0 \pmod 2\}$ of K with index 2. Thus, we have

$$\Lambda_2/P \cong F^* \cong \mathbf{Z}(v_0/2) \oplus [K + \mathbf{Z}(w/2)].$$

The expression in the right-hand side can be used to draw the Coxeter-Vinberg graph. We choose v_0 as the controlling vector. As the root basis orthogonal to v_0, the following can be chosen:

$$e_i = -v_i + v_{i+1} \quad (1 \leq i \leq 11)$$
$$e_{12} = -v_{12}.$$

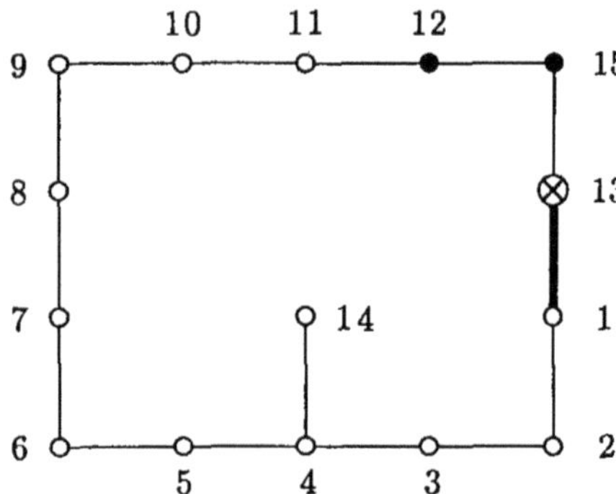

At the second step we get vectors:

$$e_{13} = v_0/2 + v_1$$
$$e_{14} = v_0 + v_1 + v_2 + v_3 + v_4$$
$$e_{15} = v_0 + (v_1 + v_2 + \cdots + v_{12})/2.$$

Drawing the graph for these 15 vectors, we get the left graph. By Proposition 8.5 this is the Coxeter-Vinberg graph of Λ_2/P in the case $Z_{1,0}$.

Corollary 9.3. *Let P be the lattice corresponding to $Z_{1,0}$. For every positive definite full root submodule $L \subset \Lambda_2/P$ there exists a primitive isotropic vector $u \in \Lambda_2/P$ in a nice position with respect to L such that the root system for $(\mathbf{Z}u)^{\perp}/\mathbf{Z}u$ is either of type $E_7 + F_4$ or of type $E_8 + BC_3$.*

Proof. We can regard that the root basis Δ_L for L is a subset of the above system of 15 vectors $e_1, \ldots, e_{15}$. The graph made from Δ_L is a Dynkin graph and it has no bold edge. Thus, either $e_1 \notin \Delta_L$ or $e_{13} \notin \Delta_L$.

Consider the case where $e_1 \notin \Delta_L$ first. Set

$$u_1 = -(e_{11} + 2e_{12} + 2e_{13} + 2e_{15}).$$

This u_1 is a primitive isotropic vector and $(\mathbf{Z}u_1)^{\perp}/\mathbf{Z}u_1$ has the root system of type $E_8 + BC_3$.

$$(u_1, e_1) = 2, \quad (u_1, e_{10}) = 1, \quad \text{and}$$
$$(u_1, e_i) = 0 \quad (1 \leq i \leq 15, \ i \neq 1, \ 10).$$

The vector e_{10} is a long root. Thus, u_1 is in a nice position.

In the case where $e_{13} \notin \Delta_L$, consider

$$u_2 = -(e_9 + 2e_{10} + 3e_{11} + 4e_{12} + 2e_{15}).$$

Also, this u_2 is a primitive isotropic vector and $(\mathbf{Z}u_2)^\perp/\mathbf{Z}u_2$ has the root system of type $E_7 + F_4$.

$$(u_2, e_8) = 1, \quad (u_2, e_{13}) = 1,$$
$$(u_2, e_i) = 0 \quad (1 \le i \le 15, \ i \ne 8, \ 13).$$

Thus, u_2 is in a nice position in this case. Q.E.D.

Corollary 9.4. *Let P be the lattice corresponding to the case $Z_{1,0}$. The root system of the quotient quasi-lattice Λ_1/P is of type G for some embedding $P \hookrightarrow \Lambda_1$ if, and only if, $G = E_7 + F_4$, $E_8 + BC_3$ or $B_{10} + BC_1$.*

The case of $Q_{2,0}$.

The quotient quasi-lattice Λ_2/P does not satisfy the equivalent conditions in Proposition 8.4. Thus, we cannot draw down the Coxeter-Vinberg graph. Instead, the decomposition of P is made use of to deal with this case.

The lattice P has the following decomposition:

$$P = P_0' \oplus T \oplus H_0, \quad P_0' \cong Q(D_4), \quad T \cong Q(A_2).$$

Here, $P' = P_0' \oplus T$. We denote $\overline{P} = P_0' \oplus H_0$. Recall that the discriminant group P'^*/P' has elements of the second kind (special elements of type B), but it has no elements of the third kind. Thus, we have to deal with only ones with length 1 as short roots. We need not consider obstruction components.

Lemma 9.5. *Assume that $N \ge 1$. We fix an embedding $P \hookrightarrow \Lambda_N$. Set $\Xi = C(T, \Lambda_N)$. T and the image of T in $\Lambda_N/\overline{P}$ are identified. By $\pi : \Lambda_N/\overline{P} \to \Lambda_N/P$ we denote the canonical surjective homomorphism.*
(1) T is primitive in $\Lambda_N/\overline{P}$.
(2) The restriction of π to $\Xi/\overline{P}$ is injective.
(3) The image $\pi(\Xi/\overline{P}) = (\Xi \oplus T)/P$ has index 3 in Λ_N/P.
(4) The root system satisfies $R(\Lambda_N/P) \subset \pi(\Xi/\overline{P})$.

Proof. Only the last (4) is not obvious. The orthogonal complement F of P in Λ_N is identified with its image in Λ_N/P. Also, we have identification $\Lambda_N/P \cong F^*$. Recall that we have a canonical group isomorphism $r : P^*/P \to F^*/F$. (Lemma 5.5.) Note that in the discriminant group $P^*/P \cong P_0'^*/P_0' \oplus T^*/T$ any element of the second kind is contained in the direct summand $P_0'^*/P_0'$. Since $F \subset \pi(\Xi/\overline{P}) \subset F^*$, and since a subgroup with index 3 in F^*/F is unique, $\pi(\Xi/\overline{P})/F = r(P_0'^*/P_0')$. Let $\alpha \in R(F^*)$ be a short root with length 1. There is a vector $x \in P^*$ with $x + \alpha \in \Lambda_N$. $\overline{x} = x \bmod P$ is an element of the second kind; thus, $\overline{x} \in P_0'^*/P_0'$. One has $\alpha \bmod F = r(\overline{x}) \in \pi(\Xi/\overline{P})/F$, and $\alpha \in \pi(\Xi/\overline{P})$. If $\beta \in R(F^*)$ is a long root, then $\beta \in F \subset \pi(\Xi/\overline{P})$. We have (4).
Q.E.D.

Lemma 9.6. *Let P be the lattice defined in the case of $Q_{2,0}$. The root system of the quotient Λ_1/P is of type G for some embedding $P \hookrightarrow \Lambda_1$ if, and only if, $G = E_6 + F_4$, $E_8 + F_2$ or B_9.*

Proof. By Lemma 9.5 we have only to consider what the root system of the orthogonal complement of T in $\Lambda_1/\overline{P}$ is. By Corollary 9.2 the root system of $\Lambda_1/\overline{P}$ is either of type E_8+F_4 or of type B_{12}.

Consider the case where $\Lambda_1/\overline{P}$ has the root system of type E_8+F_4. If the root system of T (It is of type A_2.) is contained in the component of type E_8, then $G = E_6+F_4$, while if it lies in the component of type F_4, then $G = E_8+F_2$.

Next, we consider the case where the root system of $\Lambda_1/\overline{P}$ is of type B_{12}. Let $Q \subset \Lambda_1/\overline{P}$ be the root lattice of type B_{12}. We have $T \subset Q$. Let S be the orthogonal complement of T in Q. It is easy to see that S contains a root lattice Q_1 of type B_9. Thus, $S = \mathbf{Z}\xi \oplus Q_1$ for some vector $\xi \in S$. Since $\xi^2 = d(S) = d(T) = 3$, the root system of S is of type B_9.

Conversely, we can construct an embedding $P \hookrightarrow \Lambda_1$ realizing each case $G = E_6+F_4$, E_8+F_2, B_9. Q.E.D.

Remark. By the above lemma, for some primitive embedding $P \hookrightarrow \Lambda_1$ the root system in Λ_1/P is of type B_9 and does not span $(\Lambda_1/P)\otimes\mathbf{Q}$. This is equivalent to that for some primitive isotropic vector $u \in \Lambda_2/P$ the root system of $(\mathbf{Z}u)^{\perp}/\mathbf{Z}u$ does not span $(\mathbf{Z}u)^{\perp}/\mathbf{Z}u$ over $\mathbf{Q}$. (A primitive embedding $P \hookrightarrow \Lambda_2$ is fixed.) This implies that on the fundamental polyhedron $C \subset (\Lambda_2/P)\otimes\mathbf{R}$ whose closure contains u the ray $\mathbf{R}_+u$ is not an edge of C. Thus, the conditions in Proposition 8.4 are not satisfied.

Lemma 9.7. *Let P be the lattice defined for $Q_{2,0}$. For every positive definite full root submodule $L \subset \Lambda_2/P$, we have a primitive isotropic vector $u \in \Lambda_2/P$ in a nice position with respect to L such that the root system of $(\mathbf{Z}u)^{\perp}/\mathbf{Z}u$ is either of type E_6+F_4 or of type E_8+F_2.*

Proof. We use notations in Lemma 9.5 assuming $N=2$.

By $Q(L)$ we denote the sub-quasi-lattice of L generated by roots. By Lemma 9.5 $Q(L) \subset \pi(\Xi/\overline{P})$. Let $\rho : \pi(\Xi/\overline{P}) \to \Xi/\overline{P}$ denote the inverse homomorphism of π. Set $Q = \rho(Q(L))$. Let Q' be the sub-quasi-lattice of $P(T\oplus Q, \Lambda_2/\overline{P})$ generated by roots in it. Q' is an over-quasi-lattice of $T\oplus Q$. Note that $T\oplus Q$ is generated by roots in $P(T\oplus Q, T\oplus(\Xi/\overline{P}))$. Let $Q' = \oplus_{i=1}^m Q'_i$ be the irreducible decomposition of Q'. We assume that $T \subset Q'_1$. We have $Q = (Q\cap Q'_1)\oplus(\oplus_{i=2}^m Q'_i)$. On the other hand,

$$P(Q', \Lambda_2/\overline{P})/P(T\oplus Q, T\oplus(\Xi/\overline{P})) = P(T\oplus Q, \Lambda_2/\overline{P})/P(T\oplus Q, T\oplus(\Xi/\overline{P}))$$

is isomorphic to the subgroup of $\Lambda_2/(T\oplus\Xi) \cong \mathbf{Z}/3$ and is a cyclic group. By Proposition 6.1 one knows that the root system of $T\oplus Q$ is obtained from that of Q' by one elementary transformation. Excluding common components $\oplus_{i=2}^m Q'_i$, the root basis Δ_1 of $T\oplus(Q\cap Q'_1)$ is obtained from the root basis Δ'_1 of Q'_1 by one elementary transformation. Note that Δ'_1 is irreducible, Δ'_1 and Δ_1 have the same number of elements, and Δ_1 has a component of type A_2 corresponding to T. We have never a short root with length $1/\sqrt{2}$ or $\sqrt{2/3}$ in our case, and one knows that Δ'_1 is of type either F_4, E_6, E_7, E_8 or A_2. According as the type of Δ'_1, Δ_1 is of type F_2+A_2, $3A_2$, A_5+A_2, E_6+A_2 or A_2. Let Δ'_i be a root basis for Q'_i for $2 \le i \le m$. $\Delta' = \bigcup_{i=1}^m \Delta'_i$ is a root basis of Q'.

Now, our $\overline{P}$ is isomorphic to P in the case of $J_{3,0}$. Thus, considering a conjugate one, we can assume that Δ' is a subset of the system of 15 vectors just before Corollary

9.1. We would like to show here that there exists a primitive isotropic vector $u' \in \Lambda_2/\overline{P}$ satisfying either the following (0)′ and (1)′ or the following (0)′ and (2)′:
(0)′ The root system of $(\mathbf{Z}u')^{\perp}/\mathbf{Z}u'$ is of type $E_8 + F_4$.
(1)′ u' is orthogonal to all vectors in Δ'.
(2)′ There exists a long root $\theta \in \Delta' - \Delta'_1$ such that $(\theta, u') = 1$ and $(\alpha, u') = 0$ for every $\alpha \in \Delta'$ with $\alpha \neq \theta$.

If $e_9 \notin \Delta'_1$, then $u_0 = -(e_{10} + 2e_{11} + 3e_{12} + 4e_{13} + 2e_{15})$ satisfies the desired condition.

Assume $e_9 \in \Delta'_1$. The graph of Δ'_1 is a Dynkin subgraph containing the vertex 9 in the Coxeter-Vinberg graph for $J_{3,0}$. Thus, one sees that it is of type neither F_4 nor E. Thus, $Q \cap Q'_1 = 0$, $T = Q'_1$ and either $\Delta'_1 = \{e_8, e_9\}$ or $\Delta'_1 = \{e_9, e_{10}\}$.

Consider the case $\Delta'_1 = \{e_8, e_9\}$. By assumption $e_7, e_{10} \notin \Delta'_1$. Set

$$\begin{aligned} u_1 &= -(e_7 + e_8 + e_9 + 2e_{11} + 3e_{12} + 4e_{13} + 2e_{15}) \\ &= s_{e_7} s_{e_8} s_{e_9}(u_0). \end{aligned}$$

(s_α stands for the reflection associated with α.) One can check that $(e_1, u_1) = 0$ ($i \neq 6$, 7, 10, $1 \leq i \leq 15$) and $(e_6, u_1) = 1$. Thus, u_1 satisfies the condition.

Next, we consider the case $\Delta'_1 = \{e_9, e_{10}\}$. By assumption $e_8, e_{11} \notin \Delta'$. Set

$$\begin{aligned} u_2 &= -(e_7 + 2(e_8 + e_9 + e_{10} + e_{11}) + 3e_{12} + 4e_{13} + 2e_{15}) \\ &= s_{e_8} s_{e_9} s_{e_{10}}(u_1). \end{aligned}$$

$(e_i, u_2) = 0$ ($i \neq 6$, 8, 11, $1 \leq i \leq 15$) and $(e_6, u_2) = 1$. This u_2 satisfies the desired condition.

We have shown the existence of u'.

Set $I' = \mathbf{Z}u'$. By the condition we always have $T \subset I'^{\perp}$. Set $u = \pi(u') \in \Lambda_2/P$ and $I = \mathbf{Z}u$. By Lemma 9.5, I is primitive in $\pi(\Xi/\overline{P}) = (\Xi \oplus T)/P$. If I is not primitive in Λ_2/P, we can write $u = aw$ with $a \in \mathbf{Z}$, $w \in \Lambda_2/P$ and $w \notin (\Xi \oplus T)/P$. We have $w^2 = 0$. By Lemma 5.5 (1) and Proposition 3.9 (6) one knows that w belongs to the image in Λ_2/P of the orthogonal complement F of P. In particular, we have $w \in \Xi \oplus T$, which is a contradiction. Thus, I is primitive even in Λ_2/P. By Lemma 9.5 the set of all roots in $I'^{\perp}$ orthogonal to T has one-to-one correspondence with the set of all roots in I through π. Thus, the root system of $I^{\perp}/I$ is equal to the root system consisting of all roots in $I'^{\perp}/I'$ orthogonal to T. We have here identified T and the image of T in $I'^{\perp}/I'$. The root system of $I'^{\perp}/I'$ is of type $E_8 + F_4$. Depending on which one of two components contains T, the root system of $I^{\perp}/I$ is of type either $E_6 + F_4$ or $E_8 + F_2$.

Q.E.D.

Theorem 9.8. *Consider one of $J_{3,0}$, $Z_{1,0}$, and $Q_{2,0}$ as the class X of hypersurface quadrilateral singularities. We consider the Ik-conditions depending on X. (See Section 7.) By r we denote the number of vertices in the Dynkin graph G. $Q = Q(G)$ stands for the root (quasi-)lattice of type G. $r = \operatorname{rank} Q$.*

[I] The following two conditions (a) and (b) are equivalent:
(a) $G \in PC(X)$ and the I2-condition holds for the root lattice $Q = Q(G)$ of type G.
(b) G contains no vertex corresponding to a short root and can be obtained from one of the following basic Dynkin graphs by elementary transformations repeated twice:

The basic Dynkin graphs:
The case $X = J_{3,0}$: $E_8 + F_4$, B_{12}
The case $X = Z_{1,0}$: $E_7 + F_4$, $E_8 + BC_3$, $B_{10} + BC_1$
The case $X = Q_{2,0}$: $E_6 + F_4$, $E_8 + F_2$, B_9

[II] The following two conditions (A) and (B) are also equivalent:
(A) $G \in PC(X)$ *and the I1-condition holds for the root lattice* $Q = Q(G)$ *of type* G.
(B) G *contains no vertex corresponding to a short root and can be obtained from one of the essential basic Dynkin graphs in Theorem 0.3 by one of the following 3 kinds of procedures:*
- ⟨1⟩ *elementary transformations repeated twice:*
- ⟨2⟩ *an elementary transformation following after a tie transformation*
- ⟨3⟩ *a tie transformation following after an elementary transformation.*

Proof. First, note that in our case P has no associate numbers k with $k \geq 4$.
[I] (a)⇒(b) Obviously, G has no vertex corresponding to a short root. By Theorem 2.2 and Theorem 5.11 there exists a full embedding $Q(G) \hookrightarrow \Lambda_3/P$. By Lemma 7.3 the orthogonal complement of $Q = Q(G)$ contains a 2-dimensional isotropic submodule. Thus, we can apply Theorem 6.3 (1) twice and one knows that G can be obtained from the Dynkin graph of Λ_1/P for some embedding $P \hookrightarrow \Lambda_1$ by two elementary transformations. By Corollary 9.2, Corollary 9.4 and Lemma 9.6 the Dynkin graph of Λ_1/P is one of the above basic graphs.
[I] (b)⇒(a) By Corollary 9.2, Corollary 9.4 and Lemma 9.6 for any basic graph G'' there exists a primitive embedding $P \hookrightarrow \Lambda_1$ such that the root system of Λ_1/P is of type G''.

Let G be a Dynkin graph obtained from G'' by two elementary transformations. By applying Theorem 6.3 (2) twice one can conclude that there exists a full embedding $Q(G) \hookrightarrow \Lambda_3/P$. By Theorem 6.3 (2) the orthogonal complement of $Q(G)$ contains a 2-dimensional isotropic submodule. By Lemma 7.3 the I2-condition holds. If G contains no vertex corresponding to a short root, by Theorem 5.11 and Theorem 2.2 one knows $G \in PC(X)$.
[II] (A)⇒(B) Obviously, G has no vertex corresponding to a short root. By Theorem 2.2 and Theorem 5.11 there exists a full embedding $Q(G) \hookrightarrow \Lambda_3/P$. By Lemma 7.2 the orthogonal complement of $Q(G)$ contains an isotropic vector. Thus, by Theorem 6.3 (1) for some positive definite full root submodule $M \subset \Lambda_2/P$ such that G is obtained from the Dynkin graph G' of M by one elementary transformation. By Corollary 9.1, Corollary 9.3 and Lemma 9.7 we have a primitive isotropic vector $u \in \Lambda_2/P$ in a nice position with respect to M. Here, moreover, we can assume that the Dynkin graph of $(\mathbf{Z}u)^{\perp}/\mathbf{Z}u$ is an essential basic graph. Thus, by Theorem 6.3 (1) or Theorem 6.4 (1) one can conclude (B). Note that the procedure ⟨3⟩ is dispensable.
[II] (B)⇒(A) If we apply Theorem 6.3 (2) once and Theorem 6.4 (2) once in the above proof of (b)⇒(a) instead of applying Theorem 6.3 twice, one knows $G \in PC(X)$ under (B). If we applied the procedure ⟨1⟩ or ⟨2⟩, then by Theorem 6.3 (2) for $N = 2$ one can conclude the existence of an isotropic vector. In the case of the procedure ⟨3⟩, by the latter half of Theorem 6.4 (2) for $N = 2$ and by Theorem 6.3 (2) for $N = 1$, also, one knows the existence of an isotropic vector. By Lemma 7.2 one has the I1-condition.

Q.E.D.

Note that also considering the case where we apply Theorem 6.4 twice, one knows the "if" part under the condition (2) of our main theorem Theorem 0.3 is true. Moreover, thanks to the above theorem, only the cases where the I1-condition fails are remaining. In the rest of this chapter but Section 12 and 14, we consider the "only if" part of Theorem 0.3 under the assumption that the I1-condition does not hold.

10 A transcendental isotropic cycle

Let Λ_3 be the even unimodular lattice of signature (19, 3), and P be the lattice associated with the hypersurface quadrilateral singularity. If a Dynkin graph G belongs to $PC(X)$, then we have an embedding $P \oplus Q(G) \hookrightarrow \Lambda_3$ satisfying Looijenga's conditions (L1) and (L2) and the induced embedding $Q(G) \hookrightarrow \Lambda_3/P$ is full. Also, we have an elliptic K3 surface $\Phi : Z \to C(\cong \mathbf{P}^1)$ corresponding to the embedding.

The most important part of the remaining parts of the verification of Theorem 0.3 is the following:

Proposition 10.1. *If $G \in PC(X)$, and if G is not in the exception list in Theorem 0.3, then with respect to some full embedding $Q(G) \hookrightarrow \Lambda_3/P$, there exists a primitive isotropic vector u in Λ_3/P in a nice position.*

To show this proposition we use the theory of the monodromy for elliptic surfaces and construct a certain transcendental cycle in the elliptic K3 surface $Z \to C$. The fact that the monodromy around a singular fiber of type I_0^* has a very simple form, i.e., simply multiplying -1 is essentially used for the construction.

Now, recall that we have an isomorphism $H^2(Z, \mathbf{Z}) \xrightarrow{\sim} \Lambda_3$ preserving bilinear forms up to sign. (Note that it *reverses* the sign.) Through this isomorphism we can use the geometry on the elliptic K3 surface $\Phi : Z \to C$ to show an isotropic vector in Λ_3.

We denote the fiber $\Phi^{-1}(a)$ over a point $a \in C$ by F_a for simplicity. By $\Sigma = \{c_1, c_2, \ldots, c_t\}$ we denote the set of critical values of Φ. We put $F_i = F_{c_i}$ for simplicity.

We can assume the following in our situation:

(1) For some point $c_1 \in \Sigma$ the fiber F_1 over c_1 is a singular fiber of type I_0^*.

(2) There is a section $s_0 : C \to Z$ (i.e., a morphism of varieties with $\Phi(s_0(x)) = x$ for $x \in C$) whose image is denoted by C_5.

Recall that C_5 and F_1 are contained in the curve IF at infinity. The curve IF at infinity has 6 (when $X = J_{3,0}$), 7 (when $X = Z_{1,0}$) or 8 (when $X = Q_{2,0}$) components. The lattice P has a basis associated with the dual graph of the components of IF. The union $\mathcal{E}$ of smooth rational curves on Z disjoint from IF coincides with the union of components disjoint from IF of singular fibers of $Z \to C$. The dual graph of $\mathcal{E}$ is G by definition.

We divide the case into three:

((1)) The surface $Z \to C$ has another singular fiber of type I^* apart from F_1.

((2)) $Z \to C$ has a singular fiber of type II^*, III^* or IV^*.

((3)) $Z \to C$ has no singular fiber of type I^*, II^*, III^* or IV^* apart from F_1.

In this section we show that in case ((1)) there exists a non-zero transcendental 2-cycle Ξ in Z with $\Xi^2 = 0$ orthogonal to the section C_5 and to all irreducible components of fibers of $Z \to C$. In particular, the orthogonal complement of $S = P \oplus Q(G)$ in Λ_3 contains an isotropic vector; thus, we have a desired isotropic vector in Λ_3/P.

We have the following facts (Kodaira [8], Shioda [14]):

- $C \cong \mathbf{P}^1$ (One-dimensional projective space).
- For any irreducible curve A in Z, the self-intersection number A^2 is equal to -2 if, and only if, A is smooth and rational.
- Let $e(F)$ denote the Euler number of a fiber F. Then, we have

$$\sum_{i=1}^{t} e(F_i) = 24. \tag{1}$$

- The set MW of all sections of Φ has a structure of an abelian group (the Mordell-Weil group) when we fix an element, say C_5, as the unit element. This abelian group MW is finitely generated. Let a be the rank of MW and ρ be the Picard number of Z (= rank Pic(Z)).

$$\rho = 2 + a + \sum_{i=1}^{t}(m(F_i) - 1) \tag{2}$$

where $m(F)$ denotes the number of irreducible components of a singular fiber F. (See Lemma 10.5.)
- If a singular fiber F is not of type I, then $m(F) = e(F) - 1$, while if F is of type I, then $m(F) = e(F)$.
- Let t_1 denote the number of singular fibers of type I of Φ, and t be the number of all singular fibers of Φ. By above (1) and (2) we have:

$$\rho = 26 + a - 2t + t_1. \tag{3}$$

- If the functional invariant $J : C \to \mathbf{P}^1$ of the elliptic surface Φ is not constant, then

$$20 - \rho + a \geq \nu(I_0^*) + \nu(II) + \nu(III) + \nu(IV), \tag{4}$$

where $\nu(T)$ denotes the number of singular fibers of type T. (Shioda [14]. For general elliptic surfaces 20 should be replaced by $b_2 - 2p_g$. In our case the second Betti number $b_2 = 22$, the geometric genus $p_g = 1$.)

Lemma 10.2. $\quad t \geq 3$.

Proof. Since $\rho \leq \dim H^1(Z, \Omega_Z^1) = 20$, by (3) we have $2t \geq 6 + a + t_1 \geq 6$. Q.E.D.

Recall here the concept of *parallel translation* along a path. Let $r : [0,1] \to C - \Sigma$ be a path, i.e., a continuous mapping from the closed interval between 0 and 1. We have the induced mapping

$$Z(r) = \bigcup_{0 \leq \tau \leq 1} F_{r(\tau)} \to [0,1].$$

($F_{r(\tau)} = \Phi^{-1}(r(\tau))$.) Since $[0,1]$ is contractible, this family $Z(r) \to [0,1]$ is trivial, i.e., there is a homeomorphism $\chi : F_{r(0)} \times [0,1] \to Z(r)$ such that its composition with $Z(r) \to [0,1]$ coincides with the projection $F_{r(0)} \times [0,1] \to [0,1]$. For $0 \leq \tau \leq 1$ by χ_τ :

$F_{r(0)} \to F_{r(\tau)}$ we denote the composition of the natural isomorphism $F_{r(0)} \cong F_{r(0)} \times \{\tau\}$ and the restriction of χ to $F_{r(0)} \times \{\tau\}$. The homeomorphism $\chi_\tau {\chi_{\tau'}}^{-1} : F_{r(\tau')} \to F_{r(\tau)}$ is induced for $\tau, \tau' \in [0,1]$. This is called the *parallel translation* from $r(\tau')$ to $r(\tau)$ along r. It depends on the homeomorphism χ, but the isotopy class of the parallel translation depends only on the homotopy class of the path in $C - \Sigma$ connecting $r(\tau')$ and $r(\tau)$. In particular, we can define an isomorphism of cohomology groups $r_* : H^*(F_{r(\tau')}, \mathbf{Z}) \to H^*(F_{r(\tau)}, \mathbf{Z})$ associated with the parallel translation, which depends only on the homotopy class of the path r. (Thus, we can denote it by r_*.)

Now, let b_τ be the intersection point of $C_5 = s_0(C)$ and $F_{r(\tau)}$. The section s_0 induces a section $[0,1] \to Z(r)$ whose image of τ is b_τ. Note here that we can take χ such that $\chi(b_0, \tau) = b_\tau$ for $0 \le \tau \le 1$. Then, the induced homeomorphism $F_{r(\tau')} \to F_{r(\tau)}$ sends $b_{\tau'}$ to b_τ.

A closed path r induces a homomorphism $r_* : H^*(F_{r(0)}, \mathbf{Z}) \to H^*(F_{r(0)}, \mathbf{Z})$ since $r(1) = r(0)$ and $F_{r(1)} = F_{r(0)}$. It it called the *monodromy* along r.

The fixed base point is denoted by $b \in C - \Sigma$. For $1 \le i \le t$ let l_i be a path connecting b and c_i contained in $C - \Sigma$ except the ending point c_i. Here, we can assume that if $i \ne j$, l_i and l_j have no common point except the starting point b.

By r_i we denote the closed path which starts from b, goes along l_i until a point just before c_i, then switches to a circle with a small radius with center c_i, proceeds on it in the positive direction round once, and then goes again along l_i in the opposite direction back to the base point b. We can assume that no point in Σ is inside the circular part of r_i except c_i.

Set $E = H^1(F_b, \mathbf{Z})$. For any closed path r in $C - \Sigma$ with the starting point and the ending point b, we have the associated monodromy $r_* : E \to E$. It is a linear isomorphism preserving the intersection form $\cdot$ on E.

Choosing a basis α, β of E with $\alpha \cdot \beta = 1$, we can represent the monodromy r_* by an integral 2 by 2 matrix $\begin{pmatrix} x & y \\ z & w \end{pmatrix}$ with determinant 1. This implies that α is transformed to $x\alpha + z\beta$ and β to $y\alpha + w\beta$ when we go along the closed path r.

We would like to give a remark here. Kodaira's paper on elliptic surfaces (Kodaira [8]) is a very important reference. However, we should note that in it he uses a basis α', β' of E such that $\alpha' \cdot \beta' = -1$ and $\beta' \cdot \alpha' = 1$, and, moreover, that he writes the transposed matrix of the matrix under our representation. (This is because he considers homology instead of cohomology.) If $\begin{pmatrix} x' & y' \\ z' & w' \end{pmatrix}$ is Kodaira's matrix, it implies that α' is transformed to $x'\alpha' + y'\beta'$ and β' to $z'\alpha' + w'\beta'$. Thus, Kodaira's matrix is represented by $\begin{pmatrix} x & y \\ z & w \end{pmatrix} = \begin{pmatrix} x' & -z' \\ -y' & w' \end{pmatrix}$ under our notation.

Proposition 10.3. *Assume that Φ has another singular fiber of type I^* except F_1. Then, there exists an isotropic vector $u \in H^2(Z, \mathbf{Z})$ satisfying the following conditions (1) and (2):*

(1) u is orthogonal to the cohomology class of the section $C_5 = s_0(C)$ and to all the cohomology classes of irreducible components of singular fibers.

(2) *Let $D \subset Z$ be a section of Φ. If D has finite order in the abelian group MW of sections with the unit element C_5, then u is also orthogonal to the cohomology class of D.*

Proof. We can assume that F_2 is of type I_n^*. According to Kodaira [8], the monodromy around c_1 and c_2 can be represented by $\begin{pmatrix} -1 & 0 \\ 0 & -1 \end{pmatrix}$ and $\begin{pmatrix} -1 & 0 \\ n & -1 \end{pmatrix}$ respectively. In particular, the monodromy r_{1*} equals to the multiplication of -1. The monodromy r_{2*} around c_2 has a primitive vector $\gamma \in E$ such that it is sent to $-\gamma$ by r_{2*}. We can represent the class γ by an oriented simple closed curve Γ on F_b. We can assume moreover that Γ does not pass through the intersection point b_0 of $C_5 = s_0(C)$ and F_b. Let $\overline{r} = r_2 r_1$ be the composed closed path made by connecting r_2 after r_1. The path $\overline{r}$ can be regarded as a continuous mapping $\overline{r} : [0,1] \to C - \Sigma$ with $\overline{r}(0) = \overline{r}(1) = b$. Let Γ_τ be the image of Γ by the parallel translation along $\overline{r}$ from b to $\overline{r}(\tau)$. Γ_τ is an oriented simple closed curve on the Riemann surface $F_{\overline{r}(\tau)}$ of genus 1. $\Gamma = \Gamma_0$. We can assume that Γ_τ does not pass through the intersection point b_τ of $F_{\overline{r}(\tau)}$ and C_5. Consider the 2-chain

$$\widetilde{\Gamma} = \bigcup_{0 \le \tau \le 1} \Gamma_\tau \subset Z - \Phi^{-1}(\Sigma) - C_5.$$

The boundary satisfies $\partial \widetilde{\Gamma} = -\Gamma + \Gamma_1$. We have

$$\Gamma, \Gamma_1 \subset F_b - \{b_0\}.$$

The class γ of Γ is transformed to $-\gamma$ when we go along r_1. Next, $-\gamma$ is transformed to γ when we go along r_2. Thus, the class of Γ_1 is also γ. Two curves Γ and Γ_1 are homologous in the compact Riemann surface F_b.

Note here that

$$H_1(F_b) \cong H_1(F_b - \{b_0\}).$$

Thus, there exists a 2-chain Θ in the punctured Riemann surface $F_b - \{b_0\}$ such that $\partial\Theta = \Gamma - \Gamma_1$.

Consider the chain $U = \widetilde{\Gamma} + \Theta$ of the sum. This is a 2-cycle and defines a class $[U] \in H_2(Z, \mathbf{Z})$. Let $u \in H^2(Z, \mathbf{Z})$ be the Poincare-dual class of $[U]$. By construction U intersects neither C_5 nor any component of singular fibers; thus, u is orthogonal to the cohomology classes of them.

Now, let r' be a closed path in $C - \Sigma$ homotope to $\overline{r}$. (Here, we consider homotopy without any base point.) We can repeat the above construction using r' instead of $\overline{r}$. Let U' be the resulting 2-cycle. Then, U' is homologous to U in Z. Any general r' has no intersection point with $\overline{r}$, and U' has no intersection point with U. Thus, we have

$$u^2 = [U] \cdot [U'] = 0.$$

Finally, we would like to show $u \neq 0$. Note that we have the third singular fiber by Lemma 10.2.

◇ Case 1. One of the singular fibers F_i $(3 \le i \le t)$ is not of type I.

We can assume that F_3 is not of type I. F_3 is simply connected.

We can choose a smooth path $q : [0, 1] \to C$ such that $q(0) = c_1$, $q(1) = c_3$, $q(\tau) \notin \Sigma$ for $0 < \tau < 1$, and q intersects $\overline{r} = r_1 r_0$ at only one point c_{11} in a neighborhood of c_1. We can assume further that q and $\overline{r}$ intersect at c_{11} transversally.

Let B_1 and B_3 be sufficiently small non-empty open discs on C with center c_1 and c_3 respectively. We can assume that B_1 is contained inside the circular part of the path r_1. Note that the inverse images $\Phi^{-1}(B_1)$ and $\Phi^{-1}(B_3)$ are simply connected.

Let $\gamma' \in H^1(F_{c_{11}})$ be the image of $\gamma \in H^1(F_b)$ by the parallel translation along r_1. Let $\delta' \in H^1(F_{c_{11}})$ be a primitive vector with $\gamma' \cdot \delta' \neq 0$. The class δ' can be realized by a simple closed curve Δ on the Riemann surface $F_{c_{11}}$. By Δ_τ we denote the image of Δ lying over $q(\tau)$ by the parallel translation along q. Δ_τ is a simple closed curve on the Riemann surface $F_{q(\tau)}$. Choose a sufficiently small positive real number ϵ such that $q(\epsilon) \in B_1$ and $q(1-\epsilon) \in B_3$. Set

$$\widetilde{\Delta} = \bigcup_{\epsilon \leq \tau \leq 1-\epsilon} \Delta_\tau.$$

$\widetilde{\Delta}$ is a 2-chain and the boundary satisfies the following:

$$\partial\widetilde{\Delta} = -\Delta_\epsilon + \Delta_{1-\epsilon}, \qquad \Delta_\epsilon \subset \Phi^{-1}(B_1), \qquad \Delta_{1-\epsilon} \subset \Phi^{-1}(B_3).$$

Since $\Phi^{-1}(B_1)$ and $\Phi^{-1}(B_3)$ are simply connected, we have 2-chains $\widetilde{\Delta}'$ and $\widetilde{\Delta}''$ in $\Phi^{-1}(B_1)$ and $\Phi^{-1}(B_3)$ respectively such that $\partial\widetilde{\Delta}' = \Delta_\epsilon, \partial\widetilde{\Delta}'' = -\Delta_{1-\epsilon}$. The chain $V = \widetilde{\Delta} + \widetilde{\Delta}' + \widetilde{\Delta}''$ of their sum is a 2-cycle and it defines a class $[V] \in H_2(Z, \mathbf{Z})$. Cycles U and V intersect only on the fiber $F_{c_{11}}$ and by construction their intersection number satisfies

$$[U] \cdot [V] = \pm\gamma' \cdot \delta' \neq 0.$$

Thus, we have $[U] \neq 0$ and $u \neq 0$.

◇ Case 2. Every singular fiber F_i for $3 \leq i \leq t$ is of type I.

Set $C' = \mathbf{P}^1$, and let $f : C' \to C$ be the branched double cover branching at c_1 and c_2. Let $\widetilde{Z}'$ be the normalization of the fiber product of Z and C' over C. This $\widetilde{Z}'$ is a branched double cover of Z. The branching locus is the union of 8 disjoint smooth rational curves on Z. Singular fibers F_1 and F_2 contain 4 components with multiplicity 1 respectively. The union of them is the branching locus.

Let c_1' be the unique inverse image of c_1 by f, and c_2' be that of c_2. The fiber over c_1' of the induced morphism $\widetilde{Z}' \to C'$ contains 4 disjoint smooth rational curves which are inverse images of the branching locus, and they are exceptional curves of the first kind on $\widetilde{Z}'$. Similarly the fiber over c_2' contains 4 exceptional curves of the first kind. We can contract these 8 exceptional curves and let Z' denote the resulting smooth surface. We have the induced morphism $\Phi' : Z' \to C'$.

$$\begin{array}{ccccc} Z' & \longleftarrow & \widetilde{Z}' & \longrightarrow & Z \\ \downarrow{\scriptstyle \Phi'} & & \downarrow & & \downarrow{\scriptstyle \Phi} \\ C' & = & C' & \xrightarrow{f} & C \end{array}$$

The section $s_0 : C \to Z$ induces the section $s'_0 : C' \to Z'$. Set $C'_5 = s'_0(C')$. For simplicity by F'_a we denote the fiber $\Phi'^{-1}(a)$ over $a \in C'$. For every point $a \in C'$ the fiber F'_a does not contain an exceptional curve of the first kind. $F'_{c'_1}$ is a smooth elliptic curve and $F'_{c'_2}$ is a (singular) fiber of type I_{2n}. (A fiber of type I_0 is a smooth elliptic curve.) Let Σ' denote the set of critical values for Φ'. $f^{-1}(\{c_3, \dots, c_t\}) \subset \Sigma' \subset f^{-1}(\{c_3, \dots, c_t\}) \cup \{c'_2\}$.

Let b' and b'' be the inverse images of b by f. We fix one b' of two as the base point of C'. Let c'_i be the ending point of the lifting l'_i of l_i with the starting point b' $(3 \leq i \leq t)$. Let c''_i be the ending point of the lifting l''_i of l_i with the starting point b'' $(3 \leq i \leq t)$. For $3 \leq i \leq t$ the fibers $F'_{c'_i}$ and $F'_{c''_i}$ are isomorphic to F_{c_i} and are of type I by definition.

We define paths on C'.

Let l be the lifting of r_1 with the starting point b'. The path l passes through a neighborhood of c'_1 and has the ending point b''. Let l'_2 be the lifting of l_2 with the starting point b'. The ending point of l'_2 is c'_2. We define the closed path r'_2 to be the one which starts from b', goes first along l'_2, then switches to a small circle with center c'_2 just before c'_2, proceeds round once on the circle in positive direction, and again along l'_2 comes back to b'.

For $3 \leq i \leq t$, let r'_i be the lifting of r_i with the starting point b'. Note that the ending point of r'_i is not b'' but b' and r'_i is a closed path. It goes round c'_i just once. Let $\overline{r}_i$ be the lifting of r_i with the starting point b''. Set $r''_i = l^{-1}\overline{r}_i l$. It is the composition of l, $\overline{r}_i$ and l in the inverse direction, has the starting point b', and goes round c''_i just once.

By construction $E' = H^1(F'_{b'}, \mathbf{Z})$ is identified with $E = H^1(F_b, \mathbf{Z})$ through the induced isomorphism f_*.

For each one r of $2t-3$ closed paths

$$r'_2, r'_3, \dots, r'_t, r''_3, \dots, r''_t$$

the monodromy transformation $r_* : E' \to E'$ is defined. Let G be the subgroup in the automorphism group of E' generated by these $2t-3$ monodromy transformations. We define the sheaf $\mathcal{G}$ on C' by

$$\mathcal{G} = j_* j^* R^1 \Phi'_* \mathbf{Z}_{Z'},$$

where $j : C' - \Sigma' \hookrightarrow C'$ is the inclusion morphism and $\mathbf{Z}_{Z'}$ is the constant sheaf on Z' with values in the set of integers $\mathbf{Z}$. By definition we have

$$H^0(C', \mathcal{G}) \cong E'^G = \{\, x \in E' \mid g(x) = x \text{ for every } g \in G \,\}.$$

Lemma 10.4. (Kodaira) $\qquad H^0(C', \mathcal{G}) = 0.$

Proof. Let N be the normal bundle of $C'_5 = s'_0(C')$ in Z'. By $\mathcal{F}$ we denote the pull-back of N by s'_0, which is a sheaf on C'. According to Kodaira [8], we have an injective homomorphism of sheaves $\mathcal{G} \to \mathcal{F}$. By [8] Theorem 12.3 we have $\deg \mathcal{F} = -\chi(\mathcal{O}_{Z'})$. Moreover, by [8] Theorem 12.2 $\chi(\mathcal{O}_{Z'}) > 0$. Thus, $H^0(\mathcal{F}) = 0$ and $H^0(\mathcal{G}) = 0$. Q.E.D.

Now, $\gamma \in E = E'$ is a primitive vector sent to $-\gamma$ by r_{2*}.

Since the monodromy $(r_2')_*$ equals to $(r_{2*})^2$, $\gamma \in E'$ is invariant by $(r_2')_*$. By Lemma 10.4 the following (5) or (6) holds:

$$(r_i')_*\gamma \neq \gamma \text{ for some } i \text{ with } 3 \leq i \leq t. \tag{5}$$

$$(r_i'')_*\gamma \neq \gamma \text{ for some } i \text{ with } 3 \leq i \leq t. \tag{6}$$

Assume that the case (5) takes place. If we regard $\gamma \in E$ by going down to the world of E via $f_* : E' \to E$, we have $(f_*r_i')_*\gamma \neq \gamma$. However, since $f_*r_i' = r_i$ by definition, one knows

$$r_{i*}\gamma \neq \gamma. \tag{7}$$

Assume that the case (6) takes place. Similarly $(f_*r_i'')_*\gamma \neq \gamma$. Here, by definition $f_*r_i'' = r_1^{-1}r_ir_1$. The homomorphism r_{1*} has been the multiplication of -1. Thus, $(f_*r_i'')_* = (r_1)_*^{-1}r_{i*}r_{1*} = r_{i*}$. Therefore, the above (7) holds also in this case.

In the following we fix a number i with $3 \leq i \leq t$ satisfying (7).

The monodromy r_{i*} has the matrix representation in the form $\begin{pmatrix} 1 & 0 \\ -b_i & 1 \end{pmatrix}$. Thus, we have a unique primitive vector $\delta \in E$ with $r_{i*}\delta = \delta$ up to sign. δ is a vanishing cycle of a singular point of the singular fiber F_i and it is defined associated with the path l_i with the starting point b and with the ending point c_i. By (7) $\gamma \neq \pm\delta$. Since γ is also primitive, one knows that the intersection number satisfies

$$\gamma \cdot \delta \neq 0.$$

Now, we can discuss similarly as in Case (1).

Let $q : [0,1] \to C$ be a smooth path as in the following figure:

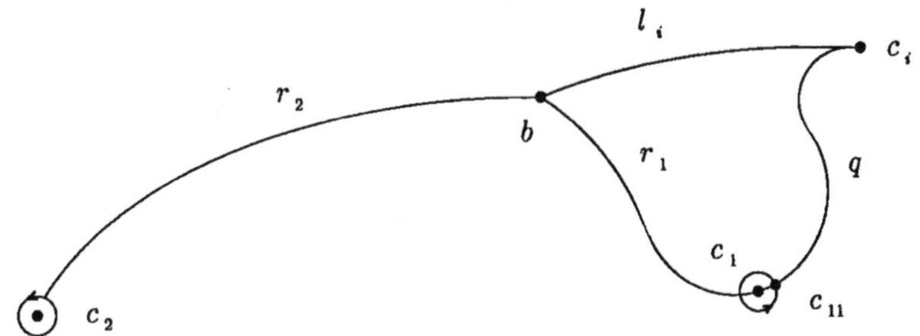

This satisfying the following conditions:

(1) The starting point c_1, the ending point c_i, i.e., $q(0) = c_1$, $q(1) = c_i$.
(2) For $0 < \tau < 1$ $q(\tau) \notin \Sigma$.
(3) The closed path $\overline{r} = r_2r_1$ and q intersect only at one point c_{11} on the circular part of r_1 in a neighborhood of c_1. They intersect at c_{11} transversally.
(4) The composition path $l_il_1^{-1}$ and q are homotope in $(C - \Sigma) \cup \{c_1, c_i\}$ with the starting point c_1 and the ending point c_i keeping fixed.
(5) For some sufficiently small positive real number ϵ_1 $q(\tau)$ and $l_i(\tau)$ coincide for $1-\epsilon_1 \leq \tau \leq 1$.

Let $\delta', \gamma' \in H^1(F_{c_{11}})$ be the image of $\delta, \gamma \in H^1(F_b)$ by the parallel translation along r_1. We can represent the class δ' by an oriented simple closed curve Δ on the Riemann surface $F_{c_{11}}$ of genus 1. Let $\Delta_\tau \subset F_{q(\tau)}$ denote the image of Δ by the parallel translation along q from c_{11} to $q(\tau)$. Let B_i be a sufficiently small non-empty open disc on C with the center c_i, and B_1 be a sufficiently small non-empty open disc on C with the center c_1. Let ϵ be a sufficiently small positive real number with $q(\epsilon) \in B_1$, $q(1-\epsilon) \in B_i$ and $0 < \epsilon < \epsilon_1$.

Set

$$\widetilde{\Delta} = \bigcup_{\epsilon \leq \tau \leq 1-\epsilon} \Delta_\tau.$$

$\widetilde{\Delta}$ is a 2-chain satisfying

$$\partial\widetilde{\Delta} = -\Delta_\epsilon + \Delta_{1-\epsilon}, \qquad \Delta_\epsilon \subset \Phi^{-1}(B_1), \qquad \Delta_{1-\epsilon} \subset \Phi^{-1}(B_i).$$

Now, since the inverse image $\Phi^{-1}(B_1)$ is simply connected, we have a 2-chain $\widetilde{\Delta}'$ in $\Phi^{-1}(B_1)$ such that $\partial\widetilde{\Delta}' = \Delta_\epsilon$.

Next, we consider $\Delta_{1-\epsilon}$. The class δ'' in $H^1(F_{q(1-\epsilon)})$ defined by $\Delta_{1-\epsilon}$ is the image of δ' by the parallel translation from c_{11} along q. Let a be the rounding number around c_1 of the closed path $l_i^{-1}q^\# r_1^\#$ where $q^\#$ denotes the part of q between c_{11} and c_i, and $r_1^\#$ denotes the part of r_1 between b and c_{11}. Since the monodromy around c_1 is the multiplication of -1, we have

$$\delta'' = (-1)^a \delta^\#,$$

where $\delta^\#$ denotes the image of $\delta \in H^1(F_b)$ by the parallel translation along l_i to $q(1-\epsilon) = l_i(1-\epsilon)$. In particular, one knows that $\Delta_{1-\epsilon}$ is a vanishing cycle along l_i. Thus, we have a 2-chain $\widetilde{\Delta}''$ in $\Phi^{-1}(B_i)$ with $\partial\widetilde{\Delta}'' = -\Delta_{1-\epsilon}$. The 2-chain $V = \widetilde{\Delta} + \widetilde{\Delta}' + \widetilde{\Delta}''$ of their sum is a 2-cycle and defines the class $[V] \in H_2(Z, \mathbf{Z})$. The intersection number satisfies

$$[U] \cdot [V] = \pm\gamma' \cdot \delta' = \pm\gamma \cdot \delta \neq 0.$$

Thus, one knows that $[U] \neq 0$ and $u \neq 0$.

Finally, we have to show (2) in Proposition 10.3.

Lemma 10.5. *Assume that an elliptic K3 surface $\Phi : Z \to C$ has a section C_5. Let $\overline{S}$ be the subgroup of the Picard group* Pic(Z) *of Z generated by C_5 and all irreducible components of singular fibers of Φ. By MW we denote the abelian group of sections of Φ with the unit element C_5. Then, we have a group isomorphism*

$$MW \xrightarrow{\sim} \mathrm{Pic}(Z)/\overline{S},$$

which is the composition of the mapping defined by associating a curve $D \in MW$ with the line bundle $\mathcal{O}_Z(D) \in$ Pic(Z) *and the canonical surjective homomorphism* $\mathrm{Pic}(Z) \to \mathrm{Pic}(Z)/\overline{S}$

Proof. Not difficult.

Under the condition in Proposition 10.3 (2) one knows by Lemma 10.5 that D can be written as a rational linear combination of the classes of C_5 and irreducible components of singular fibers. Thus, (2) follows from (1) in the same proposition. This completes the proof of Proposition 10.3. Q.E.D.

Let $S \subset H^2(Z, \mathbf{Z})$ be the subgroup generated by the classes corresponding to the irreducible components of the curve IF at infinity (See Section 2.) and rational smooth curves on Z disjoint from IF. Note that such a smooth curve on Z is a component of a singular fiber. In the case of $J_{3,0}$, $Z_{1,0}$, or $Q_{2,0}$, C_5 is the unique component of IF which is a section. The other components of IF than C_5 are components of singular fibers. Thus, in these cases, we have

$$S \subset \overline{S} \subset \mathrm{Pic}(Z) \subset H^2(Z, \mathbf{Z}),$$

where $\overline{S}$ is the group in Lemma 10.5. Under the isomorphism $H^2(Z, \mathbf{Z}) \xrightarrow{\sim} \Lambda_3$ S corresponds to $P \oplus Q(G)$, the subgroup in S generated by components of IF corresponds to P, and the subgroup in S generated by smooth rational curves on Z disjoint from IF corresponds to $Q(G)$. Reversing the sign of bilinear forms, we have the following by Proposition 10.3:

Corollary 10.6. *Consider the case of $X = J_{3,0}, Z_{1,0}$ or $Q_{2,0}$. Let $G \in PC(X)$. If the corresponding elliptic K3 surface $Z \to C$ has a singular fiber of type I^* apart from the one of type I_0^*, then the orthogonal complement of $Q(G)$ with respect to the corresponding embedding $Q(G) \hookrightarrow \Lambda_3/P$ contains an isotropic vector. In particular, then, there is an isotropic vector in a nice position.*

Note that this corollary is a stronger claim than Proposition 10.1 under the assumption ((1)) in the beginning of this section.

The following corollary is interesting in itself. Here, recall the dual graph of IF in each case $J_{3,0}$, $Z_{1,0}$, $Q_{2,0}$. (Section 2.)

Corollary 10.7. *Consider the case of $X = J_{3,0}$, $Z_{1,0}$, or $Q_{2,0}$. Assume that a Dynkin graph $G \in PC(X)$ has a component of type D_k for some $k \geq 4$. Assume moreover that if $X = Z_{1,0}$, then $k \neq 6$. Then, for any full embedding $Q(G) \hookrightarrow \Lambda_3/P$ the orthogonal complement of the image contains an isotropic vector, and, thus, in particular, the equivalent conditions [II](A) and [II](B) in Theorem 9.8 are satisfied.*

Proof. By assumption the corresponding K3 surface Z contains a combination $\mathcal{E}_1$ of k smooth rational curves disjoint from IF whose dual graph is the Dynkin graph of type D_k. $\mathcal{E}_1$ is contained in some singular fiber F_i of $Z \to C$ with $F_i \neq F_1$. If F_i is of type I^*, the claim follows from Corollary 10.6. Otherwise F_i is of type II^*, III^*, or IV^*, and it contains several components of the curve IF at infinity. Since the union of components of F_i disjoint from IF is a combination of type D_k, one knows $X = Z_{1,0}$, F_i is of type III^*, and $k = 6$. Q.E.D.

11
A transcendental cycle with a positive self-intersection number

In this section we treat the case where the corresponding elliptic K3 surface $\Phi : Z \to C$ has a singular fiber of type II^*, III^*, or IV^*. Recall that by C_5 we denote a component of the curve IF at infinity which is a section of Φ. By $[\Xi]$ we denote the homology class of a cycle Ξ or the cohomology class of Ξ.

Proposition 11.1. *Assume that the elliptic K3 surface Φ with a section C_5 has a singular fiber of type II^*, III^*, or IV^* apart from the singular fiber F_1 of type I_0^*. Then, there exists a cohomology class $\xi \in H^2(Z, \mathbf{Z})$ satisfying the following conditions (1)–(4):*

(1) $\xi^2 = +4$.

(2) The class ξ is orthogonal to the class of C_5 and to all the classes of irreducible components of singular fibers of Φ.

(3) For some two irreducible components C' and C'' of F_1 with multiplicity 1 without intersection with C_5, we can write $\xi + [C'] + [C''] = 2\eta$ for some $\eta \in H^2(Z, \mathbf{Z})$.

(4) Let $D \in MW$ be an arbitrary section of Φ. If D has finite order in the abelian group MW with the unit element C_5, then ξ is orthogonal to the cohomology class of D.

Proof. We assume that the singular fiber F_2 over $c_2 \in \Sigma$ is of type II^*, III^*, or IV^*. Consider the paths on C as in the following figure:

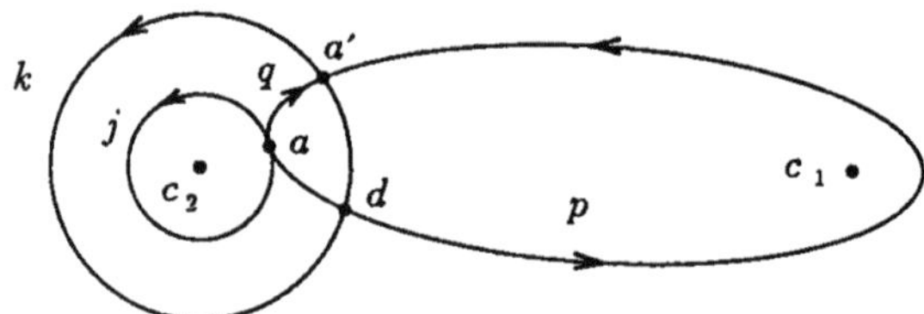

The path j and k go on circles with center c_2 with a sufficiently small radius in the positive direction. j has a shorter radius than k. The point a and a' lie on the path j and k respectively. These are regarded as the starting point and the ending point of the respective closed path. The smooth path q has the starting point a and the ending point a', and it does not intersect j and k except at a and a'. The path p is a smooth path which has also the starting point a and the ending point a', it has no intersection with j except at a, and p and k have a unique intersection point d except the ending point a'. At d they intersect transversally. The composed path $r = q^{-1}p$ goes round the point c_1 just once in the positive direction. The inner domain surrounded by k contains no points in the set Σ of critical values of Φ except c_2, and the inner domain surrounded by r contains no points in Σ except c_1.

Let α, β be a basis of $E = H^1(F_a, \mathbf{Z})$ with $\alpha \cdot \beta = 1$ such that the associated matrix of the monodromy j_* has the following form:

$$\begin{pmatrix} 0 & -1 \\ 1 & 1 \end{pmatrix} \text{ (case of type } II^*\text{)} \qquad \begin{pmatrix} 0 & -1 \\ 1 & 0 \end{pmatrix} \text{ (case of type } III^*\text{)}$$

$$\begin{pmatrix} -1 & -1 \\ 1 & 0 \end{pmatrix} \text{ (case of type } IV^*\text{)}$$

Let A be an oriented simple closed curve on the Riemann surface F_a representing the cohomology class α. We can assume that A has no intersection with C_5. The closed path $j : [0,1] \to C - \Sigma$ with $j(0) = j(1) = a$ defines the parallel translation $F_a \to F_{j(\tau)}$ for $0 \leq \tau \leq 1$. By $A_\tau \subset F_{j(\tau)}$ we denote the image of A by the parallel translation. For every τ with $0 \leq \tau \leq 1$ we can assume that A_τ has no intersection with C_5. We can define a 2-chain $\widetilde{J}$ by the following:

$$\widetilde{J} = \bigcup_{0 \leq \tau \leq 1} A_\tau \subset Z - \Phi^{-1}(\Sigma) - C_5.$$

The boundary satisfies $\partial \widetilde{J} = -A + A_1$. The cohomology class of $\partial \widetilde{J}$ in $H^1(F_a)$ is equal to $j_* \alpha - \alpha$.

Next, let $A' \subset F_{a'}$ be the image of A by the parallel translation along q. We can choose such an A' that it has no intersection with C_5. Also, the closed path $k : [0,1] \to C - \Sigma$ with $k(0) = k(1) = a'$ defines the parallel translation. By $A'_\tau \subset F_{k(\tau)}$ we denote the image of A' by the parallel translation along k. We can assume that for every τ with $0 \leq \tau \leq 1$ A'_τ does not intersect C_5. A 2-chain $\widetilde{K}$ is defined by

$$\widetilde{K} = \bigcup_{0 \leq \tau \leq 1} A'_\tau \subset Z - \Phi^{-1}(\Sigma) - C_5.$$

$\partial \widetilde{K} = -A' + A'_1$. The cohomology class of $\partial \widetilde{K}$ in $H^1(F_{a'})$ is the image of $j_* \alpha - \alpha$ by the parallel translation $q_* : H^1(F_a) \to H^1(F_{a'})$.

Now, we have an oriented simple closed curve $\Gamma \subset F_a$ representing the cohomology class $j_* \alpha - \alpha$, since $j_* \alpha - \alpha$ is primitive in E. By $\Gamma_\tau \subset F_{p(\tau)}$ we denote the image of Γ by the parallel translation along the path $p : [0,1] \to C - \Sigma$ with $p(0) = a$, $p(1) = a'$. For every $0 \leq \tau \leq 1$ we choose such a Γ_τ that it has no intersection with C_5. Set

$$\widetilde{P} = \bigcup_{0 \leq \tau \leq 1} \Gamma_\tau.$$

The boundary satisfies $\partial \widetilde{P} = -\Gamma + \Gamma_1$.

Here, the cohomology class of $-A + A_1 - \Gamma$ is zero, and the support of this cycle does not pass through the intersection point a_0 of F_a and C_5. Since $H_1(F_a) \cong H_1(F_a - \{a_0\})$, we have a 2-chain Θ in the punctured Riemann surface $F_a - \{a_0\}$ such that $\partial \Theta = A - A_1 + \Gamma$.

Consider the image $\Gamma' \subset F_a$ of $\Gamma_1 \subset F_{a'}$ by the parallel translation along the inverse of q. The homology class of Γ' coincides with the homology class of Γ applied to by the monodromy r_* around c_1; thus, it is $-(j_* \alpha - \alpha)$. It follows from this that the homology class of $-A' + A'_1 + \Gamma_1$ is zero in $F_{a'}$. There exists a 2-chain Θ' in $F_{a'}$ with $\partial \Theta' = A' - A'_1 - \Gamma_1$ such that the support of Θ' does not pass through the intersection point a'_0 of $F_{a'}$ and C_5.

The chain $\Xi = \widetilde{J} + \widetilde{K} + \widetilde{P} + \Theta + \Theta'$ of their sum is a 2-cycle, and defines the homology class $[\Xi] \in H_2(Z, \mathbf{Z})$. Let $\xi \in H^2(Z, \mathbf{Z})$ be the Poincare dual class of $[\Xi]$. By construction ξ satisfies the condition (2) obviously.

In order to see the condition (3) we need several constructions. First, let $\Gamma^*_\tau \subset F_{q(\tau)}$ be the image of Γ by the parallel translation along q. Adjusting the parallel translation along p, we can assume that Γ_1^* coincides with Γ_1 except that the orientation is opposite. Second, we choose a smooth map $T : [0,1] \times [0,1] \to C - \Sigma$ such that $T(\tau, 0) = j(\tau)$, $T(\tau, 1) = k(\tau)$, and $T(0, \sigma) = T(1, \sigma) = q(\sigma)$ for $\tau, \sigma \in [0,1]$. Denoting $\Phi^{-1}(T(\tau, \sigma)) = F_{\tau,\sigma}$, we have the parallel translation $F_a = F_{0,0} \to F_{\tau,\sigma}$ associated with T. By $A_{\tau,\sigma}$ and by $\Gamma_{\tau,\sigma}$ we denote the image of A and Γ by the translation associated with T respectively. We can assume $A_\tau = A_{\tau,0}$, and $A'_\tau = A_{\tau,1}$. Setting $\widetilde{Q} = -\bigcup_{0\leq\sigma\leq 1} A_{0,\sigma} + \bigcup_{0\leq\sigma\leq 1} A_{1,\sigma}$, and $\overline{Q} = \bigcup_{0\leq\sigma\leq 1} \Gamma^*_\sigma$, we divide Ξ into three parts. Set $\Xi_1 = \widetilde{J} + \widetilde{K} + \widetilde{Q}$, $\Xi_2 = -\widetilde{Q} + \overline{Q} + \Theta + \Theta'$, and $\Xi_3 = \widetilde{P} - \overline{Q}$. Obviously, $\Xi = \Xi_1 + \Xi_2 + \Xi_3$. However, Ξ_i $(i = 1, 2, 3)$ is not a cycle if we use $\mathbf{Z}$ as the coefficients.

Let us use $\mathbf{Z}/2$-coefficients here. We consider homology groups over $\mathbf{Z}/2$. Then, Ξ_i $(i = 1, 2, 3)$ are cycles. Besides, $\partial\Pi = \Xi_1$ for a 3-chain $\Pi = \bigcup_{0\leq\tau\leq 1, 0\leq\sigma\leq 1} A_{\tau,\sigma}$. Let Θ_σ be a continuous family of 2-chains such that $\partial\Theta_\sigma = A_{0,\sigma} - A_{1,\sigma} + \Gamma^*_\sigma$ as chains in $F_{q(\sigma)}$, and $\Theta_0 = \Theta$, $\Theta_1 = \Theta'$. Then, we have $\partial\widetilde{\Theta} = \Xi_2$ for the 3-chain $\widetilde{\Theta} = \bigcup_{0\leq\sigma\leq 1} \Theta_\sigma$. Consequently, one knows that $[\Xi] = [\Xi_3]$ in $H_2(Z, \mathbf{Z}/2)$.

Now, let $U \subset C$ be a contractible neighborhood of the point c_1 containing the path $r = q^{-1}p$ and not containing any point in Σ except c_1. The class $[\Xi_3]$ is defined in $H_2(\Phi^{-1}(U), \mathbf{Z}/2)$. Let $C_0, \ldots, C_4$ be the irreducible components of the central singular fiber F_1. We assume that C_0 has multiplicity 2, and C_4 intersects C_5. By construction the intersection $[\Xi_3] \cdot [C_i] = 0$ for $0 \leq i \leq 4$.

On the other hand, since $H_2(\Phi^{-1}(U), \mathbf{Z}/2) \cong H_2(F_1, \mathbf{Z}/2) = \sum_{i=0}^4 \mathbf{Z}/2[C_i]$, we can express $[\Xi_3]$ as the linear combination of $[C_i]$'s. Setting $u_0 = 2[C_0] + \sum_{i=1}^4 [C_i]$, we have elements $\epsilon, \delta_0, \ldots, \delta_3 \in \mathbf{Z}/2$ with $[\Xi_3] = \epsilon u_0 + \sum_{i=0}^3 \delta_i [C_i]$. Then, $\epsilon = \epsilon u_0 \cdot [C_5] = [\Xi_3] \cdot [C_5] = [\Xi] \cdot [C_5] = 0$. $\delta_0 = [\Xi_3] \cdot [C_1] = 0$, and $\delta_1 + \delta_2 + \delta_3 = [\Xi_3] \cdot [C_0] = 0$. It implies that either for some two C', C'' of C_1, C_2, C_3, $[\Xi] = [\Xi_3] = [C'] + [C'']$, or $[\Xi] = [\Xi_3] = 0$. If the first case takes place, then by the universal coefficient theorem $H^2(Z, \mathbf{Z}/2) = H^2(Z, \mathbf{Z}) \otimes \mathbf{Z}/2$, we have the condition (3). If the second case takes place, we can write $\xi = 2\eta$ for some $\eta \in H^2(Z, \mathbf{Z})$. We assume here the condition (1) $[\Xi]^2 = 4$. Then, we have $\eta^2 = 1$, which contradicts that $H^2(Z, \mathbf{Z})$ is an even lattice. Note that in the following proof of the condition (1) we do not use the condition (3). Thus, we can complete the proof of condition (3).

As for the condition (4), it follows from the condition (2) as in Proposition 1.2.

The condition (1) is remaining. To compute the self-intersection number we consider the small perturbation j', k', p' of the paths j, k, p as in the following figure:

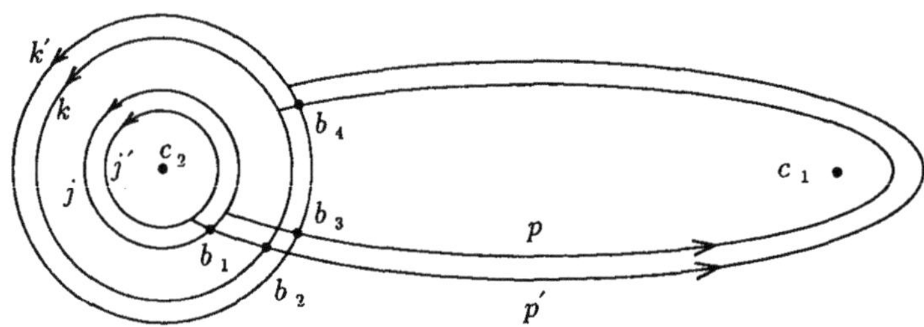

Four intersection points appear. Let b_1 be the intersection point of p' and j, b_2 be that of p' and k, and b_3, b_4 be those of k' and p. We assume that b_3 is nearer to d than b_4. Let Ξ' be the 2-cycle associated with j', k' and p', whose construction is similar to the case for Ξ. We can check that Ξ and Ξ' are homologous. Thus, the self-intersection number ξ^2 is equal to the intersection number of Ξ and Ξ'.

The intersection points of Ξ and Ξ' are contained in $\bigcup_{\nu=1}^{4} F_{b_\nu}$. After computing the local intersection number in the neighborhood of F_{b_ν}, we can take the sum.

First, we consider the neighborhood of F_{b_1}. We assume that $j(\tau_0) = b_1$ and $p'(\tau_1) = b_1$. Let B_1 be a sufficiently small neighborhood of b_1 in C. The inverse image $\Phi^{-1}(B_1)$ can be identified with the product $B_1 \times F_{b_1}$. Let p'_B be the part in B_1 of the path p', and j_B be the part in B_1 of j. Ξ can be identified locally with $j_B \times A_{\tau_0}$, while Ξ' can be identified with $p'_B \times \Gamma'_{\tau_1}$. Under the identification by the parallel translation along j $\tau_0 \leq \tau \leq 1$, the homology class of A_{τ_0} coincides with $j_*\alpha$, while that of Γ'_{τ_1} is $j_*\alpha - \alpha$. Thus,

$$\mathrm{int}(A_{\tau_0},\ \Gamma'_{\tau_1}) = j_*\alpha \cdot (j_*\alpha - \alpha) = 1.$$

Here, $\mathrm{int}(X,\ Y)$ stands for the local intersection number of X and Y. (Note that the order of two 1-cycles and the sign.) In $\Phi^{-1}(B_1)$, we have

$$\begin{aligned}\mathrm{int}(\Xi,\ \Xi') &= \mathrm{int}(j_B \times A_{\tau_0},\ p'_B \times \Gamma'_{\tau_1})\\ &= -\mathrm{int}(j_B,\ p'_B)\,\mathrm{int}(A_{\tau_0},\ \Gamma'_{\tau_1})\\ &= -(-1) \times 1\\ &= +1.\end{aligned}$$

Similarly, one knows that for each ν ($\nu = 1,\ 2,\ 3,\ 4$), Ξ and Ξ' have the local intersection number $+1$ in the neighborhood of F_{b_ν}.

Therefore, $\xi^2 = +4$. Q.E.D.

We apply Proposition 11.1 and show a nice isotropic vector.

We consider the case for $J_{3,0}$ first. Let $G \in PC(J_{2,0})$. Under our assumption in this section G contains a component of type E. Let $S = P \oplus Q(G) \hookrightarrow \Lambda_3$ be the corresponding lattice embedding. By Proposition 11.1 we have a vector $\xi \in \Lambda_3$ orthogonal to S such that $\xi^2 = -4$. (Note that when we move from $H^2(Z, \mathbf{Z})$ to Λ_3, we reverse the sign of the bilinear form.) The induced embedding $Q(G) \hookrightarrow \Lambda_3/P$ is full, and the image is contained in the orthogonal complement L of $\mathbf{Z}\xi$ in Λ_3/P. L has signature $(14, 1)$; thus, we can define the Coxeter-Vinberg graph for L. Since $Q(G)$ is full even in L, G is a subgraph of the Coxeter-Vinberg graph for L.

Therefore, we would like to draw the Coxeter-Vinberg graph for L.

Set $R = P \oplus \mathbf{Z}\xi$. The discriminant group of R is $R^*/R \cong (\mathbf{Z}/2 + \mathbf{Z}/2) \oplus \mathbf{Z}/4$. For $\overline{\alpha} = (a_1, a_2, b) \in R^*/R$, the discriminant quadratic form can be written $q_R(\overline{\alpha}) \equiv a_1^2 + a_1 a_2 + a_2^2 - (b^2/4) \bmod 2\mathbf{Z}$. Thus, $q_R \equiv 0 \Leftrightarrow \overline{\alpha} = (0,0,0), (1,0,2), (0,1,2)$, or $(1,1,2)$. One knows, in particular, that any isotropic subgroup in R^*/R has order≤ 2. It implies $[\widetilde{R} : R] \leq 2$ for the primitive hull $\widetilde{R}$ of R in Λ_3. On the other hand, the condition (3) in Proposition 11.1 implies that $[\widetilde{R} : R]$ is a multiple of 2. Consequently, one has $[\widetilde{R} : R] = 2$. After some calculation one has $\widetilde{R}^*/\widetilde{R} \cong \mathbf{Z}/4$ and $q_{\widetilde{R}}(c) \equiv 3c^2/4 \bmod 2\mathbf{Z}$. (See Lemma 4.4.)

Let M be the orthogonal complement of R in Λ_3. $M^*/M \cong \tilde{R}^*/\tilde{R} \cong \mathbf{Z}/4$ by Lemma 5.5.

Next, let K be the orthogonal complement of $\mathbf{Z}\xi$ in Λ_3. The group K^*/K has order 4. The quotient K/P can be identified with L, and we can regard M as its subgroup with finite index. Since $M \oplus P \subset K$ and since the group $(M \oplus P)^*/(M \oplus P)$ has order 16, one has $[K/P : M] = [K : M \oplus P] = 2$.

Consequently, $L = K/P$ is a unimodular lattice of signature $(14, 1)$. It is known that such a lattice is unique up to isomorphisms (Milnor-Husemoller [10]), and we can find its Coxeter-Vinberg graph in Vinberg [20], which is as in the right figure. By γ_i we denote the fundamental root in L associated with the vertex in the right graph with the attached number i.

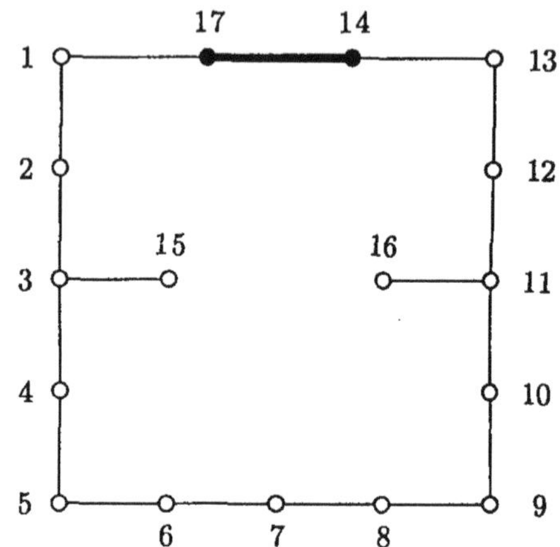

The Coxeter-Vinberg graph for L.

Proposition 11.2. *Consider the case for $J_{2,0}$. Let $G \in PC(J_{2,0})$, and G_0 be an arbitrary component of G. Assume that the corresponding elliptic K3 surface $\Phi : Z \to C$ has a singular fiber of type II^*, III^* or IV^*. We regard $Q(G)$ as a submodule of Λ_3/P by the corresponding embedding. Then, there exists an isotropic element u in Λ_3/P satisfying either the following condition (1) or (2):*

(1) u is orthogonal to $Q(G)$.

(2) For some root basis $\Delta \subset Q(G)$ there is a long root $\theta \in \Delta - \Delta_0$ such that $u \cdot \theta = 1$ and $u \cdot \alpha = 0$ for every $\alpha \in \Delta$ with $\alpha \neq \theta$, where Δ_0 denotes the component of Δ corresponding to G_0.

Proof. We can assume that the root basis $\Delta \subset Q(G)$ is contained in the set $\{\gamma_i \mid 1 \leq i \leq 17\}$ of the fundamental roots of L. Obviously, $\gamma_{14}, \gamma_{17} \notin \Delta$, since they are short roots. Moreover, either γ_6, γ_7 or γ_8 does not belong to Δ, since G contains a subgraph of type E. Thus, one can conclude that either $\gamma_5 \notin \Delta_0$ or $\gamma_9 \notin \Delta_0$, since the graph corresponding to Δ_0 is connected.

First, we consider the case where $\gamma_5 \notin \Delta_0$. Set $u_1 = -(2\gamma_1 + 2\gamma_2 + 2\gamma_3 + \gamma_4 + \gamma_{15} + 2\gamma_{17})$. We can check that u_1 is an isotropic vector in $L \subset \Lambda_3/P$. Moreover,

$$u \cdot \gamma_i = 0 \text{ for } 1 \leq i \leq 17,\ i \neq 5,\ 14$$
$$u \cdot \gamma_5 = u \cdot \gamma_{14} = 1$$

Set $\theta = \gamma_5$. If $\gamma_5 \notin \Delta$, then u_1 satisfies the above (1), while if $\gamma_5 \in \Delta$, then it satisfies (2).

The second case $\gamma_9 \notin \Delta_0$ is similar. By the symmetry of the graph it is obvious that the element $u_2 = -(\gamma_{10} + 2\gamma_{11} + 2\gamma_{12} + 2\gamma_{13} + 2\gamma_{14} + \gamma_{16})$ satisfies conditions.

Q.E.D.

We proceed to the case of $Z_{1,0}$ and $Q_{2,0}$. By X we denote either $Z_{1,0}$ or $Q_{2,0}$. Let $G \in PC(X)$. We consider the corresponding elliptic K3 surface $Z \to C$. In this case, there is a unique singular fiber containing a component of IF apart from F_1. We assume that a singular fiber F_2 contains a component of IF.

We would like to reduce our case to the case of $J_{3,0}$. Now, let $\overline{IF}$ denote the union of F_1 and $C_5 = s_0(C)$. $\overline{IF}$ is a union of some components of IF, and is same as IF in the case of $J_{3,0}$. Let $\mathcal{E}$ (resp. $\overline{\mathcal{E}}$) denote the union of irreducible components disjoint from IF (resp. $\overline{IF}$) of singular fibers F_i, $2 \leq i \leq t$. One has

$$\mathcal{E} \cap (\bigcup_{i=3}^{t} F_i) = \overline{\mathcal{E}} \cap (\bigcup_{i=3}^{t} F_i) \quad \text{and} \quad (\mathcal{E} \cap F_2) \cup T \subset (\overline{\mathcal{E}} \cap F_2) \cup B = F_2.$$

Here, B denotes the component of F_2 intersecting C_5, and T denotes the union of components of IF not contained in $\overline{IF}$. Note, in particular, that B consists of a unique component, and that it has multiplicity 1 as a component of F_2.

Assume that for some component E_1 of $\mathcal{E}$ contained in $\bigcup_{i=3}^{t} F_i$ the cohomology class $U \in H^2(Z)$ satisfies the following:

$[E_1] \cdot U = -1$, $[C_i] \cdot U = 0$ for every component C_i of $\overline{IF}$, and $[E] \cdot U = 0$ for every component E of $\overline{\mathcal{E}}$ with $E \neq E_1$.

Then, U is orthogonal to the general fiber of Φ and orthogonal to all components of F_2 except possibly one component B. However, U is orthogonal also to B since the class of B can be written as an integral linear combination of the class of the general fiber and the classes of other components of F_2. Thus, one can conclude that U is orthogonal to all components of IF and to all components of $\mathcal{E}$ with a unique exception E_1.

By the same reason, if U is orthogonal to all components of $\overline{IF}$ and to all components of $\overline{\mathcal{E}}$; then, it is orthogonal to all components of IF and to all components of $\mathcal{E}$.

By translating this fact to the lattice theory we can get the proof.

By $\overline{P}$ we denote the sublattice of P generated by the part of the basis corresponding to the components in $\overline{IF}$. (Recall that P has a basis which has one-to-one correspondence with the components of IF.) rank $\overline{P} = 6$ and $\overline{P} = P_0' \oplus H_0$ in the notation in Section 2 and 3. Moreover, $\overline{P}$ is isomorphic to P in the case of $J_{3,0}$. Let G (resp. $\overline{G}$) be the dual graph of $\mathcal{E}$ (resp. $\overline{\mathcal{E}}$), and G_0 (resp. $\overline{G}_0$) be the sub-dual-graph of G (resp. $\overline{G}$) corresponding to $\mathcal{E} \cap F_2$ (resp. $\overline{\mathcal{E}} \cap F_2$). $\overline{G}_0$ has a unique component and $G - G_0 = \overline{G} - \overline{G}_0 = G^*$. By the above we have $\overline{P} \oplus Q(\overline{G}_0) \supset P \oplus Q(G_0)$. The lattice $\overline{P} \oplus Q(\overline{G}_0) \oplus Q(G^*)$ has the embedding into Λ_3 coming from the geometric situation. By Proposition 11.2 one has an isotropic vector $u \in \Lambda_3/\overline{P}$ satisfying either (1) or (2) in Proposition 11.2 for the pair of graphs $\overline{G}$ and $\overline{G}_0$. u is orthogonal to $Q(\overline{G}_0)$ in any case. The orthogonal complement of $\overline{P}$ in Λ_3 contains an isotropic vector $\widetilde{u}$ which corresponds to u under the quotient homomorphism $\Lambda_3 \to \Lambda_3/\overline{P}$. $\widetilde{u}$ is orthogonal to $P \oplus Q(G_0)$. The image of $\widetilde{u}$ by the quotient homomorphism $\Lambda_3 \to \Lambda_3/P$ is an isotropic vector in a nice position.

We have shown Proposition 10.1 for $X = J_{3,0}$, $Z_{1,0}$ or $Q_{2,0}$ under the assumption ((2)) in the beginning of Section 10.

12 Nikulin's lattice theory

In this section we explain briefly Nikulin's theory on even lattices (Nikulin [11]). His theory is applied in Section 13, 14, etc.

By M_p we denote the p-Sylow subgroup of a finite abelian group M, and $l(M)$ stands for the minimum number of generators of M.

Let L be an integral even lattice. The discriminant quadratic form $q_L : L^*/L \to \mathbf{Q}/2\mathbf{Z}$ introduced in Section 3 plays an important role in his theory. The combination of this concept and the Hasse principle is the heart of his theory.

Let p be a prime number and $\mathbf{Z}_p$ be the ring of p-adic integers. A free module V over $\mathbf{Z}_p$ of finite rank equipped with a symmetric bilinear form with values in $\mathbf{Z}_p$ is called a *p-adic lattice*. It is even, if $x^2 \in 2\mathbf{Z}_p$ for any $x \in V$. Otherwise it is said to be *odd*. For a non-degenerate even p-adic lattice V the discriminant quadratic form $q_V : V^*/V \to \mathbf{Q}_p/2\mathbf{Z}_p$ is defined by $q_V(\xi \bmod V) \equiv \xi^2 \bmod 2\mathbf{Z}_p$. Here, $V^* = \mathrm{Hom}_{\mathbf{Z}_p}(V, \mathbf{Z}_p)$ is identified with the submodule $\{\xi \in V \otimes_{\mathbf{Z}_p} \mathbf{Q}_p \mid (\xi, \eta) \in \mathbf{Z}_p \text{ for every } \eta \in V\}$. Note that we have a natural inclusion $\mathbf{Q}_p/2\mathbf{Z}_p \subset \mathbf{Q}/2\mathbf{Z}$ such that the image is the subgroup of elements whose order is a power of p and $\mathbf{Q}/2\mathbf{Z} \cong \bigoplus_{\text{all } p} \mathbf{Q}_p/2\mathbf{Z}_p$.

The following follows from the definition:

Lemma 12.1. *Let L be a non-degenerate integral even lattice. The discriminant quadratic form of the p-adic lattice $L \otimes \mathbf{Z}_p$ is equal to the restriction of the discriminant quadratic form q_L of L to the p-Sylow subgroup $(L^*/L)_p$ of the discriminant group L^*/L of L.*

Deeper results are the following:

Lemma 12.2. (Nikulin [11] Corollary 1.9.3) *An even non-degenerate p-adic lattice V is uniquely determined up to isomorphisms by* rank V, *the discriminant of $V \otimes \mathbf{Q}_p$ defined in $\mathbf{Q}_p/\mathbf{Q}_p^{*2}$, and the discriminant quadratic form q_V.*

Lemma 12.3. (Nikulin [11] Theorem 1.9.1) *An even non-degenerate p-adic lattice V with* rank $V = l(V^*/V)$ *is uniquely determined up to isomorphisms only by the discriminant quadratic form q_V except in the case when $p = 2$ and V^*/V has the orthogonal decomposition in the form $V^*/V \cong M_1 \oplus M_2$, $M_2 \cong \mathbf{Z}/2$. (The concept of orthogonality is defined with respect to the discriminant bilinear form.)*

We have applied the following in Section 5 Proposition 5.2. Though the original version in Nikulin [11] is much more exact, we quote it in a weaker form for simplicity.

Lemma 12.4. (Nikulin [11] Theorem 1.12.2, Theorem 1.14.4) *Let L be an even non-degenerate integral lattice of signature (t_+, t_-) and Λ be an indefinite even unimodular lattice of signature (l_+, l_-).*
(1) If $l_+ \geq t_+$, $l_- \geq t_-$ and rank $\Lambda -$ rank $L \geq l(L^*/L) + 1$, *then there exists a primitive embedding $L \hookrightarrow \Lambda$.*
(2) If $l_+ > t_+$, $l_- > t_-$ and rank $\Lambda -$ rank $L \geq l(L^*/L) + 2$, *then a primitive embedding $L \hookrightarrow \Lambda$ is unique up to orthogonal transformations of Λ.*

13
Combinations of graphs of type A

In this section we assume that all singular fibers of the elliptic K3 surface $\Phi : Z \to C$ are of type I, II, III or IV except the unique exception F_1 of type I_0^*. Every component of the Dynkin graph corresponding to this surface is of type A.

We will show Proposition 10.1. However, under this assumption it is extremely difficult to construct a nice transcendental cycle uniformly applicable to all examples. We have to adopt another strategy.

We have a remark here. Assume that a Dynkin graph G with no vertex corresponding to a short root can be made from one of the essential basic graphs by elementary or tie transformations applied twice. Then, we can construct a full embedding $Q(G) \hookrightarrow \Lambda_3/P$ with a primitive isotropic vector in a nice position, and we can conclude $G \in PC(X)$. This is a consequence of the theories in Section 6 and 9. (Recall Theorem 6.3 (2) and Theorem 6.4 (2).) Note here that the constructed embedding may not be equivalent to the given embedding. However, we need not claim that *every* full embedding has a nice isotropic vector to show Proposition 10.1. It is enough to show that if $G \in PC(X)$, and if G is not an exception, then G can be made from an essential basic Dynkin graph by two of elementary or tie transformations.

By r we denote the number of vertices in the Dynkin graph under consideration.

We begin with the case $X = J_{3,0}$.

Consider the above elliptic K3 surface $\Phi : Z \to C$. Let A_{n_i} be the dual graph of the components disjoint from C_5 of the singular fiber F_i for $2 \le i \le t$. (Here, A_0 stands for the empty graph $\emptyset$.) We have $G = \sum_{i=2}^{t} A_{n_i}$. We have the following proposition.

Proposition 13.1. *We consider the case $X = J_{3,0}$ under the above assumption.*
(1) If the Picard number ρ of Z satisfies $\rho = 6 + r$, then the group MW of sections is finite.
(2) $\nu(G) \le 18 - r$, where $\nu(G)$ denotes the number of components of G.
(3) $r \le 13$.
(4) Set $N(p) = \{i \mid 2 \le i \le t,\ n_i + 1 \equiv 0 \pmod p\}$. If the corresponding embedding $P \oplus Q(G) \hookrightarrow \Lambda_3$ is not primitive, then there is a subset $M \subset N(2)$ such that $\sum_{i \in M}(n_i + 1) = 12$.
(5) For any odd prime number p $N(p)$ contains at most $15 - r$ elements.

Proof. (1) Since $r = \sum_{i=2}^{t} n_i$, we have $a + \sum_{i=2}^{t}(m(F_i) - n_i - 1) = 0$ by the equality (2) at the beginning of Section 10. By definition $m(F_i) > n_i$; thus, $a = 0$, which implies MW is finite.
(2) Without loss of generality we can assume $\rho = 6 + r$ by Theorem 2.2. In the equality (3) at the beginning of Section 10, we can substitute $a = 0$, $t - t_1 = 1 + \nu(II) + \nu(III) + \nu(IV)$, $t_1 = \nu(G) - \nu(III) - \nu(IV) + \nu(I_1)$. Thus, $\nu(G) = 18 - r - \{2\nu(II) + \nu(III) + \nu(IV) + \nu(I_1)\}$.
(3) If $r > 13$, then $r = 14$ since $20 \ge \rho \ge 6 + r$. Assuming $r = 14$, we will deduce a contradiction.

Then, first, we have $\rho = 20 = 6 + r$. Thus, $a = 0$ by (1). If the functional invariant J is not constant, then we have

$$0 = 20 - \rho + a \geq \nu(I_0^*) + \nu(II) + \nu(III) + \nu(IV) \geq \nu(I_0^*) = 1,$$

by the inequality (4) in Section 10, which is a contradiction. Thus, J is constant and all singular fibers are of type I_0^*, II, III, or IV. This implies that all components of G are of type A_1 or A_2; thus, $\nu(G) \geq 7$ since G has 14 vertices. On the other hand, by (2) we have $\nu(G) \leq 4$, which is a contradiction.
(4) Let S be the subgroup of $\mathrm{Pic}(Z)$ generated by the class of C_5 and the classes of all components of singular fibers. Since in our case under the isomorphism $H^2(Z, \mathbf{Z}) \xrightarrow{\sim} \Lambda_3$ S corresponds exactly to $P \oplus Q(G)$, the assumption in (4) implies $\widetilde{S}/S \neq 0$ for the primitive hull $\widetilde{S}$ of S in $H^2(Z, \mathbf{Z})$.

On the other hand, by Lemma 10.5 the quotient $\mathrm{Pic}(Z)/S$ is isomorphic to the group MW of sections. Here, $\widetilde{S} \subset \mathrm{Pic}(Z)$ since $\mathrm{Pic}(Z)$ is always primitive in $H^2(Z, \mathbf{Z})$, and one knows that $\widetilde{S}/S$ is isomorphic to the subgroup $\mathrm{Tor}\, MW$ of MW consisting of all elements with finite order. Thus, we have a section $C' \in \mathrm{Tor}\, MW$ with $C' \neq C_5$. $[C'] \in \widetilde{S}$.

Now, let S_i denote the subgroup of S generated by the classes of components disjoint from IF of the singular fiber F_i. We have $S = (\mathbf{Z}[F] + \mathbf{Z}[F + C_5]) \oplus \bigoplus_{i=1}^t S_i$, where F denotes a general fiber of Φ. Thus, we can write

$$[C'] = m[F] + [F + C_5] + \sum_{i=1}^{t} \chi_i$$

for some $m \in \mathbf{Z}$, $\chi_i \in S_i^*$.

Here, we recall some general facts on elliptic surfaces, which may be to be added in the beginning of section 10. (Kodaira [8], Shioda [14])

- Let $Z^{\#}$ be the set of points on Z at which the Jacobian matrix of Φ has rank 1. $Z^{\#} \to C$ has the structure of a group variety over C. In particular, for every point $a \in C$ $F_a^{\#} = \Phi^{-1}(a) \cap Z^{\#}$ has the induced structure of a complex Lie group. This group structure depends only on F_a. ($F_a^{\#}$ is the set of simple points of the fiber $F_a = \Phi^{-1}(a)$.)
- With respect to the induced group homomorphism $MW \to F_a^{\#}$, the induced homomorphism $\mathrm{Tor}\, MW \to \mathrm{Tor}\, F_a^{\#}$ is injective for every point $a \in C$. By $\mathrm{Tor}\, M$ we here denote the subgroup of an abelian group M consisting of all elements of finite order.

We consider χ_1. C' intersects a unique component of F_1 with multiplicity 1, and the component intersecting C' does not intersect C_5, since $\mathrm{Tor}\, E \to \mathrm{Tor}\, F_1^{\#}$ is injective and every component of F_1 contains at most one point in $\mathrm{Tor}\, F_1^{\#}$. It implies $\chi_1 \neq 0$, and under the isomorphism $S_1^* \cong Q(-D_4)^*$ χ_1 corresponds to a fundamental weight associated with a vertex of the Dynkin graph D_4 with only one edge. ($Q(-X)$ denotes the negative definite root lattice associated with a Dynkin graph X. The bilinear form on $Q(-X)$ is (-1) times that on $Q(X)$.) Consequently, one knows $\chi_1^2 = -1$. By injectivity one knows, moreover, that C' has order 2 in MW.

Next, we consider χ_i for $2 \leq i \leq t$. Assume $\chi_i \neq 0$. Since $\operatorname{Tor} MW \to \operatorname{Tor} F_i^{\#}$ is injective, $F_i^{\#}$ contains a point with order 2 which is not on the component intersecting C_5. It implies F_i is of type either III or I_{2k} for some k. We have $n_i + 1 \equiv 0 \pmod 2$. The injectivity also implies that under the isomorphism $S_i^* \cong Q(-A_{n_i})^*$ χ_i corresponds to the fundamental weight associated with the central vertex of the Dynkin graph A_{n_i}. In particular, one has $\chi_i^2 = -(n_i+1)/4$.

We calculate m. By injectivity for $\operatorname{Tor} MW \to \operatorname{Tor} F_a^{\#}$ $a \in C$, C' and C_5 have no intersection. Thus,

$$0 = [C'] \cdot [C_5] = m[F] \cdot [C_5] + [F + C_5] \cdot [C_5] = m - 1.$$

We have $m = 1$.

Set $M = \{i \mid 2 \leq i \leq t, \chi_i \neq 0\}$. By the above one has $M \subset N(2)$, and

$$-2 = [C']^2 = ([F] + [F + C_5])^2 + \chi_1^2 + \sum_{i=2}^{t} \chi_i^2 = 2 - 1 - \sum_{i \in M} \frac{(n_i + 1)}{4},$$

which implies the equality in (4).

(5) Assuming that for some odd prime p $N(p)$ contains a set U with $16 - r$ elements, we deduce a contradiction.

For $V = \{i \mid 2 \leq i \leq t,\ n_i \neq 0\}$ we have $V \supset U$. First, we would like to show that $V \neq U$. Indeed, if $N(p)$ contains $17 - r$ or more elements, then $V \supset N(p) \neq U$. Thus, we can consider only the case where $V = N(p)$ and V has just $16 - r$ elements. Now, since $r = \sum_{i \in V} n_i$, we have $16 = r + (16 - r) = \sum_{i \in V} (n_i + 1) \equiv 0 \pmod p$, which contradicts that p is an odd prime.

Choose an element $e \in V - U$. The singular fiber F_e over $c_e \in \Sigma$ is either of type I_m for $m = n_e + 1$, or of type III, or of type IV, since $n_e > 0$.

Next, we consider the homotopy theory.

Fixing a base point $b \in C - \Sigma$, we draw a path l_i connecting b and a point $c_i \in \Sigma$ for $1 \leq i \leq t$ as in the beginning part of Section 10. Here, by exchanging the numbering we assume, moreover, that when we go on a small circle with center b in the positive direction, we encounter l_i's in the order of the attached number i. Associated with l_i, we define the closed path r_i as in Section 10. The homotopy classes $[r_i]$ of r_i are generators of $\pi_1(C - \Sigma, b)$ and are subject to a unique relation $[r_1][r_2] \cdots [r_t] = 1$.

Let $J : C - \Sigma_1 \to \mathbf{P}^1 - \{\infty\}$ be the functional invariant of the elliptic surface Φ where we denote $\Sigma_1 = \Sigma - \{c_1\}$. By $j : \mathcal{H} \to \mathbf{P}^1 - \{\infty\}$ we denote the j-function from the upper half plane $\mathcal{H} = \{z \in \mathbf{C} \mid \operatorname{Im} z > 0\}$. The multi-valued function $j^{-1}J$ defines the monodromy representation $\overline{\chi} : \pi_1(C - \Sigma_1, b) \to \mathrm{PSL}(2, \mathbf{Z}) = \mathrm{SL}(2, \mathbf{Z})/\{+1, -1\}$. Let $f : \mathrm{SL}(2, \mathbf{Z}) \to \mathrm{PSL}(2, \mathbf{Z})$ denote the canonical surjective homomorphism, and Σ_2 be an arbitrary finite set with $\Sigma_1 \subset \Sigma_2 \subset C - \{b\}$.

Lemma 13.2. (Kodaira [8]) *The following two sets have one-to-one correspondence:*

(1) The set of an isomorphism class of an elliptic surface $W \to C$ over C with a section $s_0 : C \to W$ whose critical values are contained in Σ_2 and whose functional invariant coincides with J.

(2) The set of a representation $\chi : \pi_1(C-\Sigma_2, b) \to \mathrm{SL}(2, \mathbf{Z})$ such that the composition $f\chi$ coincides with the composition $\pi_1(C-\Sigma_2, b) \to \pi_1(C-\Sigma_1, b) \xrightarrow{\overline{\chi}} \mathrm{PSL}(2, \mathbf{Z})$.
The correspondence is given by associating the elliptic surface $W \to C$ with the monodromy representation on the first cohomology group $H^1(F_b, \mathbf{Z})$ of the fiber over b.

Lemma 13.3. (Kodaira [8]) *An elliptic surface $W \to D$ is a K3 surface if, and only if, $D \cong \mathbf{P}^1$, it has no multiple fiber, and the sum of the Euler numbers of singular fibers is equal to 24.*

Let $\chi_1 : \pi_1(C-\Sigma, b) \to \mathrm{SL}(2, \mathbf{Z})$ denote the representation associated with our elliptic K3 surface $Z \to C$. We have $\chi_1([r_1]) = -1$ and $\chi_1([r_1]) \cdot \chi_1([r_2]) \cdots \chi_1([r_t]) = 1$. We can construct another representation χ_2 by setting $\chi_2([r_1]) = 1$, $\chi_2([r_e]) = -\chi_1([r_e])$ and $\chi_2([r_i]) = \chi_1([r_i])$ for $1 \le i \le t$ with $i \ne 1, e$. Since -1 commutes with any element, $\chi_2([r_1]) \cdot \chi_2([r_2]) \cdots \chi_2([r_t]) = 1$, and it defines a representation of $\pi_1(C-\Sigma, \mathbf{Z})$ such that $f\chi_2 = f\chi_1 = \overline{\chi}$.

Let $W \to C$ be the elliptic surface corresponding to χ_2. By Kodaira [8] the type of a singular fiber is uniquely determined by the $\mathrm{SL}(2, \mathbf{Z})$-conjugacy class of the monodromy matrix around it. Thus, the fibers over c_i with $1 \le i \le t$, $i \ne 1, e$ are same as those of $Z \to C$. On the contrary, the fiber over c_1 is smooth and the fiber over c_e is of type I_m^*, III^* or II^*, according as that in Z is of type I_m, III or IV. The combination $I_0^* + I_m$ has been replaced by $I_0 + I_m^*$ in the first case. Note here that for the both pairs of singular fibers the sum of the Euler numbers is $m+6$, and they are equal. Thus, by Lemma 13.3 one can conclude that W is also a K3 surface. In the second case $I_0^* + III$ has been replaced by $I_0 + III^*$. Also, in this case, for the both pairs the sum of the Euler numbers is 9. In the third case $I_0^* + IV$ has been replaced by $I_0 + II^*$. For the both pairs the sum of them is 10. By Lemma 13.3 W is a K3 surface even in these cases.

Next, we compare Dynkin graphs. Let G_W be the dual graph associated with the set of all components of singular fibers in W disjoint from the image $s_0(C)$ of the section s_0. By construction we have $G_W = G - A_{m-1} + D_{m+4}$ in the first case. Thus, G_W has $r+5$ vertices. In the second, third case we have $G_W = G - A_1 + E_7$ and $G_W = G - A_2 + E_8$ respectively. Thus, G_W has $r+6$ vertices.

Let $S_W \subset \mathrm{Pic}(W)$ be the subgroup generated by $s_0(C)$ and all the components of singular fibers of $W \to C$. Setting $G_0 = D_{m+4}$, E_7 or E_8 according as the first, second or third case takes place, we have

$$S_W \cong H_0 \oplus Q(-G_W) \cong H_0 \oplus Q(-G_0) \oplus \bigoplus_{\substack{i=2 \\ i \ne e}}^{t} Q(-A_{n_i}),$$

where H_0 denotes a hyperbolic plane. $\mathrm{rank}\, S_W \ge r+7$. Note that the discriminant group S_W^*/S_W has at least $(16-r)$ of p-torsions corresponding to U.

Let $\widetilde{S}_W$ be the primitive hull of S_W in $H^2(W, \mathbf{Z})$. The quotient $\widetilde{S}_W/S_W$ is isomorphic to the group of all sections of $W \to C$ with finite order, and it is isomorphic to a subgroup in the group of the singular fiber over c_e. Since the fiber over c_e is either I_m^*, III^* or II^*, $\widetilde{S}/S$ is a 2-primary group. Thus, for the p-Sylow subgroup we have $(\widetilde{S}_W/S_W)_p = 0$, since p is odd. We have $l((\widetilde{S}_W^*/\widetilde{S}_W)_p) = l((S_W^*/S_W)_p) \ge 16-r$.

Let T_W be the orthogonal complement of S_W in $H^2(W, \mathbf{Z})$, which has rank 22. $\operatorname{rank} T_W \leq 22 - (r+7) = 15 - r$. Thus, $l((T_W^*/T_W)_p) \leq l(T_W^*/T_W) \leq 15 - r$. On the other hand, since $\widetilde{S}_W^*/\widetilde{S}_W \cong T_W^*/T_W$, we have $l((T_W^*/T_W)_p) \geq 16 - r$, which is a contradiction. Q.E.D.

We have a byproduct of the proof of (4).

Lemma 13.4. *Assume that $G = \sum_{i\in I} A_{k_i} \in PC(J_{3,0})$. Then, for any embedding $S = P \oplus Q(G) \hookrightarrow \Lambda_3$ satisfying Looijenga's conditions (L1) and (L2) the following hold:*
(1) For the primitive hull $\widetilde{S}$ every non-zero element in the quotient $\widetilde{S}/S$ has order 2.
(2) Any non-zero element $\overline{\alpha}$ in $\widetilde{S}/S$ can be written

$$\overline{\alpha} = \overline{\chi}_0 + \sum_{i\in M} \overline{\chi}_i,$$

for some subset $M \subset I$, where $0 \neq \overline{\chi}_0 \in P^*/P \cong Q(D_4)^*/Q(D_4)$, $\overline{\chi}_i \in Q(A_{k_i})^*/Q(A_{k_i})$ has order 2, $k_i + 1 \equiv 0 \pmod 2$ for $i \in M$ and $\sum_{i\in M}(k_i+1) = 12$.
(3) If $\widetilde{S}/S$ is not cyclic, then there are subsets $M_1, M_2 \subset N(2) = \{i \in I \mid k_i + 1 \equiv 0 \pmod 2\}$ such that $\sum_{i\in M_\nu}(k_i+1) = 12$ for $\nu = 1$ and 2, and $\sum_{i\in M_1\cap M_2}(k_i+1) = 6$.

Let us consider the case $r = 13$ further. By Proposition 13.1 (2) we can consider only the graph $G = A_{k_1} + A_{k_2} + A_{k_3} + A_{k_4} + A_{k_5}$ corresponding to the division $k_1+k_2+k_3+k_4+k_5 = 13$ of 13 into a sum of 5 non-negative integers $k_1 \geq k_2 \geq k_3 \geq k_4 \geq k_5 \geq 0$. There are 57 kinds of such divisions as follows. We omit 0.

(1) 13	(2) 12+1	(3) 11+2	(4) 10+3
(5) 9+4	(6) 8+5	(7) 7+6	(8) 11+1+1
(9) 10+2+1	(10) 9+3+1	(11) 9+2+2	(12) 8+4+1
(13) 8+3+2	(14) 7+5+1	(15) 7+4+2	(16) 7+3+3
(17) 6+6+1	(18) 6+5+2	(19) 6+4+3	(20) 5+5+3
(21) 5+4+4	(22) 10+1+1+1	(23) 9+2+1+1	(24) 8+3+1+1
(25) 8+2+2+1	(26) 7+4+1+1	(27) 7+3+2+1	(28) 7+2+2+2
(29) 6+5+1+1	(30) 6+4+2+1	(31) 6+3+3+1	(32) 6+3+2+2
(33) 5+5+2+1	(34) 5+4+3+1	(35) 5+4+2+2	(36) 5+3+3+2
(37) 4+4+4+1	(38) 4+4+3+2	(39) 4+3+3+3	(40) 9+1+1+1+1
(41) 8+2+1+1+1	(42) 7+3+1+1+1	(43) 7+2+2+1+1	(44) 6+4+1+1+1
(45) 6+3+2+1+1	(46) 6+2+2+2+1	(47) 5+5+1+1+1	(48) 5+4+2+1+1
(49) 5+3+3+1+1	(50) 5+3+2+2+1	(51) 5+2+2+2+2	(52) 4+4+3+1+1
(53) 4+4+2+2+1	(54) 4+3+3+2+1	(55) 4+3+2+2+2	(56) 3+3+3+3+1
(57) 3+3+3+2+2			

Note that in each item the number of odd numbers is 1, 3, or 5.
We divide these 57 items into 4 classes.

[1] (1) (2) (3) (5) (6) (8) (9) (10) (11) (12) (17) (18)
(20) (21) (23) (26) (27) (30) (32) (34) (43) (47) (49) (53)

[2] (14) (37)

⟦3⟧ (4) (7) (13) (15) (16) (19) (22) (24) (25) (28) (29) (31) (33) (35) (36) (38) (39) (40) (41) (42) (44) (45) (46) (48) (50) (51) (52) (54) (55) (56)

⟦4⟧ (57)

For the 24 items in the class ⟦1⟧ we can make the corresponding graph $G = A_{k_1} + A_{k_2} + A_{k_3} + A_{k_4} + A_{k_5}$ from the essential Dynkin graph $E_8 + F_4$ by tie transformations repeated twice. In particular, $G \in PC = PC(J_{3,0})$. The following table shows an example of the Dynkin graph which we can make by the first tie transformation:

$(1) \leftarrow B_{13}$	$(2) \leftarrow B_{13}$	$(3) \leftarrow A_{11} + A_1$
$(5) \leftarrow B_{13}$	$(6) \leftarrow A_5 + E_7$	$(8) \leftarrow A_{11} + A_1$
$(9) \leftarrow A_9 + A_2 + A_1$	$(10) \leftarrow A_9 + A_2 + A_1$	$(11) \leftarrow A_9 + A_2 + A_1$
$(12) \leftarrow A_8 + A_4$	$(17) \leftarrow A_6 + E_6$	$(18) \leftarrow A_6 + E_6$
$(20) \leftarrow A_5 + E_7$	$(21) \leftarrow A_4 + E_8$	$(23) \leftarrow A_9 + A_2 + A_1$
$(26) \leftarrow A_7 + D_5$	$(27) \leftarrow A_7 + D_5$	$(30) \leftarrow A_4 + E_8$
$(32) \leftarrow A_6 + E_6$	$(34) \leftarrow A_5 + E_7$	$(43) \leftarrow A_7 + A_2 + A_1 + B_2$
$(47) \leftarrow A_5 + D_5 + B_2$	$(49) \leftarrow A_5 + D_5 + B_2$	$(53) \leftarrow A_4 + E_6 + B_2$

The item (14) in the class ⟦2⟧ contains 3 odd numbers and $N(2)$ does not contain subset M satisfying the condition in Proposition 13.1 (4). Thus, the corresponding graph $G = A_7 + A_5 + A_1$ is not a member of PC.

Indeed, let T be the orthogonal complement of $S = P \oplus Q(G)$ in Λ_3 and $\widetilde{S}$ be the primitive hull of S. $\operatorname{rank} T = 3$. Since $\widetilde{S}^*/\widetilde{S} \cong T^*/T$, $l(S^*/S) - 2l(\widetilde{S}/S) \leq l(\widetilde{S}^*/\widetilde{S}) \leq \operatorname{rank} T = 3$. On the other hand, $l(S^*/S) \geq l((S^*/S)_2) = l((P^*/P)_2) + l((Q(G)^*/Q(G))_2) = 2 + \#N(2) = 5$. ($\#U$ denotes the number of elements in a set U.) Consequently, one has $l(\widetilde{S}/S) \geq 1$. By Proposition 13.1 (4) $G \notin PC$.

For item (37) in the class ⟦2⟧, the corresponding $N(5) = \{i \in I \mid k_i + 1 \equiv 0 \pmod 5\}$ contains 3 elements. Thus, $G = 3A_4 + A_1 \notin PC$ by Proposition 13.1 (5).

As for the 30 items in the class ⟦3⟧ we treat them after the case $r = 12$. The unique item (57) in the class ⟦4⟧ corresponds to the exception $3A_3 + 2A_2$ in Theorem 0.3, and will be discussed in the next section Section 14.

Let us proceed to the case $r = 12$.

By Proposition 13.1 (2), this case corresponds to the division of $12 = \sum_{i=1}^{6} k_i$ into a sum of 6 non-negative integers $k_1 \geq k_2 \geq \cdots \geq k_6 \geq 0$. There are 58 kinds of such divisions as in the following table. We omit 0:

[1] 12	[2] 11+1	[3] 10+2	[4] 9+3
[5] 8+4	[6] 7+5	[7] 6+6	[8] 10+1+1
[9] 9+2+1	[10] 8+3+1	[11] 8+2+2	[12] 7+4+1
[13] 7+3+2	[14] 6+5+1	[15] 6+4+2	[16] 6+3+3
[17] 5+5+2	[18] 5+4+3	[19] 4+4+4	[20] 9+1+1+1
[21] 8+2+1+1	[22] 7+3+1+1	[23] 7+2+2+1	[24] 6+4+1+1
[25] 6+3+2+1	[26] 6+2+2+2	[27] 5+5+1+1	[28] 5+4+2+1
[29] 5+3+3+1	[30] 5+3+2+2	[31] 4+4+3+1	[32] 4+4+2+2
[33] 4+3+3+2	[34] 3+3+3+3	[35] 8+1+1+1+1	[36] 7+2+1+1+1
[37] 6+3+1+1+1	[38] 6+2+2+1+1	[39] 5+4+1+1+1	[40] 5+3+2+1+1

[41] 5+2+2+2+1 [42] 4+4+2+1+1 [43] 4+3+3+1+1 [44] 4+3+2+2+1
[45] 4+2+2+2+2 [46] 3+3+3+2+1 [47] 3+3+2+2+2 [48] 7+1+1+1+1+1
[49] 6+2+1+1+1+1 [50] 5+3+1+1+1+1 [51] 5+2+2+1+1+1 [52] 4+4+1+1+1+1
[53] 4+3+2+1+1+1 [54] 4+2+2+2+1+1 [55] 3+3+3+1+1+1 [56] 3+3+2+2+1+1
[57] 3+2+2+2+2+1 [58] 2+2+2+2+2+2

For simplicity we would like to use the following proposition effectively in what follows. Claim (1) is a direct consequence of our theory of elementary and tie transformations. (See Theorem 6.3 and 6.4.) Claim (2) follows easily from Theorem 2.2 or Theorem 5.11.

Proposition 13.5. *(1) If a Dynkin graph G can be obtained from a basic Dynkin graph G_0 by elementary or tie transformations applied twice, then any subgraph G' of G can be obtained from G_0 by elementary or tie transformations applied twice.*
(2) If $G \in PC(X)$, then any subgraph G' belongs to $PC(X)$.

A. For each item $[a]$ among the above 58 items except the following 13 we can find an item (b) in the class $[\![1]\!]$ of the case $r = 13$ such that the corresponding graph $G(b)$ to (b) contains the corresponding graph $G[a]$ to $[a]$. (Thus, $G[a] \subset G(b) \in PC = PC(J_{3,0})$.):

[16], [33], [34], [35], [37], [41], [45], [48], [49], [52], [53], [57], [58].

By Proposition 13.5 (1) for all items except the above 13, we can construct the corresponding graph from $E_8 + F_4$ by elementary or tie transformations applied twice.

We can discuss only the above 13 items in what follows. It turns out that each of the above 13 does not belong to PC.

B. In the case [48] the division of 12 contains 6 odd numbers, but there are no M_1, M_2 as in Lemma 13.4 (3). If the corresponding graph G is in PC, then $l(\widetilde{S}/S) \geq 2$, and by Lemma 13.4 (3) we have a contradiction. Thus, $G \notin PC$.

C. For the case [35], [37], [49], [52], and [53], the division of 12 contains 4 odd numbers, but there is no subset M satisfying the condition in Proposition 13.1 (4). If $G \in PC$, then $l(\widetilde{S}/S) \geq 1$, and we have a contradiction by Proposition 13.1 (4). $G \notin PC$.

D. Next, we consider the case [41], [45], [57], [58]. In these cases the set $N(3)$ contains 4 or more elements. Thus, by Proposition 13.1 (5) the corresponding graph is not a member of PC.

E. The remaining items are the following three; [16], [33], [34].

For the former 2 cases [16], [33], the division contains 2 odd numbers, and there is no subset M as in Proposition 13.1 (4). Thus, we can consider only the primitive embedding $P \oplus Q(G) \hookrightarrow \Lambda_3$.

We consider [16] 6+3+3 $G = A_6 + 2A_3$.

Assume that $G \in PC$. Then, we have an embedding $S = P \oplus Q(G) \hookrightarrow \Lambda_3$ satisfying (L1), (L2). It is primitive. The discriminant group $S^*/S \cong (\mathbf{Z}/2 + \mathbf{Z}/2) \oplus \mathbf{Z}/4 \oplus \mathbf{Z}/4 \oplus \mathbf{Z}/7$. Here, the first and the second direct summand $\mathbf{Z}/2 + \mathbf{Z}/2$ correspond to P^*/P. The third $\mathbf{Z}/4$ and the fourth $\mathbf{Z}/4$ correspond to $Q(A_3)^*/Q(A_3)$ respectively, and the fifth $\mathbf{Z}/7$ to $Q(A_6)^*/Q(A_6)$. The discriminant quadratic form q_S on S^*/S can be written $q_S(a_1, a_2, b_1, b_2, c) \equiv a_1^2 + a_1a_2 + a_2^2 + 3(b_1^2 + b_2^2)/4 + 6c^2/7 \bmod 2\mathbf{Z}$ for $(a_1, a_2, b_1, b_2, c) \in (\mathbf{Z}/2 + \mathbf{Z}/2) \oplus \mathbf{Z}/4 \oplus \mathbf{Z}/4 \oplus \mathbf{Z}/7$. Let T denote the orthogonal complement of S. T has signature $(2,2)$, and the discriminant D of T is positive. We have an isomorphism $T^*/T \cong S^*/S$. By this isomorphism the discriminant quadratic form q_T of T satisfies $q_T \equiv -q_S$. One has $D = |D| = \#T^*/T = 7 \cdot 2^6$.

Next, we consider the lattice $T_2 = T \otimes \mathbf{Z}_2$ over 2-adic integers $\mathbf{Z}_2$. By Lemma 12.1 $T_2^*/T_2 \cong (T^*/T)_2 \cong (\mathbf{Z}/2 + \mathbf{Z}/2) \oplus \mathbf{Z}/4 \oplus \mathbf{Z}/4$, and the discriminant quadratic form q on T_2 satisfies $q(\overline{\alpha}) \equiv (a_1^2 + a_1 a_2 + a_2^2) - 3(b_1^2 + b_2^2)/4 \bmod 2\mathbf{Z}$ for an element $\overline{\alpha} = (a_1, a_2, b_1, b_2)$. By Lemma 12.3 T_2 is equivalent over $\mathbf{Z}_2$ to the lattice defined by the matrix $A = \begin{pmatrix} 4 & 2 & 0 & 0 \\ 2 & 4 & 0 & 0 \\ 0 & 0 & -3\cdot 2^2 & 0 \\ 0 & 0 & 0 & -3\cdot 2^2 \end{pmatrix}$. Thus, $D \equiv \det A = 3^3 \cdot 2^6 \bmod \mathbf{Z}_2^{*2}$. By the 2 expressions of D one knows that $7 = 3\mu^2$ for some $\mu \in \mathbf{Z}_2^* = \mathbf{Z}_2 - 2\mathbf{Z}_2$, which is equivalent to $7 \equiv 3 \pmod 8$. It is a contradiction. Thus, $G \notin PC$.

For the case [33] 4+3+3+2 $G = A_4 + 2A_3 + A_2$, the reasoning is similar to that in [16]. We can conclude $G \notin PC$.

Now, we consider the last case [34] 3+3+3+3 $G = 4A_3$.

Assuming $G \in PC$, we define a lattice S, an embedding $S \hookrightarrow \Lambda_3$, the orthogonal complement T and its discriminant D similarly to the above case [16]. In this case, $I = \widetilde{S}/S$ is not zero, since $N(2)$ contains 4 elements. On the other hand, we have no subsets M_1, M_2 as in Lemma 13.4. Thus, I is cyclic of order 2.

Now, $S^*/S \cong (\mathbf{Z}/2 + \mathbf{Z}/2) \oplus \mathbf{Z}/4 \oplus \mathbf{Z}/4 \oplus \mathbf{Z}/4 \oplus \mathbf{Z}/4$. The discriminant quadratic form q can be written $q(\overline{\alpha}) \equiv x_1^2 + x_1 x_2 + x_2^2 + 3(y_1^2 + y_2^2 + y_3^2 + y_4^2)/4 \bmod 2\mathbf{Z}$ for $\overline{\alpha} = (x_1, x_2, y_1, y_2, y_3, y_4) \in S^*/S$. Note here that $P \cong H_0 \oplus Q(D_4)$ and $Q(D_4)$ has an action of the symmetric group of degree 3 associated with the symmetry of the Dynkin graph D_4 and that we can exchange the order of 4 A_3-components. Thus, by Lemma 13.4 (2) we can assume without loss of generality that I is generated by the element $\overline{\alpha} = (1,0,2,2,2,0)$. One can check that the orthogonal complement $I^\perp$ of I with respect to the discriminant bilinear form b on S^*/S is generated by $\overline{\alpha}$, and $\overline{\beta}_1 = (0,1,1,0,0,0)$, $\overline{\beta}_2 = (0,1,0,1,0,0)$, $\overline{\beta}_3 = (0,1,0,0,1,0)$, $\overline{\gamma} = (0,0,0,0,0,1)$. Note that $\overline{\beta}_i$'s $(i = 1,2,3)$ and $\overline{\gamma}$ are mutually orthogonal with respect to b, and $q(\overline{\beta}_1) \equiv q(\overline{\beta}_2) \equiv q(\overline{\beta}_3) \equiv -1/4$, $q(\overline{\gamma}) \equiv 3/4 \pmod{2\mathbf{Z}}$. Thus, the discriminant quadratic form q_1 on $\widetilde{S}^*/\widetilde{S} \cong I^\perp/I \cong \mathbf{Z}/4 \oplus \mathbf{Z}/4 \oplus \mathbf{Z}/4 \oplus \mathbf{Z}/4$ can be written $q_1(a_1, a_2, a_3, b) \equiv q(\sum a_i \overline{\beta}_i + b\overline{\gamma}) \equiv -(a_1^2 + a_2^2 + a_3^2 + a_4^2)/4 + 3b^2/4 \bmod 2\mathbf{Z}$ for $(a_1, a_2, a_3, b) \in \mathbf{Z}/4 \oplus \mathbf{Z}/4 \oplus \mathbf{Z}/4 \oplus \mathbf{Z}/4$. One has $2^8 = D \equiv -3 \cdot 2^8 \bmod \mathbf{Z}_2^{*2}$, which is equivalent to $1 \equiv -3 \pmod 8$. It is a contradiction. We conclude $4A_3 \notin PC$.

Here we consider the class ⟦3⟧ in the case of $r = 13$. Indeed, for each item (a) in the class ⟦3⟧ we can find an item [b] belonging to the 13 exceptions in paragraph **A** such that the corresponding graph $G(a)$ to (a) contains the corresponding graph $G[b]$ to [b]. By Proposition 13.5 (2) we conclude that any item in ⟦3⟧ does not belong PC.

If $r \leq 11$, we have nothing to verify thanks to Theorem 9.8 ⟦II⟧, since then the I1-condition is automatically satisfied. (See Section 7.) We can complete the study in the case $J_{3,0}$.

Let us proceed to the case $Z_{1,0}$. Z carries the curve IF at infinity associated with $Z_{1,0}$. IF has 7 components. Five of 7 are components of the singular fiber F_1 of type I_0^* of Φ. One of the remaining two is the section C_5. The last remaining component

C_6 is a component of another singular fiber. We assume that F_2 contains C_6. The dual graph of the set of components disjoint from IF of singular fibers is the graph G.

Recall here that in this section all singular fibers of Φ except F_1 are of type I, II, III or IV. Note that we can assume further that the Picard number ρ of Z is equal to $r+7$. (If $\rho > r+7$, then we consider the deformation of Z under the condition that it keeps the union of the curve IF and the combination of curves corresponding to the graph G. The general deformation of Z under this condition has the Picard number $r+7$, it has the structure of the elliptic K3 surface, and all the singular fibers are of type I, II, III, or IV except the unique exception of type I_0^*.)

Lemma 13.6. *Under the above assumptions the singular fiber F_2 is of type III or I_2.*

Proof. For $2 \le i \le t$ let $n(F_i)$ denote the number of components of F_i disjoint from IF. By definition we have $\sum_{i=2}^{t} n(F_i) = r$. By the equality (2) in the beginning of Section 10, we have

$$a + m(F_2) - n(F_2) - 2 + \sum_{i=3}^{t} (m(F_i) - n(F_i) - 1) = 0.$$

Since $m(F_2) \ge n(F_2) + 2$ and $m(F_i) \ge n(F_i) + 1$ for $3 \le i \le t$, we have, in particular, $m(F_2) = n(F_2) + 2$. It implies that a component of F_2 intersecting C_6 is unique except C_6. If F_2 is of type either IV or I_n with $n \ge 3$, it has never this property. Q.E.D.

Proposition 13.7. *The following two conditions are equivalent:*

(1) $G \in PC(Z_{1,0})$, and there exists a K3 surface Z containing the curve IF associated with $Z_{1,0}$ such that with respect to the associated structure $\Phi : Z \to C$ of the elliptic surface every singular fiber is of type I, II, III or IV and such that the dual graph of the set $\mathcal{E}$ of smooth rational curves on Z disjoint from IF is equal to G.

(2) $G + A_1 \in PC(J_{3,0})$ and every component of G is of type A.

Proof. Assume that there is a K3 surface Z with the above mentioned properties. We can assume moreover $\rho = r + 7$. By Lemma 13.6 F_2 contains only one component C' not contained in IF. Let $\overline{IF}$ denote the union of F_1 and C_5. $\overline{IF}$ is the curve at infinity in the case of $J_{3,0}$. Obviously, the set $\overline{\mathcal{E}}$ of smooth rational curves disjoint from $\overline{IF}$ coincides with $\mathcal{E} \cup \{C'\}$. Thus, the dual graph of $\overline{\mathcal{E}}$ is $G + A_1$ and it belongs to $PC(J_{3,0})$. Obviously, every component of G is of type A under the assumption.

Conversely, assume that a Dynkin graph G with only components of type A satisfies $G + A_1 \in PC(J_{3,0})$. Let $\Phi : Z \to C$ be the associated elliptic K3 surface. Φ has a singular fiber F_1 of type I_0^* and Z has a curve C_5 which is a section. The dual graph of the union of components disjoint from $\overline{IF} = F_1 \cup C_5$ of singular fibers coincides with $G + A_1$ by definition.

In the case $J_{3,0}$ singular fibers of Φ with 2 or more components other than F_1 have one-to-one correspondence with components of the Dynkin graph, and the type of the singular fiber is uniquely determined by the type of the corresponding irreducible Dynkin graph except for A_1 and A_2. Both type III and type I_2 correspond to A_1, and both type IV and type I_3 correspond to A_2. A smooth fiber, type I_1 and type II correspond to the empty graph.

Thus, our $Z \to C$ has a singular fiber F_2 of type III or I_2 corresponding to the component A_1 of $G + A_1$. Every singular fiber except F_1 is of type I, II, III, or IV. Anyway, F_2 has 2 components and only one C'' of two intersects C_5. The curve $IF = \overline{IF} \cup C''$ is the curve at infinity in the case $Z_{1,0}$. The dual graph of all components of singular fibers disjoint from IF coincides with G. Thus, $G \in PC(Z_{1,0})$. Q.E.D.

By Proposition 13.7 and by Proposition 13.1 (3) we can assume $r \leq 12$. If $r \leq 10$, thanks to Theorem 9.8, we have nothing to verify.

Let us consider the case $r = 12$ first. Thanks to Proposition 13.7, by the following method we can give the verification of our case: First, we collect items $(k_1, \ldots, k_5)$ with $\sum_{i=1}^{5} k_i = 13$ such that $k_i = 1$ for some i from the class [[1]] and [[4]] in the case $r = 13$. Second, we check whether for each item in the collection the graph G defined by $G + A_1 = \sum_{i=1}^{5} A_{k_i}$ can be made from $E_7 + F_4$ or $E_8 + BC_3$ by tie transformations applied twice.

Of course, for every item in the collection the answer is affirmative. In the following we show the graph $G = \sum_{i=1}^{5} A_{k_i} - A_1$ (Note that this is different from $\sum_{i=1}^{5} A_{k_i}$.) and an example of the Dynkin graph G' which can be made after the first tie transformation. As the basic graph, $E_7 + F_4$ can be used for the items (2) through (34). For (43), (47), (49), (53) $E_8 + BC_3$ can be used. There are 15 items in the collection.

G	$\leftarrow G'$		
(2) A_{12}	$\leftarrow A_{11}$	(8) $A_{11} + A_1$	$\leftarrow A_{11}$
(9) $A_{10} + A_2$	$\leftarrow A_9 + A_2$	(10) $A_9 + A_3$	$\leftarrow A_9 + A_2$
(12) $A_8 + A_4$	$\leftarrow A_8 + F_4$	(17) $2A_6$	$\leftarrow A_6 + A_5$
(23) $A_9 + A_2 + A_1$	$\leftarrow A_9 + A_2$	(26) $A_7 + A_4 + A_1$	$\leftarrow A_7 + A_3 + A_1$
(27) $A_7 + A_3 + A_2$	$\leftarrow A_7 + A_3 + A_1$	(30) $A_6 + A_4 + A_2$	$\leftarrow E_7 + A_4$
(34) $A_5 + A_4 + A_3$	$\leftarrow E_7 + A_4$	(43) $A_7 + 2A_2 + A_1$	$\leftarrow A_7 + A_2 + A_1 + BC_1$
(47) $2A_5 + 2A_1$	$\leftarrow D_5 + A_5 + BC_1$	(49) $A_5 + 2A_3 + A_1$	$\leftarrow D_5 + A_5 + BC_1$
(53) $2A_4 + 2A_2$	$\leftarrow E_6 + A_4 + BC_1$		

Next, we consider the case $r = 11$. First, we can collect items $(k_1, \ldots, k_6)$ with $\sum_{i=1}^{6} k_i = 12$ such that $k_i = 1$ for some i from items [1]–[58] in the above list. Then, it is not difficult to see that every item in the collection is either one of the 13 exceptions treated in paragraph **A**, or the corresponding graph G is a subgraph of a graph in the above 15 just discussed. We can complete the proof by Proposition 13.5 (1).

Remark. Among the 13 exceptions in **A** there are 8 items with $k_i = 1$ for some i. The corresponding graph G is as follows. (Note that G does not have 12 vertices but 11 ones.)

[35] $A_8 + 3A_1$	[37] $A_6 + A_3 + 2A_1$	[41] $A_5 + 3A_2$
[48] $A_7 + 4A_1$	[49] $A_6 + A_2 + 3A_1$	[52] $2A_4 + 3A_1$
[53] $A_4 + A_3 + A_2 + 2A_1$	[57] $A_3 + 4A_2$	

We can show that $G \notin PC(Z_{1,0})$ for three items [41], [49], [57]. However, for the other 5 items we can make the corresponding graph from $E_7 + F_4$ by tie transformations applied twice.

Note that this fact does not contradict Proposition 13.7, because in the case of these 5 graphs, in the corresponding elliptic K3 surface the singular fiber F_2 containing the component C_6 of IF is of type either I^*, II^*, III^* or IV^*.

We complete the case $Z_{1,0}$.

The third case is $Q_{2,0}$. We can show the following in this case:

Proposition 13.8. *The following two conditions are equivalent:*

(1) $G \in PC(Q_{2,0})$, and there exists a K3 surface Z containing the curve IF associated with $Q_{2,0}$ such that with respect to the associated structure $\Phi : Z \to C$ of the elliptic surface every singular fiber is of type I, II, III or IV and such that the dual graph of the set $\mathcal{E}$ of smooth rational curves disjoint from IF on Z is equal to G.

(2) $G + A_2 \in PC(J_{3,0})$ and every component of G is of type A.

By Theorem 9.8 [II], Proposition 13.1 (3) and Proposition 13.8 we can consider only the case $r = 11$ or $r = 10$.

First, we treat the case $r = 11$. By Proposition 13.8 we can consider the division $(k_1, \dots, k_5)$ of 13 with $k_i = 2$ for some i.

Note that the case (57) 3+3+3+2+2 in the class ⟦4⟧ corresponding to the exception $3A_3 + 2A_2$ in Theorem 0.3 satisfies $k_5 = 2$. We consider this case separately in Section 14.

Excluding (57), by the following method we can give the verification of our case: First, we collect items $(k_1, \dots, k_5)$ with $k_i = 2$ for some i from the class ⟦1⟧ in the case $J_{3,0}$ $r = 13$. Second, we check for each item in the collection whether the graph G defined by $G + A_2 = \sum_{i=1}^{5} A_{k_i}$ can be made from $E_6 + F_4$ or $E_8 + F_2$ by tie transformations applied twice.

We can check this affirmatively for every item in the collection. The following list of this collection shows the numbering of the item, the corresponding graph G, and an example of the Dynkin graph G' which we can make after the first tie transformation. For every item we can use $E_6 + F_4$ at the start. The list contains 10 items:

G	$\leftarrow G'$		
(3) A_{11}	$\leftarrow A_9 + A_1$	(9) $A_{10} + A_1$	$\leftarrow A_9 + A_1$
(11) $A_9 + A_2$	$\leftarrow A_9 + A_1$	(18) $A_6 + A_5$	$\leftarrow 2A_5$
(23) $A_9 + 2A_1$	$\leftarrow A_9 + A_1$	(27) $A_7 + A_3 + A_1$	$\leftarrow A_7 + A_1 + B_2$
(30) $A_6 + A_4 + A_1$	$\leftarrow A_6 + A_1 + F_4$	(32) $A_6 + A_3 + A_2$	$\leftarrow A_6 + 2A_2$
(43) $A_7 + A_2 + 2A_1$	$\leftarrow A_7 + A_1 + B_2$	(53) $2A_4 + A_2 + A_1$	$\leftarrow A_4 + 2A_2 + B_2$

We proceed to the case $r = 10$. In this case, our problem is reduced to the analysis of the decomposition $(k_1, \dots, k_6)$ of 12 into 6 integers. It is not difficult to show one of the following three conditions always holds for each item in the above [1]–[58]:

(1) $k_i \neq 2$ for $1 \le i \le 6$.

(2) $k_i = 2$ for some $1 \le i \le 6$ and the graph G_0 defined by $G_0 + A_2 = \sum_{i=1}^{6} A_{k_i}$ is a subgraph of one of the 10 graphs G in the list just above.

(3) It is one of the 13 items discussed in paragraph **A**.

By Proposition 13.5 and Proposition 13.8 we can complete the proof.

In this section we have shown Proposition 10.1 for $J_{3,0}$, $Z_{1,0}$, and $Q_{2,0}$ under the assumption ((3)) in the beginning of Section 10.

14
The exceptions in Theorem 0.3

In this section we study the exception in Theorem 0.3.

First, we consider $G = 3A_3 + 2A_2$ in the case $J_{3,0}$.

In this case, $P = H_0 \oplus P_0'$, $P_0' \cong Q(D_4)$. H_0 is a hyperbolic plane. Set $S = P \oplus Q(G)$. Consider the discriminant group $S^*/S \cong (\mathbf{Z}/4)^3 \oplus (\mathbf{Z}/2)^2 \oplus (\mathbf{Z}/3)^2$. Each of three $\mathbf{Z}/4$-components corresponds to A_3, $(\mathbf{Z}/2)^2$ corresponds to P, and $(\mathbf{Z}/3)^2$ to $2A_2$. The discriminant quadratic form can be written $q(a_1, a_2, a_3, b_1, b_2, x_1, x_2) \equiv 3(a_1^2 + a_2^2 + a_3^2)/4 + b_1^2 + b_1b_2 + b_2^2 + 2(x_1^2 + x_2^2)/3 \bmod 2\mathbf{Z}$.

Assume $G \in PC(J_{3,0})$. We have the corresponding elliptic K3 surface $\Phi : Z \to C$ and the corresponding embedding $S \hookrightarrow \Lambda_3$. Every singular fiber except one F_1 of type I_0^* is of type I, II, III, or IV, since every component of G is of type A. Thus, we can apply the theory in Section 13.

Since $l((S^*/S)_2) = 5 > \operatorname{rank} \Lambda_3 - \operatorname{rank} S = 3$, there is no primitive embedding $S \hookrightarrow \Lambda_3$. Let $\widetilde{S}$ be the primitive hull of S in Λ_3. Every non-zero element in the quotient $I = \widetilde{S}/S$ has order 2 by Lemma 13.4 (1). We have $I \cong \mathbf{Z}/2$, since there are no M_1, M_2 satisfying the condition in Lemma 13.4 (3). By Lemma 13.4 (2) the generator of I is either $(2,2,2,1,0,0,0)$, $(2,2,2,0,1,0,0)$ or $(2,2,2,1,1,0,0)$.

Note that these three elements are conjugate with respect to the action of the symmetric group of degree 3 on $P_0' \cong Q(D_4)$ induced by the symmetry of the Dynkin graph D_4.

Let S_1 be the inverse image by $S^* \to S^*/S$ of the subgroup in S^*/S generated by the first element $(2,2,2,1,0,0,0)$. It is an even overlattice of S with index 2.

Proposition 14.1. *Any embedding $S = H_0 \oplus P_0' \oplus Q(3A_3 + 2A_2) \hookrightarrow \Lambda_3$ satisfying Looijenga's conditions (L1) and (L2) is the composition of an isomorphism $S \xrightarrow{\sim} S$ induced by an isomorphism $P_0' \xrightarrow{\sim} P_0'$ of the direct summand and a primitive embedding $S_1 \hookrightarrow \Lambda_3$.*

Next, we compute the discriminant quadratic form q_1 of S_1. Let $\overline{\sigma}_1$, $\overline{\sigma}_2$, $\overline{\sigma}_3 \in S^*/S$ be the elements of order 4 corresponding to $(1,0,0,0,1,0,0)$, $(0,1,0,0,1,0,0)$, $(0,0,1,0,1,0,0) \in (\mathbf{Z}/4)^3 \oplus (\mathbf{Z}/2)^2 \oplus (\mathbf{Z}/3)^2$ respectively. Let $\overline{\tau}_1$, $\overline{\tau}_2 \in S^*/S$ be the elements of order 3 corresponding to $(0,0,0,0,0,1,0)$ and $(0,0,0,0,0,0,1)$ respectively. We can check that the orthogonal complement $I^\perp$ of I with respect to the discriminant bilinear form b on S^*/S is the direct sum of I and the 5 cyclic groups generated by $\overline{\sigma}_1$, $\overline{\sigma}_2$, $\overline{\sigma}_3$, $\overline{\tau}_1$, $\overline{\tau}_2$. Thus, we have $S_1^*/S_1 \cong I^\perp/I \cong (\mathbf{Z}/4)^3 \oplus (\mathbf{Z}/3)^2$. Note that any two of $\overline{\sigma}_1, \overline{\sigma}_2, \overline{\sigma}_3, \overline{\tau}_1, \overline{\tau}_2$ are orthogonal with respect to b, $q(\overline{\sigma}_\nu) \equiv -1/4 \bmod 2\mathbf{Z}$ $\quad(\nu = 1,2,3)$, and $q(\overline{\tau}_\nu) \equiv 2/3 \bmod 2\mathbf{Z}$ $\quad(\nu = 1,2)$. Thus, the discriminant quadratic form q_1 of S_1 can be written

$$q_1(\overline{\sigma}) \equiv -\frac{1}{4}(a_1^2 + a_2^2 + a_3^2) + \frac{2}{3}(b_1^2 + b_2^2) \bmod 2\mathbf{Z},$$

for an element $\overline{\sigma} \in S_1^*/S_1$ corresponding to $(a_1, a_2, a_3, b_1, b_2) \in (\mathbf{Z}/4)^3 \oplus (\mathbf{Z}/3)^2$.

In what follows we consider $S_1^*/S_1 \cong (\mathbf{Z}/4)^3 \oplus (\mathbf{Z}/3)^2$. Let $\overline{\kappa}_1$ and $\overline{\kappa}_2$ be the elements of order 12 in S_1^*/S_1 corresponding to $(1,2,0,1,1)$ and $(2,1,0,1,-1)$ respectively. Let $\overline{\lambda} \in S_1^*/S_1$ be the element of order 4 corresponding to $(0,0,1,0,0)$. We can check that

S_1^*/S_1 is the direct sum of three cyclic groups generated by $\overline{\kappa}_1$, $\overline{\kappa}_2$ and $\overline{\lambda}$. After some calculation one has

$$q_1(a_1\overline{\kappa}_1 + a_2\overline{\kappa}_2 + b\overline{\lambda}) \equiv \frac{1}{12}(a_1^2 + a_2^2) - \frac{1}{4}b^2 \bmod 2\mathbf{Z},$$

for $a_1, a_2 \in \mathbf{Z}/12$, $b \in \mathbf{Z}/4$.

Proposition 14.2. *(1) S_1 has a primitive embedding into the even unimodular lattice Λ_3 of signature $(19,3)$.*
(2) If $\eta \in S_1$, $\eta \notin S$ and $\eta \cdot u_0 = 0$, then $\eta^2 \geq 4$.

Proof. (1) Let T be the lattice of rank 3 defined by the diagonal matrix whose diagonal entries are -12, -12 and 4. T is an even lattice and we can define the discriminant quadratic form q_T on T^*/T. By the above calculation we have an isomorphism $\phi : S_1^*/S_1 \xrightarrow{\sim} T^*/T$ such that $-q_T\phi$ coincides with the discriminant quadratic form q_1 of S_1. By Lemma 5.5 (4) S_1 has a primitive embedding into Λ_3 such that the orthogonal complement of the image is isomorphic to T.
(2) Let $\omega_0 \in (P_0')^*$ be the vector corresponding under the isomorphism $(P_0')^* \cong Q(D_4)^*$ to the fundamental weight associated with one of three vertices at the end of the Dynkin graph D_4. For $\nu = 1, 2, 3$ let $\chi_\nu \in Q(G)^*$ be the fundamental weight associated with the central vertex of the ν-th component of G of type A_3. Set $\xi = \omega_0 + \chi_1 + \chi_2 + \chi_3$. We can assume $S_1 = S \cup (S + \xi)$.

Now, by assumption we can write $\eta = \xi + \zeta$ for some $\zeta \in S$. We have $0 = \eta \cdot u_0 = \xi \cdot u_0 + \zeta \cdot u_0 = \zeta \cdot u_0$. Thus, $\zeta = ku_0 + \zeta_0$ for some $k \in \mathbf{Z}$, $\zeta_0 \in P_0' \oplus Q(G)$. Setting $\eta_0 = \xi + \zeta_0$, one has $\eta^2 = \eta_0^2$, since $\eta = \eta_0 + ku_0$ and $\eta_0 \cdot u_0 = 0$. Our problem is reduced to showing $\eta_0^2 \geq 4$.

Recall here the concept of the expected minimum square length. (Section 3 Lemma 3.1, etc.) $S_0 = P_0' \oplus Q(G)$ is an even positive definite lattice, and we can define the expected minimum square length $m(\overline{x})$ for each element $\overline{x} \in S_0^*/S_0$. For an element $x \in S_0^*$ we write $\overline{x} = x \bmod S_0 \in S_0^*/S_0$. By definition $\eta_0^2 \geq m(\overline{\eta}_0) = m(\overline{\xi}) = m(\overline{\omega}_0 + \overline{\chi}_1 + \overline{\chi}_2 + \overline{\chi}_3) = 4$. Q.E.D.

Corollary 14.3. *In the case of $J_{3,0}$ the lattice $P \oplus Q(3A_3 + 2A_2)$ has an embedding into Λ_3 satisfying Looijenga's conditions (L1) and (L2). In particular, $3A_3 + 2A_2 \in PC(J_{3,0})$.*

Proposition 14.4. *In case $J_{3,0}$ with respect to any full embedding $Q(3A_3 + 2A_2) \hookrightarrow \Lambda_3/P$ there is no isotropic vector in a nice position.*

Proof. Assume that we have a primitive embedding $P \hookrightarrow \Lambda_3$, a full embedding $Q(G) \hookrightarrow \Lambda_3/P$ and a primitive isotropic vector $\overline{u} \in \Lambda_3/P$ in a nice position. We will deduce a contradiction.

Let $u \in \Lambda_3$ be the vector in the orthogonal complement of P whose image under the canonical surjective homomorphism $\Lambda_3 \to \Lambda_3/P$ is $\overline{u}$. Such a u exists by Proposition 5.6 (2). By the definition of a nice position we have a root basis $\Delta \subset Q(G)$ and a long root $\theta \in \Delta$ such that $\alpha \cdot u = 0$ for every $\alpha \in \Delta$ with $\alpha \neq \theta$ and $\theta \cdot u = 1$ or 0. The induced embedding $S = P \oplus Q(G) \hookrightarrow \Lambda_3$ satisfies Looijenga's (L1) and (L2).

Assume that $\theta \cdot u = 0$. Then, we have the I1-condition, i.e., $\epsilon_p(Q(3A_3 + 2A_2)) = 1$ for every prime p, since the orthogonal complement of $P \oplus Q(G)$ contains an isotropic vector u. However, $\epsilon_3(Q(3A_3 + 2A_2)) = (3,3)_3 = -1$, which is a contradiction. One knows $\theta \cdot u = 1$.

Let T be the orthogonal complement of S in Λ_3. Since $P \oplus Q(G) \oplus T \subset \Lambda_3 \subset P^* \oplus Q(G)^* \oplus T^*$ and since u is orthogonal to P, we can write $u = \omega + \tau$ for $\omega \in Q(G)^*$, $\tau \in T^*$. The vector $\omega \in Q(G)^*$ is the fundamental weight associated with θ and Δ, i.e., $\omega \cdot \theta = 1$ and $\omega \cdot \alpha = 0$ for $\alpha \in \Delta$ with $\alpha \neq \theta$.

Note here that considering $w(u)$ instead of u for an element w of the Weyl group of $Q(G)$, we can assume moreover that the root basis Δ coincides with the one given beforehand.

We have two cases.

(a) The long root θ lies on a component of type A_3.

(b) θ lies on a component of type A_2.

We consider case (a) first. We use the notations ξ, ω_0, χ_ν in the verification of Proposition 14.2 (2). By Proposition 14.1 there is an isomorphism $\sigma : S \to S$ keeping every element in $Q(G)$ fixed such that $\sigma(\xi) \in \Lambda_3$. We have $\mathbf{Z} \ni u \cdot \sigma(\xi) = \omega \cdot \xi = \omega \cdot \chi_\nu$. Here, we assumed that θ lies on the ν-th A_3-component of Δ. If θ corresponds to a vertex of the Dynkin graph A_3 at the end, then $\omega \cdot \chi_\nu = 1/2$. Thus, θ corresponds to the central vertex of A_3 and $\omega = \chi_\nu$. In particular, $\omega^2 = 1$.

Since $0 = u^2 = \omega^2 + \tau^2$, one has $\tau^2 = -1$ for $\tau \in T^*$.

Next, we consider the 2-adic lattice $T^* \otimes \mathbf{Z}_2 = (T \otimes \mathbf{Z}_2)^*$. By Proposition 14.1 the discriminant quadratic form of $T_2 = T \otimes \mathbf{Z}_2$ coincides with $-(q_1)_2$, where $(q_1)_2$ denotes the restriction to the 2-Sylow subgroup of the discriminant quadratic form q_1 of S_1. The discriminant D of T is equal to that of T_2, and it satisfies $D = 2^6 \cdot 3^2 \equiv 2^6 \mod \mathbf{Z}_2^{*2}$.

Let $T' = (\mathbf{Z}_2)^3$ be the 2-adic lattice whose intersection matrix is the diagonal matrix with diagonal entries 4, 4, 4. By the calculation of q_1 before Proposition 14.2 one knows that T_2 and T' have the same rank, the same discriminant quadratic form, and the same discriminant modulo $\mathbf{Z}_2^{*2}$. By Lemma 12.2 they are isomorphic as 2-adic lattices. Thus, there is an isomorphism $T_2^* \cong (\mathbf{Z}_2)^3$ such that the quadratic form is given by $x^2 = (x_1^2 + x_2^2 + x_3^2)/4$ for any element $x \in T_2^*$ corresponding to $(x_1, x_2, x_3) \in (\mathbf{Z}_2)^3$.

Assume $\tau \in T_2^*$ corresponds to $(x_1, x_2, x_3) \in (\mathbf{Z}_2)^3$. One has $x_1^2 + x_2^2 + x_3^2 = -4$. Let k_1 be the number of x_ν's ($\nu = 1, 2, 3.$) with $x_\nu \notin 2\mathbf{Z}_2$. One has $-4 \equiv k_1 \pmod 4$. Thus, $k_1 = 0$. We can write $x_\nu = 2y_\nu$ with $y_\nu \in \mathbf{Z}_2$ for $\nu = 1, 2, 3$. Let k_2 be the number of y_ν's ($\nu = 1, 2, 3.$) with $y_\nu \notin 2\mathbf{Z}_2$. One has $-1 = y_1^2 + y_2^2 + y_3^2 \equiv k_2 \pmod 4$, and $k_2 = 3$. Then, one has congruent relations $y_\nu^2 \equiv 1 \pmod 8$ ($\nu = 1, 2, 3$) modulo 8. We have $-1 = y_1^2 + y_2^2 + y_3^2 \equiv 3 \pmod 8$, which is a contradiction. The case (a) never takes place.

We proceed to the case (b). Since ω is a fundamental weight of $Q(A_2)$, $3\omega \in Q(G) \subset \widetilde{S}$ and $\omega^2 = 2/3$. The integer 3 is invertible in $\mathbf{Z}_2$ and we have $\omega \in \widetilde{S} \otimes \mathbf{Z}_2$. Since $\omega \bmod \widetilde{S} \otimes \mathbf{Z}_2 = 0$ corresponds to $\tau \bmod T_2$ under the canonical isomorphism $(\widetilde{S} \otimes \mathbf{Z}_2)^*/(\widetilde{S} \otimes \mathbf{Z}_2) \cong T_2^*/T_2$, $\tau \in T_2 = T \otimes \mathbf{Z}_2$. We can identify the quadratic form on T_2 with $2^2(x_1^2 + x_2^2 + x_3^2)$ and $\tau^2 \in \mathbf{Z}_2$ is a multiple of 2^2.

On the other hand, $\tau^2 = u^2 - \omega^2 = -2/3$, which is not a multiple of 2^2. We have a contradiction. Thus, the case (b) never takes place, either. Q.E.D.

Corollary 14.5. *The Dynkin graph $3A_3 + 2A_2$ can never be made from $E_8 + F_4$ or B_{12} by applying elementary or tie transformations twice.*

The case $G = 3A_3 + A_2$ for $X = Q_{2,0}$ follows easily from the above case. By Proposition 13.8 one has $G \in PC(Q_{2,0})$ since $G + A_2 = 3A_3 + 2A_2 \in PC(J_{3,0})$.

Next, assume that there are an embedding $P \oplus Q(G) \hookrightarrow \Lambda_3$ satisfying (L1) and (L2) and a primitive isotropic vector $\overline{u} \in \Lambda_3/P$ in a nice position. Let $u \in \Lambda_3$ be the vector orthogonal to P and mapped to $\overline{u}$ by $\Lambda_3 \to \Lambda_3/P$. u is also a primitive isotropic vector.

Note here that the lattice P in our case $Q_{2,0}$ has the decomposition $P = \overline{P} \oplus T$ where $\overline{P}$ is isomorphic to P defined for $J_{3,0}$ and $T \cong Q(A_2)$. Thus, the given embedding induces the embedding $\overline{P} \oplus Q(G + A_2) \hookrightarrow \Lambda_3$. Regarding this as the one defined for $J_{3,0}$, we would like to show that this also satisfies (L1) and (L2).

Let us consider the associated elliptic surface $\Phi : Z \to C$ with $\rho = r + 8 = 19$. Let F_1 be the singular fiber of Φ contained in the curve at infinity IF. F_1 is of type I_0^*. Let F_2 be another singular fiber containing 2 components of IF. The other singular fibers than F_1 and F_2 are of type I, II, III, or IV, since G has only components of type A.

If F_2 is of type I^*, then the orthogonal complement of $P \oplus Q(G)$ contains an isotropic vector by Proposition 10.3. Thus, $\epsilon_p(Q(G)) = (3, -d(Q(G)))_p$ for every prime p. However, $\epsilon_3(Q(G)) = (-1, 3)_3 = -1$ and $(3, -d(Q(G)))_3 = (3, -2^6 \cdot 3)_3 = (3, -3)_3 = +1$, which is a contradiction.

Assume that F_2 is of type either II^*, III^* or IV^*. Let G_0 be the dual graph of the set of components disjoint from IF in F_2. One knows G_0 is of type either E_6, A_5 or $2A_2$. Neither of them is contained in $G = 3A_3 + A_2$, which is a contradiction.

Consequently, F_2 is also of type I or IV, and by the verification of Proposition 13.7 one knows that the embedding $\overline{P} \oplus Q(G + A_2) \hookrightarrow \Lambda_3$ also satisfies (L1) and (L2).

The image of u by the surjective homomorphism $\Lambda_3 \to \Lambda_3/\overline{P}$ is a primitive isotropic vector in a nice position with respect to the embedding $Q(3A_3 + 2A_2) = Q(G + A_2) \hookrightarrow \Lambda_3/\overline{P}$, which contradicts Proposition 14.4.

Proposition 14.6. *(1) $3A_3 + A_2 \in PC(Q_{2,0})$.*
(2) In case $Q_{2,0}$ with respect to any full embedding $Q(3A_3 + A_2) \hookrightarrow \Lambda_3/P$ there is no isotropic vector in a nice position.
(3) The Dynkin graph $3A_3 + A_2$ can never be made from any one of $E_6 + F_4$, $E_8 + F_2$, B_9 by applying elementary or tie transformations twice.

We complete the verification of Theorem 0.3 here.

Chapter 3

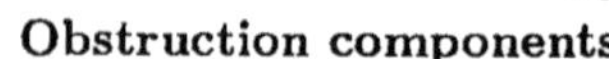

15
Obstruction components

The theme of Chapter 3 is $W_{1,0}$ and $S_{1,0}$. This section is devoted to providing theoretical foundation of the concept of obstruction components, which appeared in Theorem 0.5 for the first time and is indispensable to discuss $W_{1,0}$, $S_{1,0}$ and $U_{1,0}$.

In Definition 5.9 of Section 5 the concept of obstruction components was defined. In this section we show that they behave under an elementary transformation and a tie transformation just as in Definition 0.4. (Indeed, Definition 0.4 is not a definition but a theorem under the definition in Section 5. However, it is easier to understand it if we call it a definition in the Introduction.)

In the following we assume that $k \geq 4$ and $H = \mathbf{Z}u + \mathbf{Z}v$ ($u^2 = v^2 = 0$, $u \cdot v = 1$) denotes a hyperbolic plane.

We consider the behavior under elementary transformations first.

Let G' be a Dynkin graph. Consider the case where a full embedding

$$Q(G') \hookrightarrow (\Lambda_N/P) \oplus H$$

into the orthogonal complement of $u \in H$ is given. Moreover, we assume that G' has an obstruction component G'_0 of type A_k. By definition

$$[P(Q(G'_0), (\Lambda_N/P) \oplus H) : Q(G'_0)] = k + 1.$$

By $p : (\Lambda_N/P) \oplus H \to \Lambda_N/P$ we denote the projection to Λ_N/P. Let $M \subset \Lambda_N/P$ be a positive definite full root submodule containing the image $p(Q(G'))$, and $Q(M)$ be the root quasi-lattice of M. Recall that $M_0 = p(Q(G'))$ has the property in Theorem 6.3 (1).

Lemma 15.1. *The component Q_1 of $Q(M)$ containing the image $p(Q(G'_0))$ is also of type A_k and $[P(Q_1, \Lambda_N/P) : Q_1] = k + 1$.*

Proof. Let $\Delta' \subset Q(G')$ be a root basis, and $\Delta' = \bigcup_{i=1}^{m'} \Delta'_i$ be the decomposition into irreducible components. We assume that the component G'_0 corresponds to Δ'_1. Set $\Delta'_1 = \{\alpha'_1, \alpha'_2, \ldots, \alpha'_k\}$. We assign numbers to roots α'_i in Δ'_1 from the end of the Dynkin graph in order. Let

$$\omega'_1 = \{k\alpha'_1 + (k-1)\alpha'_2 + \cdots + 2\alpha'_{k-1} + \alpha'_k\}/(k+1)$$

be the first fundamental weight. By definition $\omega'_1 \cdot \alpha'_1 = 1$ and $\omega'_1 \cdot \alpha'_i = 0$ for $2 \leq i \leq k$. By assumption $\omega'_1 \in P(Q(G'_0), (\Lambda_N/P) \oplus H)$.

Set $\alpha_i = p(\alpha'_i)$ and $\omega_1 = p(\omega'_1)$. Let $\Delta \subset Q(M)$ be a root basis and $\Delta = \bigcup_{i=1}^{m} \Delta_i$ be the irreducible decomposition. By the theory of elementary transformations we can assume that $p(\Delta')$ is a subset of the extended root basis $\Delta^+ = \bigcup_{i=1}^{m} \Delta_i^+$. Let Δ_1^+ be the component containing $p(\Delta'_1)$. One has $\{\alpha_1, \ldots, \alpha_k\} \subset \Delta_1^+$.

Assume that Δ_1 is not of type A_k. Since $k \geq 4$, by the classification of finite irreducible root systems with a long root (Proposition 5.4 (1)), one knows that there exist number i with $1 \leq i \leq k$ and a long root $\beta \in \Delta_1^+$ such that $\beta \cdot \alpha_i = -1$ and $\beta \cdot \alpha_j = 0$ for $j \neq i$, $1 \leq j \leq k$. Then, we have

$$\omega_1 \cdot \beta = -(k+1-i)/(k+1) \notin \mathbf{Z}.$$

This contradicts Proposition 5.6 (1), since $\beta \in F$. Thus, Δ_1^+ is of type A_k and the index is $k+1$, since $\omega_1 \in P(Q_1, \Lambda_N/P)$, where Q_1 is the lattice generated by $p(\Delta_1')$. Q.E.D.

Next, we consider the situation when we go up from Λ_N/P to Λ_{N+1}/P.

Let R be a finite root system and $R = \bigoplus_{i=1}^m R_i$ be the irreducible decomposition. We assume that R_1 is of type A_k with $k \geq 4$. Assume that a full embedding $Q(R) \subset \Lambda_N/P$ is given and it satisfies $[\widetilde{Q}_1 : Q_1] = k+1$ for $Q_1 = Q(R_1)$ and $\widetilde{Q}_1 = P(Q_1, \Lambda_N/P)$. Let R' be a root subsystem of R obtained from R by one elementary transformation. We assume moreover that $R' \cap R_1 = R_1$.

Lemma 15.2. *Under the above situation there exists a full embedding*

$$\phi : Q(R') \hookrightarrow (\Lambda_N/P) \oplus H$$

satisfying the following conditions (1), (2) and (3). Moreover, there exists also a full embedding satisfying (1), (2) and (4). We denote here $Q_1' = \phi(Q_1)$ *and* $\widetilde{Q}_1' = P(Q_1', (\Lambda_N/P) \oplus H)$:

(1) The image $\phi(Q(R'))$ *of* ϕ *is orthogonal to* $u \in H$.

(2) The composition of ϕ *and the projection* $(\Lambda_N/P) \oplus H \to \Lambda_N/P$ *coincides with the given embedding* $Q(R') \hookrightarrow Q(R) \hookrightarrow \Lambda_N/P$.

(3) $[\widetilde{Q}_1' : Q_1'] < k+1$

(4) $[\widetilde{Q}_1' : Q_1'] = k+1$.

Proof. Let $\Delta \subset R$ be a root basis, $\Delta^+ = \bigcup_{i=1}^m \Delta_i^+$ be the extended root basis. By η_i we denote the maximal root for Δ_i. By definition $\Delta_i^+ = \Delta_i \cup \{-\eta_i\}$.

For every i with $1 \leq i \leq m$ we have a proper subset $\Delta_i' \subset \Delta_i^+$ and R' is the root system generated by $\bigcup_{i=1}^m \Delta_i'$. Now, Δ_1 is of type A_k and Δ_1^+ consists of $k+1$ elements. We have $k+1$ ways of choosing Δ_1', and under any choice the basis Δ_1' generates the same root system $R' \cap R_1$.

In order to define the embedding ϕ satisfying (3), we choose Δ_1' in such a way that $-\eta_1 \in \Delta_1'$. To define ϕ satisfying (4) we choose Δ_1' with $-\eta_1 \notin \Delta_1'$. $\Delta' = \bigcup_{i=1}^m \Delta_i'$ is a free basis of $Q(R')$. We define the embedding ϕ by setting for $\alpha \in \Delta_i'$

$$\phi(\alpha) = \begin{cases} \alpha \oplus 0 & (\text{if } \alpha \neq -\eta_i) \\ \alpha \oplus u & (\text{if } \alpha = -\eta_i). \end{cases}$$

Obviously, it defines an embedding of quasi-lattices satisfying (1) and (2). Moreover, fullness follows from Proposition 4.2 in Urabe [16].

We show the condition (3). Set $\Delta_1' = \{\alpha_1', \alpha_2', \ldots, \alpha_k'\}$. The numbers are assigned from the end of the Dynkin graph in order. If $-\eta_1 \in \Delta_1'$, then there is a number j $(1 \leq j \leq k)$ with $\alpha_j' = -\eta_1$. Now, if $[\widetilde{Q}_1' : Q_1'] \geq k+1$, then for $\omega_1 = \{k\alpha_1' + (k-1)\alpha_2' + \cdots + \alpha_k'\}/(k+1)$ we have $\phi(\omega_1) \in (\Lambda_N/P) \oplus H$. However, $\phi(\omega_1) = \omega_1 \oplus (k+1-j)u/(k+1) \notin (\Lambda_N/P) \oplus H$, which is a contradiction. We have (3).

When $-\eta_1 \notin \Delta_1'$, we have $\phi(\omega_1) = \omega_1 \oplus 0$, $\phi(\omega_1) \in \widetilde{Q}_1'$, and we have (4). Q.E.D.

By Lemma 15.1 and 15.2, one knows that the obstruction components behave following the rules in Definition 0.4 under elementary transformations.

We proceed to the relation between obstruction components and tie transformations.

Let G' be a Dynkin graph and $Q(G') \hookrightarrow \Lambda_{N+1}/P$ be a full embedding. We assume the following assumptions (O1) and (O2):

(O1) G' has a component G'_0 of type A_k with $k \geq 4$ such that

$$[P(Q(G'_0), \Lambda_{N+1}/P) : Q(G'_0)] = k+1.$$

(O2) For some root basis $\Delta' \subset Q(G')$, for some long root $\theta \in \Delta'$ and for an isotropic vector u belonging to F, $\theta \cdot u = 1$ and $\alpha \cdot u = 0$ for every $\alpha \in \Delta$ with $\alpha \neq \theta$.

Now, set $v = u - \alpha$, We have $u^2 = v^2 = 0$ and $u \cdot v = 1$. Setting $H = \mathbf{Z}u + \mathbf{Z}v$, $J = C(H, \Lambda_{N+1}/P)$, one has $\Lambda_{N+1}/P = J \oplus H$. Let $p : J \oplus H \to J$ denote the projection. Let $\Delta' = \bigcup_{i=1}^{m'} \Delta'_i$ be the irreducible decomposition. We assume that the component G'_0 corresponds to Δ'_1. By T we denote the submodule of $Q(G')$ generated by $\Delta' - \{\theta\}$.

Note here that $M_0 = p(T)$ has the property in Theorem 6.4 (1).

Let M be a positive definite full root submodule of $J \cong \Lambda_N/P$ containing $p(T)$, and G be the Dynkin graph of M. A root basis Δ of M is decomposed $\Delta = \bigcup_{i=1}^{m} \Delta_i$ into irreducible components. By $\Delta^+ = \bigcup_{i=1}^{m} \Delta_i^+$ we denote the extended root basis. By the theory of elementary transformations, we can assume that $p(\Delta' - \{\theta\}) \subset \Delta^+$.

Lemma 15.3. *(1) The vector u is necessarily orthogonal to $Q(G'_0)$. In particular, $\Delta'_1 \subset T$.*
(2) Let Δ_1^+ be the component of Δ^+ containing $p(\Delta'_1)$. Δ_1 is also of type A_k and for some unique root $\gamma \in \Delta_1^+$,

$$p(\Delta'_1) = \Delta_1^+ - \{\gamma\}.$$

Proof. (1) Set $\Delta'_1 = \{\beta_1, \beta_2, \ldots, \beta_k\}$. We assign numbers of β_i's from the end of the Dynkin graph in order. Let

$$\omega_1 = \{k\beta_1 + (k-1)\beta_2 + \cdots + \beta_k\}/(k+1)$$

be the first fundamental weight. $\omega_1 \cdot \beta_1 = 1$ and $\omega_1 \cdot \beta_i = 0$ for $2 \leq i \leq k$. By (O1) $\omega_1 \in P(Q(G'_0), \Lambda_{N+1}/P) = Q(G'_0)^*$. Assume that u is *not* orthogonal to $Q(G'_0)$. The root θ in (O2) belongs to Δ'_1. We have a number j with $\theta = \beta_j$, $1 \leq j \leq k$. We have

$$u \cdot \omega_1 = (k+1-j)/(k+1) \notin \mathbf{Z},$$

which is a contradiction.
(2) First, note that $\alpha \cdot x \in \mathbf{Z}$ for every long root $\alpha \in J$ and for every vector $x \in J$ by Proposition 5.6.

Assume that Δ_1 is *not* of type A_k. By the classification of root systems, one has a long root $\alpha \in \Delta_1^+$ and a root $\beta_i \in \Delta_1'$ $(1 \le i \le k)$ such that $\alpha \cdot p(\beta_i) = -1$ and $\beta \cdot p(\beta_j) = 0$ for $j \neq i$. By (O1) $p(\omega_1) \in J$. However,

$$\alpha \cdot p(\omega_1) = -(k+1-i)/(k+1) \notin \mathbf{Z},$$

which is a contradiction. The latter half is obvious, since Δ_1' has only $k+1$ elements. Q.E.D.

Finally, we consider the case when we go up from Λ_N to Λ_{N+1} by a tie transformation.

Let G be a Dynkin graph and $Q(G) \hookrightarrow \Lambda_N/P$ be a full embedding. By Δ we denote a root basis of $Q(G)$ and Δ^+ is the extended root basis.

We here assume that G' is a Dynkin graph obtained from G by one tie transformation and $Q(G') \hookrightarrow (\Lambda_N/P) \oplus H$ be the full embedding obtained by the transformation (Urabe [18] Section 1).

Corresponding to the procedure of the tie transformation, we have subsets A, $B \subset \Delta^+$ with $A \cap B = \emptyset$ satisfying the condition on G.C.D. with respect to coefficients of the maximal roots. We have

$$Q(G') = \sum_{\alpha \in \Delta^+ - (A \cup B)} \mathbf{Z}\alpha + \sum_{\alpha \in B} \mathbf{Z}(\alpha - u) + \mathbf{Z}(u+v),$$

and $\Delta' = [\Delta^+ - (A \cup B)] \cup \{\alpha - u \mid \alpha \in B\} \cup \{u+v\}$ is a root basis for $Q(G')$.

Here, assume moreover that G has an obstruction component G_0 of type A_k in Λ_N/P. We assume that the component Δ_1 and Δ_1^+ corrspond to G_0.

Lemma 15.4. *The following (1) and (2) are equivalent:*
(1) In $(\Lambda_N/P) \oplus H$, G' has an obstruction component G_0' of type A_k containing a vertex corresponding to a root in $\Delta_1^+ - A$.
(2) $\Delta_1^+ \cap B = \emptyset$ and $\Delta_1^+ \cap A$ consists of a unique element.

Proof. (1)⇒(2) It follows from Lemma 15.3.
(2)⇒(1) Under the assumption $\Delta_1' = \Delta_1^+ - A$ is a root basis of type A_k and is an irreducible component of Δ'. Let G_0' be the component of G' corresponding to Δ_1'. Then, we have

$$Q(G_0') = \sum_{\alpha \in \Delta_1'} \mathbf{Z}\alpha = \sum_{\beta \in \Delta_1} \mathbf{Z}\beta.$$

This implies that the embedding $Q(G_0') \hookrightarrow (\Lambda_N/P) \oplus H$ coincides with the composition of the identification $Q(G_0') = Q(G_0)$, the given embedding $Q(G_0) \hookrightarrow \Lambda_N/P$ and the embedding into the direct summand $\Lambda_N/P \hookrightarrow (\Lambda_N/P) \oplus H$. Thus, $[P(Q(G_0'), (\Lambda_N/P) \oplus H) : Q(G_0')] = k+1$, since we have the same number even if we erase the prime symbol and $\oplus H$. Q.E.D.

By Lemma 15.3 and 15.4 one knows that obstruction components behave following the rule in Definition 0.4 under a tie transformation.

16 Theorems with the Ik-conditions for $W_{1,0}$

In Section 16 through 19 we consider $W_{1,0}$. In this section Section 16 we would like to draw the Coxeter-Vinberg graph for Λ_2/P and to deduce some results from the graph.

For $W_{1,0}$ the discriminant group $P^*/P \cong P'^*/P'$ of $P = P' \oplus H_0$ has elements of the second kind. Elements of the third kind have the associated number 2 or 11 only. Thus, it is enough to consider only roots with length 1 or $\sqrt{2/3}$ as short roots. In addition, we have to count obstruction components of type A_{11}.

The discriminant group $P^*/P \cong P'^*/P'$ is a cyclic group of order 12. Let ξ be a generator. We can assume that for the discriminant quadratic form q_P, $q_P(\xi) \equiv 13/12 \bmod 2\mathbf{Z}$.

On the other hand, for the root lattice $Q = Q(A_{11})$ of type A_{11}, $(Q \oplus H)^*/(Q \oplus H) \cong Q^*/Q$ is also a cyclic group of order 12. It has a generator η with $q_{Q\oplus H}(\eta) \equiv 11/12 \bmod 2\mathbf{Z}$, where $q_{Q\oplus H}$ denotes the discriminant quadratic form. Thus, $q_P = -q_{Q\oplus H}$.

By Lemma 5.5 there exists a primitive embedding $P \hookrightarrow \Lambda_2$ such that the orthogonal complement F of P in Λ_2 is isomorphic to $Q \oplus H$. Thus, by Proposition 5.2 (2), for every primitive embedding $P \hookrightarrow \Lambda_2$, the orthogonal complement F of P in Λ_2 is isomorphic to $Q \oplus H$.

Now, let $K = \sum_{i=0}^{13} \mathbf{Z} v_i$ be the odd unimodular lattice with signature (13, 1). We assume that the basis satisfies $v_0^2 = -1$, $v_i^2 = +1$ $(1 \leq i \leq 13)$, $(v_i, v_j) = 0$ $(i \neq j)$. Setting $w = v_0 + v_1 + \cdots + v_{13}$, we define M to be the orthogonal complement of $\mathbf{Z}w$ in K. Set

$$g = v_0 + v_{13}, \qquad h = -(v_0 + v_{12}),$$
$$f_i = -v_i + v_{i+1} \quad (1 \leq i \leq 10), \qquad f_{11} = v_0 - v_{11} + v_{12} + v_{13}.$$

Vectors g, h, $f_1, \ldots, f_{11}$ are a basis for M. $H = \mathbf{Z}g + \mathbf{Z}h$ is a hyperbolic plane, since $g^2 = h^2 = 0$ and $(g, h) = 1$. Vectors $f_1, \ldots, f_{11}$ form a root basis of type A_{11}, which is orthogonal to g and h. Thus, one knows $M \cong Q(A_{11}) \oplus H$.

Let $p : K \otimes \mathbf{Q} \to M \otimes \mathbf{Q}$ be the orthogonal projection. By definition $p(x) = x - (x, w)w/12$, and $p(x) = x$ if, and only if, $x \in M \otimes \mathbf{Q}$. For every x, $y \in K \otimes \mathbf{Q}$, $(p(x), p(y)) = (x, p(y)) = (p(x), y)$.

Set

$$r = p(v_0) = (13v_0 + v_1 + \cdots + v_{13})/12.$$

$r^2 = -13/12$ and $r \in M \otimes \mathbf{Q}$.

Note here that g and $e_i = -v_i + v_{i+1}$ $(1 \leq i \leq 12)$ are also a basis of M. $(r, g) = (v_0, g) = -1$. $(r, e_i) = (v_0, e_i) = 0$ $(1 \leq i \leq 12)$. Thus, $M^* \supset M + \mathbf{Z}r$. On the other hand, $[M^* : M] = |d(M)| = 12$ and $[M + \mathbf{Z}r : M] = 12$. One knows $M^* = M + \mathbf{Z}r$. Since $\Lambda_2/P \cong F^*$ and $F \cong Q(A_{11}) \oplus H \cong M$, one has an isomorphism of quasi-lattices

$$\Lambda_2/P \cong M + \mathbf{Z}r.$$

We apply Vinberg's algorithm with the controlling vector r to the right-hand side above. As the root basis orthogonal to r, we take

$$e_i = -v_i + v_{i+1} \qquad (1 \leq i \leq 12).$$

By the algorithm we get successively

$$\begin{aligned} e_{13} &= v_0 + v_1 + v_2 - v_{13} \\ e_{14} &= \{3v_0 + (v_1 + \cdots + v_8) - (v_9 + \cdots + v_{13})\}/2 \\ e_{15} &= \{4v_0 + (v_1 + \cdots + v_{10}) - 2(v_{11} + v_{12} + v_{13})\}/3 \\ e_{16} &= \{5v_0 + 2(v_1 + \cdots + v_6) - (v_7 + \cdots + v_{13})\}/3 \\ e_{17} &= \{5v_0 + (v_1 + \cdots + v_{11}) - 3(v_{12} + v_{13})\}/2 \\ e_{18} &= \{7v_0 + 3(v_1 + \cdots + v_5) - (v_6 + \cdots + v_{13})\}/2. \end{aligned}$$

Drawing the graph for this system of 18 vectors, we get the following:

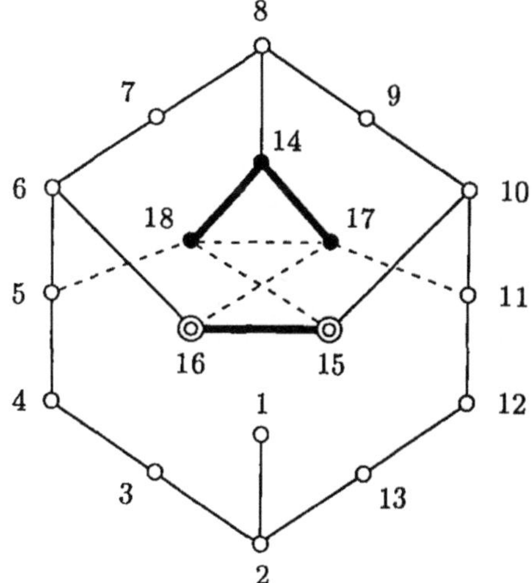

We would like to apply Proposition 8.5 to this graph. The condition $\langle a \rangle$ in Proposition 8.5 is easily checked. However, it has many dotted edges and we have to check the condition $\langle b \rangle$. Note that any Lannér subgraph contained in the above graph has only two vertices and a dotted edge.

Lemma 8.6 can be applied. Let Γ be the subgraph of the above consisting the vertices 17 and 18. The unique edge in Γ is dotted, and Γ is a Lannér subgraph. The corresponding subgraph Ξ consists of the vertices 6, 7, 8, 9, 10 and 1, 2, 3, 4, 12, 13, and Ξ is a Dynkin graph of type $E_6 + A_5$. In this case, $S(\Gamma) \cup S(\Xi)$ consists of 13 linearly independent vectors, and $13 = \text{rank}\ \Lambda_2/P$. Thus, the condition $(y, \alpha) = 0$ for $\alpha \in S(\Gamma) \cup S(\Xi)$ implies $y = 0$. By Lemma 8.6, Proposition 8.5 $\langle b \rangle$ holds for Γ.

Note that because of the symmetry of the graph we can check $\langle b \rangle$ for only two other Lannér subgraphs. The same method can be applied to them. By Proposition 8.5 the above is the Coxeter-Vinberg graph of Λ_2/P for $W_{1,0}$.

Corollary 16.1. *Let P be the lattice defined for $W_{1,0}$ and $L \subset \Lambda_2/P$ be a positive definite full root submodule. Assume that the Dynkin graph of L does not contain a component of type B_1. Then, there is a primitive isotropic vector $u \in \Lambda_2/P$ in a nice position with respect to L such that the root system of $(\mathbf{Z}u)^{\perp}/\mathbf{Z}u$ is of type $E_8+G_2+B_1$, B_9+G_2, $E_7+B_3+G_1$ or A_{11}.*

Proof. We can assume that a root basis Δ_L of L is a subset of the above $\{e_1, \dots, e_{18}\}$. The graph of Δ_L is a Dynkin graph and does not contain a dotted edge, a bold edge, or an extended Dynkin graph. In particular, either $e_{17} \notin \Delta_L$ or $e_{18} \notin \Delta_L$. By symmetry of the graph we can consider only the case $e_{18} \notin \Delta_L$. If $e_{18} \notin \Delta_L$ and $e_{17} \in \Delta_L$, then e_{11}, e_{14}, $e_{16} \notin \Delta_L$, since the graph has no dotted edge and no bold edge. If e_{18}, $e_{17} \notin \Delta_L$, then either $e_{15} \notin \Delta_L$ or $e_{16} \notin \Delta_L$, since the graph has no bold edge. It suffices to consider only the case $e_{16} \notin \Delta_L$ by symmetry. Next, consider vectors e_{10}, e_{11}, e_{15}. These form an extended Dynkin graph of type G_2 and at least one of them does not belong to Δ_L. We have 4 cases to be considered.

Case (1) : $e_{11}, e_{14}, e_{16}, e_{18} \notin \Delta_L$
Case (2) : $e_{11}, e_{16}, e_{17}, e_{18} \notin \Delta_L$
Case (3) : $e_{10}, e_{16}, e_{17}, e_{18} \notin \Delta_L$
Case (4) : $e_{15}, e_{16}, e_{17}, e_{18} \notin \Delta_L$.

Case (1) and (2).

In these cases $e_{11}, e_{16}, e_{18} \notin \Delta_L$. Consider $u_1 = -(e_{14}+e_{17})$. $u_1^2 = 0$ and $(u_1, e_i) = 0$ for $i \neq 8, 11, 16, 18$, $1 \le i \le 18$. This u_1 is a primitive isotropic vector with $(u_1, e_8) = 1$ and $(u_1, \alpha) = 0$ for $\alpha \in \Delta_L - \{e_8\}$.

Consider the subgraph in the above Coxeter-Vinberg graph consisting of all the vertices connected to neither the vertex 14 nor 17, plus the vertices 14 and 17. This subgraph is the extended Dynkin graph of type $E_8+G_2+B_1$. Thus, the root system of $(\mathbf{Z}u_1)^{\perp}/\mathbf{Z}u_1$ is of type $E_8+G_2+B_1$.

Case (3).

Consider $u_2 = -(e_{15}+e_{16})$. $u_2^2 = 0$. $(u_2, e_i) = 0$ for $i \neq 6, 10, 17, 18$, $1 \le i \le 18$, and, thus, $(u_2, \alpha) = 0$ for every $\alpha \in \Delta_L - \{e_6\}$. Moreover, $(u_2, e_6) = 1$. This u_2 is a primitive isotropic vector in a nice position with respect to L and the root system of $(\mathbf{Z}u_2)^{\perp}/\mathbf{Z}u_2$ is of type $E_7+B_3+G_1$.

Case (4).

Four vectors e_7, e_8, e_9, e_{14} form the extended Dynkin graph of type B_3. Thus, one of them does not belong to Δ_L. Depending one of the four, we have four subcases in Case (4).

Case (4.1). $e_7, e_{15}, e_{16}, e_{17}, e_{18} \notin \Delta_L$.

Consider $u_3 = -(e_5+2e_6+3e_{16})$. $u_3^2 = 0$. $(u_3, e_i) = 0$ for $i \neq 4, 7, 15, 17, 18$, $1 \le i \le 18$. Thus, for every $\alpha \in \Delta_L - \{e_4\}$, $(u_3, \alpha) = 0$. Moreover, $(u_3, e_4) = 1$. This u_3 is primitive and the root system of $(\mathbf{Z}u_3)^{\perp}/\mathbf{Z}u_3$ is of type B_9+G_2.

Case (4.2). $e_9, e_{15}, e_{16}, e_{17}, e_{18} \notin \Delta_L$.

By symmetry of the graph, the reasoning is the same as in (4.1). We can consider the vector $u_3' = -(2e_{10}+e_{11}+3e_{15})$.

Case (4.3). $e_{14}, e_{15}, e_{16}, e_{17}, e_{18} \notin \Delta_L$.

Consider $u_4 = -(e_2 + e_3 + \cdots + e_{13})$. $u_4^2 = 0$. $(u_4, e_i) = 0$ $(2 \le i \le 13)$; thus, for every $\alpha \in \Delta_L - \{e_1\}$, $(u_4, \alpha) = 0$. Moreover, $(u_4, e_1) = 1$. This u_4 is primitive and the root system of $(\mathbf{Z}u_4)^{\perp}/\mathbf{Z}u_4$ is of type A_{11}.

Case (4.4). $e_8, e_{15}, e_{16}, e_{17}, e_{18} \notin \Delta_L$.

This is the last remaining case. If $e_{14} \in \Delta_L$ in this case, then $\{e_{14}\}$ is an irreducible component of Δ_L of type B_1, which contradicts the assumption. Thus, $e_{14} \notin \Delta_L$, and Case (4.4) is reduced to the above case (4.3). Q.E.D.

Corollary 16.2. *Let P be the lattice associated with the case $W_{1,0}$. The root system of the quotient quasi-lattice Λ_1/P is of type G for some primitive embedding $P \hookrightarrow \Lambda_1$ if, and only if, $G = E_8 + G_2 + B_1$, $E_7 + B_3 + G_1$, $B_9 + G_2$, or A_{11}.*

Moreover, if the root system of Λ_1/P is of type A_{11}, then Λ_1/P is isomorphic to the dual quasi-lattice Q^* of the root lattice $Q = Q(A_{11})$ of type A_{11}. In particular, for any full embedding $Q(A_{11}) \hookrightarrow \Lambda_1/P$, the component A_{11} is an obstruction component.

Proof. The former half is obvious. We have the latter half since the extended Dynkin graph of type A_{11} is unique in the above Coxeter-Vinberg graph. Q.E.D.

Lemma 16.3. *Let $G' + B_1$ be a Dynkin graph containing a component of type B_1. Let G be a Dynkin graph obtained from $G' + B_1$ by one tie or elementary transformation. If G contains no vertex corresponding to a short root, then G can be obtained even from G' by the same transformation.*

The Theorem 16.4 below follows from Corollary 16.1 and 16.2. Note here that Corollary 16.1 contains an additional condition "L contains no component of type B_1." However, the influence of this condition does not appear in Theorem 0.5 and Theorem 16.4. This is because Lemma 16.3 above holds.

Theorem 16.4. *We consider the Ik-conditions in the case $W_{1,0}$. (See Section 7.) By r we denote the number of vertices in the Dynkin graph G. $Q = Q(G)$ stands for the root (quasi-)lattice of type G. $r = \operatorname{rank} Q$.*

[I] The following two conditions (a) and (b) are equivalent:

(a) $G \in PC(W_{1,0})$ and the I2-condition holds for the root lattice $Q = Q(G)$ of type G.

(b) G contains no vertex corresponding to a short root and can be obtained from one of the 4 basic Dynkin graphs in Theorem 0.5 by elementary transformations repeated twice.

[II] The following two conditions (A) and (B) are also equivalent:

(A) $G \in PC(W_{1,0})$ and the I1-condition holds for the root lattice $Q = Q(G)$ of type G.

(B) G contains no vertex corresponding to a short root and can be obtained from one of the 4 basic Dynkin graphs in Theorem 0.5 by one of the following 3 kinds of procedures:

The procedures:

⟨1⟩ elementary transformations repeated twice

⟨2⟩ an elementary transformation following after a tie transformation

⟨3⟩ a tie transformation following after an elementary transformation.

17
Two singular fibers of type I^* in the case $W_{1,0}$

By the results up to Section 16 the remaining part of the verification of our Theorem 0.5 is to show the following:

Proposition 17.1. *Assume that $G \in PC(W_{1,0})$ and that $Q = Q(G)$ does not satisfy the I1-condition. Then, with respect to some full embedding $Q(G) \hookrightarrow \Lambda_3/P$ without an obstruction component A_{11}, there exists a primitive isotropic vector u in Λ_3/P in a nice position, i.e., such that either u is orthogonal to $Q(G)$, or there is a root basis $\Delta \subset Q(G)$ and a long root $\theta \in \Delta$ such that $\alpha \cdot u = 0$ for every $\alpha \in \Delta$ with $\alpha \neq \theta$ and $\theta \cdot u = 1$.*

To show this we apply the theory developed in Chapter 2.

Now, if $G \in PC = PC(W_{1,0})$, then we have an embedding $S = P \oplus Q(G) \hookrightarrow \Lambda_3$ come from an actual deformation fiber Y. The embedding satisfies Looijenga's conditions (L1) and (L2) and the induced embedding $Q(G) \hookrightarrow \Lambda_3/P$ is full.

Also, we have an elliptic K3 surface $\Phi : Z \to C$ corresponding to the embedding. By $\Sigma = \{c_1, \dots, c_t\}$ we denote the set of critical values of Φ. By F_i with $1 \leq i \leq t$ we denote the singular fiber $\Phi^{-1}(c_i)$. The elliptic surface $Z \to C$ has a singular fiber F_1 of type I_0^* in our situation by definition. Moreover, it has two sections $s_0, s_1 : C \to Z$ whose images $C_5 = s_0(C)$ and $C_6 = s_1(C)$ are disjoint. The union $IF = F_1 \cup C_5 \cup C_6$ is called the curve at infinity. The lattice P has signature $(6, 1)$ and it is defined associated with the dual graph of components of IF. The union $\mathcal{E}$ of smooth rational curves on Z disjoint from IF coincides with the union of components disjoint from IF of singular fibers of $Z \to C$. The dual graph of $\mathcal{E}$ is G by definition.

We use the same division of the case into three subcases as in Chapter 2.

((1)) The surface $Z \to C$ has another singular fiber of type I^* apart from F_1.

((2)) $Z \to C$ has a singular fiber of type II^*, III^* or IV^*.

((3)) $Z \to C$ has no singular fiber of type I^*, II^*, III^* or IV^* apart from F_1.

For each case we apply the theory in Chapter 2. However, for $W_{1,0}$ IF contains 2 sections, and the theory in Chapter 2 are not sufficient to treat $W_{1,0}$. Thus, in the first step of the verification we write down the list of possible Dynkin graphs, and then we check each item G in the list case by case. We show either G can be made from one of the basic graphs by two transformations such that the resulting graph has no obstruction component, or $G \notin PC$. To show $G \notin PC$ we apply the theory of symmetric bilinear forms, the theory of elliptic surfaces, and the theory of K3 surfaces, etc.

The case ((i)) is discussed in Section 16+i. In this section we treat the case ((1)).

In the following by G we denote a Dynkin graph belonging to $PC(W_{1,0})$ with the number of vertices r. We assume that the I1-condition fails for $Q = Q(G)$.

We assume that apart from the singular fiber F_1 of type I_0^*, F_2 is a singular fiber of $\Phi : Z \to C$ of type I^* in this section.

We have an embedding $S = P \oplus Q(G) \hookrightarrow \Lambda_3$ satisfying (L1) and (L2). Recall that the lattice P has a basis $e_0, e_1, \dots, e_6$ whose mutual intersection numbers are described by the dual graph in Section 4 "The case of $W_{1,0}(2, 2, 3, 3)$".

We set

$$u_0 = 2e_0 + e_1 + e_2 + e_3 + e_4, \quad v_0 = -u_0 - e_5, \quad f = e_6 - e_5 - 2u_0.$$

$u_0^2 = v_0^2 = 0$, $u_0 \cdot v_0 = 1$, $P = P' \oplus (\mathbf{Z}u_0 + \mathbf{Z}v_0)$, and P' has a basis e_0, e_1, e_2, e_3, f.

By Proposition 10.3 in Chapter 2 there exists a vector $u \in \Lambda_3$ satisfying the following conditions:
(1) u is isotropic.
(2) u is orthogonal to $Q(G)$.
(3) u is orthogonal to e_0, e_1, e_2, e_3, e_4, and e_5.
(4) $u \cdot e_6 > 0$.

Set $m = u \cdot e_6 = u \cdot f$. (Note that $m \neq 0$ since the I1-condition fails.)

The vector u is orthogonal to $\mathbf{Z}u_0 + \mathbf{Z}v_0$. Set $M = P' + \mathbf{Z}u$. M has signature $(5, 1)$.

Lemma 17.2. *Assume that M is not primitive in Λ_3. Then, the primitive hull $\widetilde{M}$ of M in Λ_3 contains an element u' satisfying the above conditions (1)–(4) such that $u \cdot e_6 > u' \cdot e_6 > 0$.*

Proof. By $w_0, \ldots, w_3, z$ we denote the dual basis of $e_0, \ldots, e_3, f$. In particular, $z = (2e_0 + e_1 + e_2 + 2e_3 + 2f)/6$. Set $\xi = m(2e_0 + e_1 + e_2 + 2e_3 + 2f) - 6u$. It satisfies $\xi \in M$, $\xi \cdot e_i = 0$ $(0 \le i \le 3)$, $\xi \cdot f = 0$, $\xi \cdot u = 2m^2$ and $\xi^2 = -12m^2$. The element $y_0 = \xi/12m^2$ satisfies $y_0 \cdot \xi = -1$ and $u = mz - 2m^2 y_0 = m(z - 2my_0)$. Set $N = P' \oplus \mathbf{Z}\xi$. We have $N \subset M \subset \widetilde{M} \subset \widetilde{M}^* \subset M^* \subset N^*$. (Recall that by * we denote the dual module.) Consider the discriminant group $N^*/N = P'^*/P' \oplus \mathbf{Z}y_0/\mathbf{Z}\xi$. On this group we can define the discriminant bilinear form b and the discriminant quadratic form q. For $x \in N^*$ we denote $\overline{x} = x \bmod N^* \in N^*/N$. Set $I = M/N$. $I^\perp = M^*/N$ is the orthogonal complement of I with respect to b. I is generated by a unique element $\overline{u} = m\overline{z} - 2m^2\overline{y}_0$. On the other hand, P'^*/P' is a cyclic group of order 12 generated by $\overline{w}_1$, and $\overline{z} = 2\overline{w}_1$. (Recall $w_1^2 = 13/12$; thus, $\overline{w}_1^2 \equiv 13/12 \bmod 2\mathbf{Z}$). Since

$$b(2m\overline{w}_1 - 2m^2\overline{y}_0, a\overline{w}_1 + b\overline{y}_0) \equiv \frac{1}{6}ma + \frac{1}{6}b \bmod \mathbf{Z},$$

$a\overline{w}_1 + b\overline{y}_0 \in I^\perp \Leftrightarrow ma + b \equiv 0 \pmod 6$. Choose an element $\overline{x}_0 \in \widetilde{M}/N$ with $\overline{x}_0 \notin I = M/N$. Since it is contained in $I^\perp$, we can write $\overline{x}_0 = a(\overline{w}_1 - m\overline{y}_0) + 6c\overline{y}_0$. Moreover,

$$0 \equiv q(\overline{x}_0) \equiv a^2 + \frac{acm - 3c^2}{m^2} \bmod 2\mathbf{Z}.$$

In particular, $c^2 \equiv am(am + c) \pmod 2$, and c is even. We set $c = 2d$.

First, we would like to show that there is $\overline{x}_1 \in \widetilde{M}/N$ with $\overline{x}_1 \notin I$ in the form either $\overline{x}_1 = 2A(\overline{w}_1 - m\overline{y}_0)$ (In this case, no restriction on m.), or $\overline{x}_1 = 2A(\overline{w}_1 + m\overline{y}_0)$ and $m \equiv 0 \pmod 3$.

◇ Case 1. $am - 12d \not\equiv 0 \pmod{2m}$.

Set $\overline{x}_1 = m\overline{x}_0$. Since $12m\overline{y}_0 = -12(\overline{w}_1 - m\overline{y}_0)$, we have $\overline{x}_1 = am(\overline{w}_1 - m\overline{y}_0) + d(12m\overline{y}_0) = a'(\overline{w}_1 - m\overline{y}_0)$ for $a' = am - 12d$. The assumption $a' \not\equiv 0 \pmod{2m}$ implies $\overline{x}_1 \notin I$. Since $0 \equiv q(\overline{x}_1) \equiv a'^2 \bmod 2\mathbf{Z}$, we can write $a' = 2A$. This $\overline{x}_1$ satisfies the condition.

◇ Case 2. $am - 12d \equiv 0 \pmod{2m}$.

Set $12d = am + 2me$. $\overline{x}_0 = a\overline{w}_1 + 2me\overline{y}_0$. Then, $q(\overline{x}_0) \equiv 0 \bmod 2\mathbf{Z} \Leftrightarrow 13a^2 - 4e^2 \equiv 0 \pmod{24}$. We can write $a = 2A'$ and we have $(A'+e)(A'-e) \equiv 0 \pmod 6$.

If $A'+e \equiv 0 \pmod 3$, set $3C = A'+e$. We have $C^2 \equiv C(2A'-3C) \equiv 0 \pmod 2$. C is even. Thus, $\overline{x}_1 = \overline{x}_0 = 2A'(\overline{w}_1 - m\overline{y}_0) + 6mC\overline{y}_0 = 2A(\overline{w}_1 - m\overline{y}_0)$ for $A = A' - 3C$.

If $A' - e \equiv 0 \pmod 3$, set $3C = -A' + e$. We have $C^2 \equiv 0 \pmod 2$. C is even. Thus, $\overline{x}_1 = \overline{x}_0 = 2A'(\overline{w}_1 + m\overline{y}_0) + 6mC\overline{y}_0 = 2A(\overline{w}_1 + m\overline{y}_0)$ for $A = A' + 3C$. If $m \equiv 0 \pmod 3$ we have done. Thus, we can assume $m \not\equiv 0 \pmod 3$. Since $\overline{x}_1 \in I^{\perp}$, $Am \equiv 0 \pmod 3$; thus, $A \equiv 0 \pmod 3$. Then, $\overline{x}_1 = 2A(-\overline{w}_1 + m\overline{y}_0) = 2(-A)(\overline{w}_1 - m\overline{y}_0)$, since $6\overline{w}_1 = -6\overline{w}_1$.

Next, we consider the case $\overline{x}_1 = 2A(\overline{w}_1 - m\overline{y}_0)$. If we write $A = Cm + B$ $(0 \le B < m)$, then $B \ne 0$, since $\overline{x}_1 \notin I$. The element $\overline{x}_2 = \overline{x}_1 - 2Cm(\overline{w}_1 - m\overline{y}_0) = 2B(\overline{w}_1 - m\overline{y}_0)$ satisfies $\overline{x}_2 \in \widetilde{M/N}$ and $\overline{x}_2 \notin I$.

On the other hand, we can check that $u' = Bu/m$ is a vector in N^*, $u'^2 = 0, u' \ne 0$, $u' \cdot e_i = 0$ for $0 \le i \le 5$ and $m = u \cdot e_6 > B = u' \cdot e_6$. Moreover, $u' \in \widetilde{M}$ since $\overline{u}' = \overline{x}_2$. We have the desired element.

The case $\overline{x}_1 = 2A(\overline{w}_1 + m\overline{y}_0)$, $m \equiv 0 \pmod 3$ is remaining. In this case, $2m\overline{w}_1 = -2m\overline{w}_1$ since $\overline{w}_1$ has order 12. If A is a multiple of m, then $\overline{x}_1 = 2A(\overline{w}_1 + m\overline{y}_0) = (-A)\cdot 2(\overline{w}_1 - m\overline{y}_0) \in I$, which is a contradiction. Thus, we can write $A = Cm + B$ $(0 < B < m)$. Setting $\overline{x}_2 = 2B(\overline{w}_1 + m\overline{y}_0) = \overline{x}_1 - 2Cm(\overline{w}_1 + m\overline{y}_0) = \overline{x}_1 + 2Cm(\overline{w}_1 - m\overline{y}_0)$, we have $\overline{x}_2 \in \widetilde{M/N}$, $\overline{x}_2 \notin I$.

On the other hand, setting $u'' = B(z + 2my_0) = B(2z - (u/m))$, we can check $u''^2 = 0$, $u'' \ne 0$, $u'' \cdot e_i = 0$ $(0 \le i \le 5)$, $u'' \cdot u_0 = u'' \cdot v_0 = 0$, and $u'' \cdot f = u'' \cdot e_6 = B$. Since $u'' \in N^*$ and since $\overline{u}'' = \overline{x}_2$, one knows $u'' \in \widetilde{M}$. This u'' is the desired vector.

Q.E.D.

By induction on $u \cdot e_6$ we can assume that u satisfies the following (5) in addition to (1)–(4):

(5) $M = P' + \mathbf{Z}u$ is primitive in Λ_3.

Then, of course, $M \oplus (\mathbf{Z}u_0 + \mathbf{Z}v_0) = P + \mathbf{Z}u$ is also primitive in Λ_3.

Proposition 17.3. *Assume that we have a vector $u \in \Lambda_3$ satisfying above (1)–(4) and (5).*

1. *If $u \cdot e_6 = 1$, then G is a subgraph of the Coxeter-Vinberg graph Γ of the lattice $Q(D_{12}) \oplus H$. (H denotes a hyperbolic plane.)*
2. *If $u \cdot e_6 \ge 2$, then G can be obtained from a subgraph of the above Γ by one elementary transformation.*

Proof. **1.** Setting $v = f - 2u$, one has

$$M = (\mathbf{Z}e_0 + \mathbf{Z}e_1 + \mathbf{Z}e_2 + \mathbf{Z}(e_3 + u)) \oplus (\mathbf{Z}u + \mathbf{Z}v),$$

and $u^2 = v^2 = 0$, $u \cdot v = 1$. Thus, $P + \mathbf{Z}u \cong Q(D_4) \oplus H \oplus H$. For their discriminant quadratic forms we know $q_{P+\mathbf{Z}u} = q_{Q(D_4)}$.

Let L be the orthogonal complement of $P + \mathbf{Z}u$ in Λ_3. L has signature (13, 1). The discriminant quadratic form of L is $-q_{P+\mathbf{Z}u} = -q_{Q(D_4)} = q_{Q(D_4)}$. Since $Q(D_{12}) \oplus H$

and L have the same signature and the same discriminant quadratic form, they are isomorphic by Lemma 5.5 (4) and Lemma 12.4. In particular, the Coxeter-Vinberg graph of L coincides with Γ. Since $Q(G)$ is full in L, G is a subgraph of Γ.

2. Assume that $m = u \cdot e_6 \geq 2$.

By L we denote the orthogonal complement of $R = P + \mathbf{Z}u$ in Λ_3. We have a natural isomorphism $R^*/R \cong L^*/L$ preserving discriminant quadratic forms up to sign.

Set $u_1 = u/m$ and $R_1 = R + \mathbf{Z}u_1$. R_1 is an even overlattice of R with index m, and is isomorphic to $P + \mathbf{Z}u$ in the case of $m = u \cdot e_6 = 1$. In particular, the discriminant quadratic form of R_1 is the same as that of $Q(D_4)$.

By the above isomorphism one knows that L has an overlattice L_1 with index m whose discriminant quadratic form is $q_{Q(D_4)} = -q_{Q(D_4)}$. By reasoning in **1** $L_1 \cong Q(D_{12}) \oplus H$.

Let $\tilde{Q}_1$ (resp. $\tilde{Q}$) be the primitive hull of $Q(G)$ in L_1 (resp. L). The Dynkin graph of $\tilde{Q}_1$ is a subgraph of the Coxeter-Vinberg graph Γ of L_1. Since $\tilde{Q}_1/\tilde{Q} \subset L_1/L \cong \mathbf{Z}/m$ is cyclic, the Dynkin graph of $\tilde{Q}$ is obtained from that of $\tilde{Q}_1$ by one elementary transformation. Besides, by the fullness the Dynkin graph of $\tilde{Q}$ is G. Q.E.D.

Now, we would like to draw the Coxeter-Vinberg graph Γ of $Q(D_{12}) \oplus H$.

Set $K = \sum_{i=0}^{13} \mathbf{Z}v_i$ where $v_0, \ldots, v_{13}$ is a free basis with $v_0^2 = -1, v_i^2 = 1$ $(1 \leq i \leq 13), v_i \cdot v_j = 0$ $(i \neq j)$. The sublattice $L = \left\{\sum_{i=0}^{13} x_i v_i \in K \mid \sum_{i=0}^{13} x_i \in 2\mathbf{Z}\right\}$ is isomorphic to $Q(D_{12}) \oplus H$. We use v_0 as the controlling vector. We can take

$$\gamma_i = -v_i + v_{i+1} \quad (1 \leq i \leq 12)$$
$$\gamma_{13} = -(v_{12} + v_{13})$$

as a root basis for the orthogonal complement of v_0 in L. By Vinberg's algorithm we get successively,

$$\gamma_{14} = v_0 + v_1 + v_2 + v_3$$
$$\gamma_{15} = 3v_0 + v_1 + v_2 + \cdots + v_{11}.$$

Drawing the graph for these 15 vectors, we get the following graph. This has no dotted edges, no Lannér subgraph and any extended Dynkin subgraph is a component of an extended Dynkin subgraph of rank 12. Thus, this is Γ:

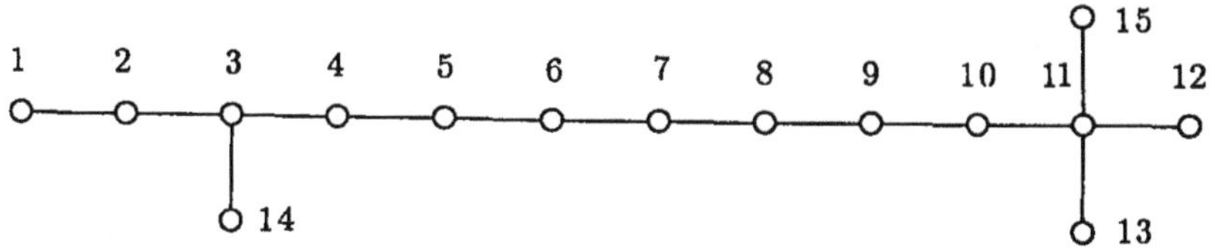

The Coxeter-Vinberg graph Γ for $Q(D_{12}) \oplus H$

Lemma 17.4. *There are 27 kinds of maximal Dynkin subgraphs of Γ. All of them but the last one (27) have 13 vertices. The following is the list:*

(1) D_{13}	(2) $D_{12}+A_1$	(3) $D_{10}+A_2+A_1$
(4) D_9+A_4	(5) D_8+D_5	(6) D_7+E_6
(7) E_7+D_6	(8) E_8+D_5	(9) $A_{10}+3A_1$
(10) $D_{10}+3A_1$	(11) A_9+4A_1	(12) $A_7+A_2+4A_1$
(13) $A_6+A_4+3A_1$	(14) $D_5+A_5+3A_1$	(15) $E_6+A_4+3A_1$
(16) $E_7+A_3+3A_1$	(17) $E_8+A_2+3A_1$	(18) A_9+D_4
(19) D_9+D_4	(20) $A_8+D_4+A_1$	(21) $A_6+D_4+A_2+A_1$
(22) $A_5+A_4+D_4$	(23) $D_5+D_4+A_4$	(24) $E_6+D_4+A_3$
(25) $E_7+D_4+A_2$	(26) $E_8+D_4+A_1$	(27) D_8+D_4

Lemma 17.5. *(1) If $\rho=r+7$, then the group MW of sections of Φ has rank 1 and only one component of the singular fiber F_i intersects IF for every $2\le i\le t$.*
(2) G has a component of type D, has 11, 12, or 13 vertices, and can be obtained from one of the 26 Dynkin graphs with 13 vertices in Lemma 17.4 by one elementary transformation.

Proof. (1) For $2\le i\le t$ by $n(F_i)$ we denote the number of components of F_i disjoint from the curve IF at infinity. The equality $\sum_{i=2}^{t} n(F_i)=r$ holds. Thus, by the equality (2) in the beginning of Section 10 and by assumption we have

$$1+\sum_{i=2}^{t} n(F_i)=a+\sum_{i=2}^{t}(m(F_i)-1),$$

since $m(F_1)=5$.

Assume $a=0$. Then, all elements in MW have finite order. In particular, the vector u in Proposition 10.3 is orthogonal also to the class of $C_6=s_1(C)$. Thus, u is orthogonal to the subgroup S generated by the classes in the union of the set of components of IF and the set of components disjoint from IF of singular fibers. Since u is isotropic, this contradicts the assumption that the I1-condition fails. Thus, $a>0$.

Since $n(F_i)\le m(F_i)-1$, by the above equality we have $a=1$ and $n(F_i)=m(F_i)-1$ for $2\le i\le t$.

(2) Recall here that by Theorem 2.2 we can assume further that the Picard number $\rho=r+7$. ($7=\text{rank } P$.) Thus, by (1), in particular, only one component of F_2 intersects IF. Note that the intersecting component has multiplicity 1. The dual graph of components of F_2 without intersection with IF is of type D and it is a component of G.

By Proposition 17.3 and by the theory of elementary transformations G can be made from a maximal Dynkin subgraph G' of Γ by an elementary transformation. By Lemma 8.7 G' has either 13 or 12 vertices. If it has 12 vertices, $C(Q(G),\ Q(D_{12})\oplus H)$ contains an isotropic vector by Lemma 8.7. Since $Q(D_{12})\oplus H\cong C(P+\mathbf{Z}u,\ \Lambda_3)$, $C(Q(G)\oplus P,\ \Lambda_3)$ contains an isotropic vector, which contradicts the assumption that the I1-condition fails. Thus, G' has 13 vertices.

Since the I1-condition fails, $r=11$, 12 or 13. Q.E.D.

In what follows we assume $\rho = r+7$. By Theorem 2.2 we do not lose any generality even if we assume this condition. For every singular fiber F_i with $2 \le i \le t$, let G_i be the Dynkin graph defined as the dual graph of components of F_i disjoint from C_5. Note that by Lemma 17.5 (1) $G = \sum G_i$.

Lemma 17.6. *(1)* $\nu(A) + 2\nu(D) + 2\nu(E) \le 18 - r$, *where* $\nu(T)$ *denotes the number of components of* G *of type* T.
(2) If $r = 13$, *then* G *has no component of type* D_4.

Proof. (1) In our situation we can substitute $\rho = 7 + r$, and $a = 1$ into the equality (3) in the beginning of Section 10. Moreover, by the note just above $t - t_1 = 1 + \nu(D) + \nu(E) + \nu(II) + \nu(III) + \nu(IV)$, $t_1 = \nu(A) - \nu(III) - \nu(IV) + \nu(I_1)$. The inequality (1) follows from these.
(2) First, assume that the functional invariant J is constant. Then, $t_1 = 0$ in the equality (3) Section 10, since J has never poles. Since $\rho = 20$, and $a = 1$ under our assumptions, we have $2t = 7$, which is a contradiction. Thus, J is not constant. We can apply the inequality (4) in the beginning of Section 10. We have the claim, since F_1 is of type I_0^* and $\nu(I_0^*) \ge 1$. Q.E.D.

Lemma 17.7. *(1)* $Q(G)$ *is primitive in* Λ_3.
(2) Let $\widetilde{S}$ *be the primitive hull of* $S = P \oplus Q(G)$ *in* Λ_3. *Then, the restriction to* $\widetilde{S}/S$ *of the projection* $S^*/S = P^*/P \oplus Q(G)^*/Q(G) \to P^*/P$ *is injective.*
(3) $(\widetilde{S}/S)_p = 0$ *for any prime* $p \ge 5$.
(4) $l((\widetilde{S}/S)_p) \le 1$ *for* $p = 2, 3$.

Proof. (1) Let $\widetilde{Q}(G)$ be the primitive hull of $Q(G)$ in Λ_3. We will deduce a contradiction, assuming that $I_Q = \widetilde{Q}(G)/Q(G) \ne 0$.

Let $\overline{P}$ be the sublattice of rank 6 in P generated by $e_0, e_1, \dots, e_5$. Set $\overline{S} = \overline{P} \oplus Q(G)$ and $\widetilde{\overline{S}}$ be the primitive hull of $\overline{S}$ in Λ_3. Note that $J = \widetilde{\overline{S}}/\overline{S}$ can be identified with the group of sections of finite order. Since $I_Q \subset J$, we have a section C' corresponding to a non-zero element $\overline{\alpha}$ in I_Q. In $J = \overline{P}^*/\overline{P} \oplus Q(G)^*/Q(G)$, $\overline{\alpha}$ is contained in the direct summand $Q(G)^*/Q(G)$. It implies that C' and C_5 intersect F_1 on the same component. Thus, $C' = C_5$ since the homomorphism from $\widetilde{\overline{S}}/\overline{S}$ to the group $F_1^{\#}$ of the singular fiber F_1 is injective, and every component of F_1 contains at most one point of finite order. We have $\overline{\alpha} = 0$, a contradiction.
(2) If it is not injective, then $(\widetilde{S}/S) \cap (Q(G)^*/Q(G)) = \widetilde{Q}(G)/Q(G)$ is not zero, which contradicts (1).
(3), (4) By (2) $I = \widetilde{S}/S$ is isomorphic to a subgroup of $P^*/P \cong \mathbf{Z}/12$. Since $l(I_p) \le l((P^*/P)_p)$, $l(I_p) \le 1$ if $p = 2$ or 3, and it is zero if $p \ge 5$. Q.E.D.

From here, for a while, we consider the case where G has 13 vertices. We consider the following conditions on a Dynkin graph F:
(A) F has 13 vertices and every component of it is of type A, D or E.
(B) F can be made from one of Dynkin graphs in Lemma 17.4 by one elementary transformation.
(C) F has a component of type D.

(D) F has no component of type D_4.
(E) $\nu(A)+2\nu(D)+2\nu(E)\le 5$ for F.

By Lemma 17.5 (2) and by Lemma 17.6 one knows that $G \in PC(W_{1,0})$ satisfies the above conditions.

Lemma 17.8. *(1) Let F' be one of Dynkin graphs in Lemma 17.4. Assume that F' satisfies one of the following three conditions:*
⟨1⟩ All components are of type A.
⟨2⟩ It has the form $F''+A_k+A_1$.
⟨3⟩ It has a component of type D_4.
Then, any graph F with 13 vertices obtained from F' by one elementary transformation does not satisfy one of the above conditions (C), (D), (E).
(2) Under the condition (C), (D) and (E), the above condition (B) is equivalent to the following (B′):

(B′) F can be made from one of nine Dynkin graphs (1)-(8), (10) in Lemma 17.4 by one elementary transformation.

The following is the list of graphs satisfying (A), (B′), (C), (D) and (E):

(1) D_{13}
(1.1) D_{13} (1.2) $D_{11}+2A_1$ (1.3) $D_{10}+A_3$
(1.4) D_8+D_5 (1.5) D_7+D_6
(2) $D_{12}+A_1$
(2.1) $D_{12}+A_1$ (2.2) $D_{10}+3A_1$ (2.3) $D_9+A_3+A_1$
(2.4) $D_7+D_5+A_1$ (2.5) $2D_6+A_1$
(3) $D_{10}+A_2+A_1$
(3.1) $D_{10}+A_2+A_1$ (3.2) $D_7+A_3+A_2+A_1$
(4) D_9+A_4
(4.1) D_9+A_4 (4.2) $D_7+A_4+2A_1$ (4.3) $D_6+A_4+A_3$
(5) D_8+D_5
(5.1) D_8+D_5 (5.2) $2D_5+A_3$ (5.3) $D_8+A_3+2A_1$
(6) D_7+E_6
(6.1) D_7+E_6 (6.2) $D_7+A_5+A_1$ (6.3) D_7+3A_2
(7) E_7+D_6
(7.1) E_7+D_6 (7.2) $2D_6+A_1$ (7.3) $D_6+2A_3+A_1$
(7.4) $D_6+A_5+A_2$ (7.5) D_6+A_7
(8) E_8+D_5
(8.1) E_8+D_5 (8.2) $E_7+D_5+A_1$ (8.3) $E_6+D_5+A_2$
(8.4) $2D_5+A_3$ (8.5) D_5+2A_4 (8.6) $D_5+A_5+A_2+A_1$
(8.7) $A_7+D_5+A_1$ (8.8) D_8+D_5 (8.9) D_5+A_8
(8.10) $D_5+2A_3+2A_1$ (8.11) $D_8+A_3+2A_1$
(10) $D_{10}+3A_1$
(10.1) $D_{10}+3A_1$

Note that (1.4) = (5.1) = (8.8), (2.2) = (10.1). (5.2) = (8.4), (5.3) = (8.11) and (5.6) = (8.10).

We divide these items into 3 classes:

MPD13: (9 items.)
(1.1), (1.4), (2.1), (3.1), (4.1), (6.1), (7.1), (8.1), (8.2).
MID13: (4 items.)
(3.2), (6.2), (7.5), (8.5).
NID13: (19 items.)
(1.2), (1.3), (1.5), (2.2), (2.3), (2.4), (2.5), (4.2), (4.3), (5.2), (5.3), (6.3), (7.2), (7.3), (7.4), (8.3), (8.6), (8.7), (8.9).

Lemma 17.9. (MPD13) *Every graph in the class MPD13 except (8.2)* $E_7+D_5+A_1$ *can be made from a basic graph* A_{11} *by two tie transformations. Also, (8.2) can be made from a basic graph* $E_8+G_2+B_1$. *In particular, they belong to* $PC(W_{1,0})$.

Lemma 17.10. (MID13) *Graphs in the class MID13 are not members of* $PC(W_{1,0})$.

Proof. Consider the case (3.2). Let $F = D_7 + A_3 + A_2 + A_1$. Assume that we have a full embedding $Q = Q(F) \hookrightarrow \Lambda_3/P$. We will deduce a contradiction.

We have the induced embedding $S = P \oplus Q(F) \hookrightarrow \Lambda_3$. By $\widetilde{S}$ we denote the primitive hull of S in Λ_3, and by T the orthogonal complement of S in Λ_3. The discriminant group of S can be written $S^*/S \cong (S^*/S)_2 \oplus (S^*/S)_3$. ($M_p$ denotes the p-Sylow subgroup of an abelian group M.) $(S^*/S)_2 \cong (\mathbf{Z}/4)^{\oplus 3} \oplus \mathbf{Z}/2$, and $(S^*/S)_3 \cong \mathbf{Z}/3 \oplus \mathbf{Z}/3$, since $P^*/P \cong \mathbf{Z}/4 \oplus \mathbf{Z}/3$, $Q(D_7)^*/Q(D_7) \cong \mathbf{Z}/4$, and $Q(A_k)^*/Q(A_k) \cong \mathbf{Z}/(k+1)$. The discriminant quadratic form on $(S^*/S)_2$ can be written

$$q(a,b,c,d) \equiv -\frac{1}{4}(a^2+b^2) + \frac{3}{4}b^2 + \frac{1}{2}c^2 \bmod 2\mathbf{Z}$$

for an element $(a,b,c,d) \in (\mathbf{Z}/4)^{\oplus 3} \oplus \mathbf{Z}/2 \cong (S^*/S)_2$. (The first component $\mathbf{Z}/4$ corresponds to P, the second to D_7, the third to A_3, and the last $\mathbf{Z}/2$ to A_1.)

$l((S^*/S)_2) = 4$. Since $\widetilde{S}^*/\widetilde{S} \cong T^*/T$, $2 = \operatorname{rank} T \geq l(T^*/T) = l(\widetilde{S}^*/\widetilde{S}) \geq l((\widetilde{S}^*/\widetilde{S})_2)$. Thus, we have a non-zero element $\overline{\alpha} = (a,b,c,d) \in (\widetilde{S}/S)_2$. It satisfies $q(\overline{\alpha}) \equiv 0 \bmod 2\mathbf{Z}$.

By solving the congruence equation, one knows that $\overline{\alpha}$ is one of the following: $\overline{\alpha}_1 = (2,2,0,0)$, $\overline{\alpha}_2 = (2,0,2,0)$, $\overline{\alpha}_3 = (0,2,2,0)$, $\overline{\beta}_1 = (\pm1,\pm1,0,\pm1)$, $\overline{\beta}_2 = (\pm1,0,\pm1,0)$, $\overline{\beta}_3 = (0,\pm1,\pm1,0)$. If $\overline{\alpha}_1 \in I_2$, then $Q(D_7) \subset Q(B_7) \subset \widetilde{Q} \subset \Lambda_3/P$. If $\overline{\alpha}_2 \in I_2$, then $Q(A_3) \subset Q(B_3) \subset \widetilde{Q} \subset \Lambda_3/P$. If $\overline{\alpha}_3 \in I_2$, then $Q(D_7+A_3) \subset Q(D_{10}) \subset \widetilde{Q} \subset \Lambda_3/P$. These cases are prohibited because of the fullness. If $\overline{\beta}_i \in I_2$, then we can reduce to the above cases, since $2\overline{\beta}_i = \overline{\alpha}_i$. One concludes $D_7 + A_3 + A_2 + A_1 \notin PC(W_{1,0})$.

Next, we consider (6.2) $G = D_7 + A_5 + A_1$. To treat (6.2) we have to apply a p-adic method. Set $S = P \oplus Q(D_7 + A_5 + A_1)$. Assume that we have an embedding $S \hookrightarrow \Lambda_3$ satisfying Looijenga's conditions (L1) and (L2). We will deduce a contradiction. We define $\widetilde{S}$ and T as above. Consider the discriminant group $S^*/S \cong \mathbf{Z}/4 \oplus \mathbf{Z}/4 \oplus \mathbf{Z}/2 \oplus \mathbf{Z}/2 \oplus \mathbf{Z}/3 \oplus \mathbf{Z}/3$. The first component $\mathbf{Z}/4$ and the fifth component $\mathbf{Z}/3$ are associated with the lattice P. The second $\mathbf{Z}/4$ is associated with the component D_7. The third $\mathbf{Z}/2$ and the last $\mathbf{Z}/3$ are associated with A_5, and the fourth $\mathbf{Z}/2$ with A_1. We have a non-zero element $\overline{\alpha} = (a, b, c, d, x, y)$ in $I = \widetilde{S}/S$. For the discriminant quadratic form q,

$$q(\overline{\alpha}) \equiv -\frac{1}{4}(a^2 + b^2) - \frac{1}{2}c^2 + \frac{1}{2}d^2 - \frac{2}{3}(x^2 + y^2) \equiv 0 \bmod 2\mathbf{Z}.$$

$\overline{\alpha}$ is one of the following: $\overline{\alpha}_1 = (0, 0, 1, 1, 0, 0)$, $\overline{\alpha}_2 = (2, 2, 0, 0, 0, 0)$, $\overline{\alpha}_3 = (2, 2, 1, 1, 0, 0)$, $\overline{\alpha}_4 = (\pm 1, \pm 1, 0, 1, 0, 0)$. If $\overline{\alpha} = \overline{\alpha}_1$, then $\widetilde{S}$ contains a long root orthogonal to P such that it is not in S. It contradicts the assumption. If $\overline{\alpha} = \overline{\alpha}_2$, then $\widetilde{S}/P$ contains a short root with length 1, which is a contradiction. If $\overline{\alpha} = \overline{\alpha}_4$, then I contains $2\overline{\alpha}_4 = \overline{\alpha}_2$ and we can reduce the problem to the second case. Thus, we can assume that I is a cyclic group of order 2 generated by $\overline{\alpha}_3$. Set $\overline{\beta}_1 = (2, 1, 0, 1, 0, 0)$, $\overline{\beta}_2 = (1, 0, 0, 1, 0, 0)$, $\overline{\gamma}_1 = (0, 0, 0, 0, 1, 0)$, and $\overline{\gamma}_2 = (0, 0, 0, 0, 0, 1) \in S^*/S$. We can check that the orthogonal complement $I^{\perp}$ of I with respect to b is a direct sum of I and 4 cyclic groups generated by these 4 elements. $\overline{\beta}_i$ $(i = 1, 2)$ generates a cyclic group of order 4, and $\overline{\gamma}_i$ $(i = 1, 2)$ generates a cyclic group of order 3. We have $\widetilde{S}^*/\widetilde{S} \cong I^{\perp}/I \cong \mathbf{Z}/4 \oplus \mathbf{Z}/4 \oplus \mathbf{Z}/3 \oplus \mathbf{Z}/3$. Note here that $\overline{\beta}_1$, $\overline{\beta}_2$, $\overline{\gamma}_1$, and $\overline{\gamma}_2$ are mutually orthogonal with respect to b, and $q(\overline{\beta}_1) \equiv -3/4$, $q(\overline{\beta}_2) \equiv 1/4$, $q(\overline{\gamma}_i) \equiv -2/3 \bmod 2\mathbf{Z}$ $(i = 1, 2)$. Thus, we can compute the discriminant form of $\widetilde{S}$. Reversing the sign of the discriminant quadratic form on $\widetilde{S}^*/\widetilde{S}$ we get the discriminant form q_T on T^*/T. We have

$$q_T(a, b, x, y) \equiv \frac{3}{4}a^2 - \frac{1}{4}b^2 + \frac{2}{3}(x^2 + y^2) \bmod 2\mathbf{Z}$$

for $(a, b, x, y) \in \mathbf{Z}/4 \oplus \mathbf{Z}/4 \oplus \mathbf{Z}/3 \oplus \mathbf{Z}/3$.

Now we consider the 2-adic lattice $T_2 = T \otimes \mathbf{Z}_2$. Note that $T_2^*/T_2 = (T^*/T)_2$. Thus, the discriminant quadratic form of T_2 has the form $3a^2/4 - b^2/4$. This implies that T_2 is equivalent over $\mathbf{Z}_2$ to the lattice whose intersection form is defined by the matrix $\begin{pmatrix} 3 \cdot 2^2 & 0 \\ 0 & -2^2 \end{pmatrix}$. Therefore, we can conclude that the discriminant D of T satisfies $D \equiv -3 \cdot 2^4 \bmod \mathbf{Z}_2^{*2}$. However, on the other hand, $|D|$ = the order of T^*/T = the order of $\widetilde{S}^*/\widetilde{S} = 3^2 \cdot 2^4$, moreover, $D = |D|$ since T has signature $(0, 2)$. One knows that there exists an element $\xi \in \mathbf{Z}_2^* = \mathbf{Z}_2 - 2\mathbf{Z}_2$ with $3 = -\xi^2$. It implies $3 \equiv -1 \pmod 8$, which is a contradiction.

To (7.5) $G = D_6 + A_7$ we can apply the 2-adic method and deduce a contradiction similarly as in the case (6.2).

Here, we discuss (8.5) $D_5 + 2A_4$. Assume that there is an embedding $S = P \oplus Q(G) \hookrightarrow \Lambda_3$ satisfying Looijenga's (L1) and (L2) for $G = D_5 + A_8$. We will deduce a contradiction. The induced embedding $Q(G) \hookrightarrow \Lambda_3/P$ is full. We define $\widetilde{S}$ and T as above. $\operatorname{rank} T = 2$.

By Lemma 17.7 (3) $m = [\widetilde{S} : S]$ is prime to 5. Thus, $(T^*/T)_5 \cong (\widetilde{S}^*/\widetilde{S})_5 \cong (S^*/S)_5$. Note here that $(S^*/S)_5 \cong \mathbf{Z}/5 \oplus \mathbf{Z}/5$. Each $\mathbf{Z}/5$-component of $(S^*/S)_5$ corresponds to an A_4-component of G. The discriminant quadratic form on $(S^*/S)_5$ can be written

$$q_5 \equiv \frac{4}{5}(x^2 + y^2) \bmod 2\mathbf{Z}.$$

The discriminant quadratic form on $(T^*/T)_5$ is $-q_5$.

We consider the lattice $T_5 = T \otimes \mathbf{Z}_5$ over 5-adic integers $\mathbf{Z}_5$. By Lemma 12.1, the discriminant quadratic form of T_5 coincides with $-q_5$. By Lemma 12.3 T_5 is equivalent over $\mathbf{Z}_5$ to the lattice defined by the diagonal matrix whose diagonal entries are $-4 \cdot 5$, $-4 \cdot 5$. Thus, the discriminant D of T satisfies $D \equiv 2^4 \cdot 5^2 \bmod \mathbf{Z}_5^{*2}$. On the other hand, $|D| = \#(T^*/T) = \#(\widetilde{S}^*/\widetilde{S}) = \#(S^*/S)/m^2 = 2^4 \cdot 3 \cdot 5^2/m^2$ (By $\#M$ we denote the order of an abelian group M.), and $D > 0$ since T has signature $(0,2)$. In conclusion, we have $2^4 \cdot 5^2 \equiv 2^4 \cdot 3 \cdot 5^2/m^2 \bmod \mathbf{Z}_5^{*2}$. It implies that $x^2 \equiv 3 \pmod 5$ has an integral solution, which is a contradiction.

Thus, we can conclude $D_5 + 2A_4 \notin PC$. Q.E.D.

The class NID13 will be considered later. We here complete the case where the number of vertices is 13 except the class NID13.

We would like to proceed to the case of 12 vertices.

Lemma 17.11. *In addition to our assumptions on the elliptic K3 surfaces we assume that the number r of vertices of $G \in PC(W_{1,0})$ is 12. Then, G has at most only one component of type D_4.*

Proof. If G has 2 or more components of type D_4, then the functional invariant must be constant. (See Lemma 17.6 (2).) One has $\rho = 19$, $a = 1$, $t_1 = 0$, and $t = 4$ by assumption and by the equality (3) in the beginning of Section 10. Thus, the combination of singular fibers must be $4I_0^*$ and $G = 3D_4$. However, then, $\epsilon_p(3D_4) = (3, d(3D_4))_p$ for every p, which contradicts the assumption. Q.E.D.

We consider the following conditions on a Dynkin graph F:

(a) F has components of type A, D, or E only, and it has 12 vertices.
(b) F has a component of type D.
(c) F has at most only one component of type D_4.
(d) The I1-condition fails for the root lattice $Q(F)$ of type F.
(e) $\nu(A) + 2\nu(D) + 2\nu(E) \leq 6$ for F.

By Lemma 17.5 (2) and Lemma 17.11 $G \in PC(W_{1,0})$ satisfies the above conditions under our assumption.

Lemma 17.12. *Let F be a Dynkin graph satisfying above (a) and (b). Let F' be the sum of components of F of type A or E. The condition (d) is equivalent to (d') below.*

(d') $$\epsilon_p(F') \neq (3, d(F'))_p \text{ for } p = 3, 5, \text{or } 7.$$

Proof. For simplicity by $\epsilon_p(F)$ and by $d(F)$ we denote the Hasse invariant $\epsilon_p(Q(F))$ and the discriminant $d(Q(F))$ of the root lattice $Q(F)$ of type F respectively. By the condition (d) $\epsilon_p(F) \neq (3, d(F))_p$ for some prime number p.

Note that the number of vertices of F' is at most 8. Let $F_D = F - F'$ be the sum of components of type D. Note that $d(F_D) = 4^m$ for some m, and $\epsilon_p(F_D) = 1$ for every prime p. We have $\epsilon_p(F) = \epsilon_p(F')\epsilon_p(F_D) \cdot (d(F'), 4^m)_p = \epsilon_p(F')$, and $(3, d(F))_p = (3, 4^m)_p(3, d(F'))_p = (3, d(F'))_p$.

Thus, $\epsilon_p(F') \neq (3, d(F'))_p$ for some prime p.

Assume $p \geq 11$. By Lemma 7.5 (3) $p \nmid d(F')$. Thus, $(3, d(F'))_p = 1$. On the other hand, by Lemma 7.5 (4) we have $\epsilon_p(F') = 1 = (3, d(F'))_p$. Consequently, we can assume $p = 2, 3, 5$ or 7.

Finally, we can omit $p = 2$ further, because of the product formula. $\big(\epsilon_\infty(F') = 1$ since $Q(F')$ is positive definite. $(3, d(F'))_\infty = 1$ since $d(F') > 0.\big)$ Q.E.D.

Proposition 17.13. *A Dynkin graph F satisfies the above conditions (a), (b), (c), (d) and (e) if, and only if, F belongs one of the following three classes MPD12, MID12, NID12:*

MPD12: (14 items.)

(1) $D_{11} + A_1$ (2) $D_8 + A_4$ (3) $D_8 + A_3 + A_1$
(4) $D_7 + A_5$ (5) $D_6 + D_5 + A_1$ (6) $D_6 + E_6$
(7) $D_6 + A_5 + A_1$ (8) $D_5 + A_7$ (9) $D_5 + E_7$
(10) $D_5 + A_6 + A_1$ (11) $D_5 + E_6 + A_1$ (12) $D_5 + A_5 + 2A_1$
(13) $D_5 + A_4 + A_3$ (14) $D_5 + A_4 + A_2 + A_1$

MID12: (11 items.)

(15) $D_9 + 3A_1$ (16) $D_7 + D_4 + A_1$ (17) $D_6 + A_4 + 2A_1$
(18) $D_6 + 2A_2 + 2A_1$ (19) $D_4 + E_6 + 2A_1$ (20) $D_4 + A_5 + A_3$
(21) $D_6 + A_3 + 3A_1$ (22) $D_4 + A_5 + 3A_1$ (23) $D_5 + 2A_3 + A_1$
(24) $D_4 + A_6 + A_2$ (25) $D_4 + A_4 + 2A_2$

NID12: (5 items.)

(26) $D_6 + 3A_2$ (27) $D_5 + 3A_2 + A_1$ (28) $D_8 + 2A_2$
(29) $D_5 + A_5 + A_2$ (30) $D_5 + A_3 + 2A_2$

Proof. By Lemma 17.12 we can consider the condition (d′) instead of (d). Assume that $\epsilon_7(F') \neq (3, d(F'))_7$. (We omit the lower index $p = 7$ in the following.) We can write $F' = A_6 + F''$ since $7 \mid d(F')$. $F'' = A_2, 2A_1, A_1$ or $\emptyset$. In any case, $7 \nmid d(F'')$; thus, $(3, d(F')) = (3, d(A_6))(3, d(F'')) = (3, 7) = -1$. Calculating $\epsilon(F')$ for the four possible cases, one has $F' = A_6 + A_2$.

Assume that $\epsilon_5(F') \neq (3, d(F'))_5$. (We omit the lower index $p = 5$ in the following.) We can write $F' = A_4 + F''$. F'' has at most 4 vertices.

Case 1. $5 \mid d(F'')$

$F' = 2A_4$. However, in this case, $\epsilon(F') = 1 = (3, d(F'))$.

Case 2. $5 \nmid d(F'')$

$(3, d(F')) = -1$. We have only to check whether $\epsilon(F') = \epsilon(A_4 + F'') = (5, d(F''))$ is equal to 1. This is equivalent to $d(F'') \equiv \pm 1 \pmod 5$. Among the 11 possibilities only the following 6 graphs satisfy $\epsilon(F') = 1$:

$$A_4 + 2A_2, \quad A_4 + 4A_1, \quad A_4 + A_3, \quad A_4 + A_2 + A_1, \quad A_4 + 2A_1, \quad A_4.$$

Assume that $\epsilon_3(F') \neq (3, d(F'))_3$. (We omit the lower index $p = 3$ in the following.) Note that in this case, we cannot conclude $3 \mid d(F')$.

Case 1. $3 \nmid d(F')$

In this case, $\epsilon(F') = 1$. Thus, the assumption $\iff (3, d(F')) = -1 \iff d(F') \equiv -1 \pmod 3$. F' has at most 8 vertices and its component is either A_1, A_3, A_4, A_6, A_7, E_7, or E_8. We can pick up the following 15 graphs satisfying the assumptions from 34 possibilities:

E_7, A_7, A_6+A_1, A_4+4A_1, A_4+A_3, A_4+2A_1, A_4, $2A_3+A_1$, A_3+5A_1, A_3+3A_1, A_3+A_1, $7A_1$, $5A_1$, $3A_1$, A_1.

Case 2. F' contains A_8.

$F'=A_8$. $\epsilon(A_8)=(-1,9)=1$. $(3,d(A_8))=(3,9)=1$. This does not satisfy the assumption.

Case 3. F' contains E_6.

There are only 4 possibilities for F'. Only the following three satisfy the assumption:

$$E_6+2A_1,\quad E_6+A_1,\quad E_6.$$

Case 4. F' contains A_5.

There are only 7 possibilities for F'. Among them only the following six satisfy the assumption:

$$A_5+A_3,\quad A_5+A_2,\quad A_5+3A_1,\quad A_5+2A_1,\quad A_5+A_1,\quad A_5.$$

Case 5. F' contains just 4 of A_2.

$F'=4A_2$. In this case, $\epsilon=1=(3,d)$. It does not satisfy the assumption.

Case 6. F contains just 3 of A_2.

We can write $F'=3A_2+F''$ with $3\not| \, d(F'')$. Then, we have $\epsilon(F')=(3,d(F''))$, and $(3,d(F'))=-(3,d(F''))$. Thus, every possibility automatically satisfies the assumption. There are three possibilities:

$$3A_2+2A_1,\quad 3A_2+A_1,\quad 3A_2.$$

Case 7. F contains just 2 of A_2.

Writing $F'=2A_2+F''$, one knows $\epsilon\neq(3,d) \iff d(F'')\equiv 1 \pmod 3$. Only the following four among possibilities satisfy the assumption:

$$2A_2+4A_1,\quad 2A_2+A_3,\quad 2A_2+2A_1,\quad 2A_2.$$

Case 8. F contains just one of A_2 and it does not contain A_5 and E_6.

We can write $F'=A_2+F''$ with $3\not| \, d(F'')$. We have $\epsilon(F')=-(3,d(F''))$ and $(3,d(F'))=-(3,d(F''))$. Thus, the assumption is never satisfied in this case.

We have obtained the list of Dynkin graph satisfying (a), (b) and (d). It is easy to exclude ones not satisfying (c) or (e). Q.E.D.

Proposition 17.14. (MPD12) *Any graph in the class MPD12 can be made from one of the essential basic Dynkin graphs by elementary or tie transformations repeated twice.*

Proof. For example we can make D_{12} from a basic graph A_{11} by one tie transformation. From D_{12} we can make any graph in MPD12 by the same transformation. Q.E.D.

Proposition 17.15. (MID12) *The eleven graphs (15)-(25) in the class MID12 do not belong to $PC(W_{1,0})$.*

Proof. For (15)–(20) we can apply a similar argument to that in the case (3.3), $r = 13$ in the above. Only by solving the congruence equation $q \equiv 0$, we can conclude that they are not in PC.

For (21) and (22) we can use the method applying Lemma 17.7 (4).

We explain case (21) only. Set $G = D_6 + A_3 + 3A_1$. By S, $\widetilde{S}$ we denote the same lattice as above. Now, we have $l((S^*/S)_2) = 7$. Note that $3 \geq l((\widetilde{S}^*/\widetilde{S})_2) \geq l((S^*/S)_2) - 2l((\widetilde{S}/S)_2) = 7 - 2l((\widetilde{S}/S)_2)$. We have $l((\widetilde{S}/S)_2) \geq 2$, which contradicts Lemma 17.7.

We here explain case (23) $G = D_5 + 2A_3 + A_1$. Set $S = P \oplus Q(G)$. We can consider only the 2-Sylow subgroup of the discriminant group $S^*/S \cong \mathbf{Z}/4 \oplus \mathbf{Z}/4 \oplus \mathbf{Z}/4 \oplus \mathbf{Z}/4 \oplus \mathbf{Z}/2 \oplus \mathbf{Z}/3$. For an element $(a, b, c_1, c_2, d) \in (\mathbf{Z}/4)^{\oplus 4} \oplus \mathbf{Z}/2 = (S^*/S)_2$ the discriminant quadratic form can be written

$$q(a, b, c_1, c_2, d) \equiv -\frac{1}{4}a^2 - \frac{3}{4}b^2 + \frac{3}{4}(c_1^2 + c_2^2) + \frac{1}{2}d^2 \bmod 2\mathbf{Z}.$$

A non-zero solution of $q \equiv 0$ is one of the following;

$\overline{\alpha}_1 = (2, 2, 0, 0, 0)$, $\overline{\alpha}_2 = (0, 0, 2, 2, 0)$, $\overline{\alpha}_3 = (2, 0, \{2, 0\}, 0)$,
$\overline{\alpha}_4 = (0, 2, \{2, 0\}, 0)$, $\overline{\beta} = (2, 2, 2, 2, 0)$, $\overline{\gamma}_1 = (0, 0, \pm1, \pm1, 1)$,
$\overline{\gamma}_2 = (0, \pm1, \{\pm1, 0\}, 1)$, $\overline{\gamma}_3 = (\pm1, 0, \{\pm1, 2\}, 1)$, $\overline{\gamma}_4 = (\pm1, \pm1, \{2, 0\}, 1)$,
$\overline{\gamma}_5 = (\pm1, 2, \{\pm1, 0\}, 1)$, $\overline{\gamma}_6 = (2, \pm1, \{\pm1, 2\}, 1)$, $\overline{\gamma}_7 = (2, 2, \pm1, \pm1, 1)$.

Here, $\{x_1, x_2, \ldots, x_k\}$ stands for $x_{\sigma(1)}, x_{\sigma(2)}, \ldots, x_{\sigma(k)}$ for some permutation σ. (This is very convenient abbreviation.) Among these, $\overline{\gamma}_i$'s have order four. The group $I = \widetilde{S}/S$ contains one of $\overline{\alpha}_i, i = 1, 2, 3, 4$ and $\overline{\beta}$. We can see that if the contained element is $\overline{\alpha}_i \in I$, the induced embedding $Q(G) \hookrightarrow \Lambda_3/P$ is not full. Thus, I contains $\overline{\beta} = (2, 2, 2, 2, 0)$. On the other hand, since $2\overline{\gamma}_i \neq \overline{\beta}$, one knows that I is generated by $(2, 2, 2, 2, 0)$. Let $I^\perp$ be the orthogonal complement of I with respect to the discriminant bilinear form b on S^*/S. By easy calculation one knows $l(I^\perp) \geq 5$. Thus, $4 \leq l(I^\perp/I) = l(\widetilde{S}^*/\widetilde{S}) = l(T^*/T)$, where T is the orthogonal complement of S in Λ_3. However, $l(T^*/T) \leq \operatorname{rank} T = 3$, which is a contradiction. Thus, $D_5 + 2A_3 + A_1 \notin PC$.

To (24) and (25) we can apply similar method to that in (8.5). Namely, by the p-adic method, (in these cases $p = 3$.) we show first that $S = P \oplus Q(G)$ has no primitive embedding into Λ_3. Next, assuming $\widetilde{S}/S \neq 0$ by solving $q \equiv 0$, we can deduce a contradiction from the fullness. (Note that we cannot apply Lemma 17.7 (3), since $p = 3$.) Q.E.D.

The class NID12 are treated after the case $r = 11$.

We complete the case of 12 vertices except the class NID12.

Now, the last remaining case is the case of 11 vertices. Also, in this case, we use the abbreviations $d(F)$ and $\epsilon_p(F)$.

Proposition 17.16. *Let F be a Dynkin graph with 11 vertices with components of type A, D, or E only. Assume that F has a component of type D and satisfies the following condition* ♡:

For some prime number p, $3d(F) \in \mathbf{Q}_p^{*2}$ and $\epsilon_p(F) \neq (-1, 3)_p$. ♡

Then, $F = D_5 + E_6$, $D_5 + A_5 + A_1$ or $D_5 + 3A_2$.

Proof. Let F' be the sum of components of F of type A or E. F' has at most 7 vertices. One knows easily that $\heartsuit$ is equivalent to the following $\heartsuit'$:

$$\text{For some prime number } p,\ 3d(F') \in \mathbf{Q}_p^{*2} \text{ and } \epsilon_p(F') \neq (-1,3)_p. \qquad \heartsuit'$$

We denote $d = d(F')$ and $\epsilon_p = \epsilon(F')$. Assume that $\heartsuit'$ holds for $p \geq 5$. We have $\epsilon_p = -1$. By Lemma 7.5 (4) p divides d. Since $3d \in \mathbf{Q}_p^{*2}$, p^2 divides d. By Lemma 7.5 (3) $8 \leq 2(p-1) \leq 7$, a contradiction.

Assume that $\heartsuit'$ holds for $p = 3$. $\epsilon_3 = +1$. Since $3d \in \mathbf{Q}_3^{*2}$, 3 divides d. F' contains a component E_6, A_5 or A_2. It is easy to see that if F' contains E_6 or A_5, then only $F' = E_6$ or $A_5 + A_1$ satisfies the condition. If F' contains A_2 but does not contain A_5, then the number s of A_2 components is odd; thus, $s = 3$ or 1. If $s = 3$, only the case $F' = 3A_2$ satisfies the condition. Assume $s = 1$. We can write $F' = F'' + A_2$. $d'' = d(F'')$ is not a multiple of 3. Since $3d = 3^2 d'' \in \mathbf{Q}_3^{*2}$, $d'' \equiv 1 \pmod 3$. On the other hand, $1 = \epsilon_3 = (-1,3)_3(3,d'')_3 = \left(\frac{-d''}{3}\right)$. Thus, $d'' \equiv -1 \pmod 3$, a contradiction.

Finally, assume that $\heartsuit'$ is satisfied for $p = 2$. By Lemma 7.7 one sees easily that if $3d \in \mathbf{Q}_2^{*2}$, then $F' = E_6$, $A_5 + A_1$, $3A_2$, $A_3 + A_2 + 2A_1$, $A_3 + A_2$, $A_2 + 4A_1$, $A_2 + 2A_1$, or A_2. Calculating ϵ_2 for each case, one sees that only the former 3 satisfy the condition $\epsilon_2 = 1$. Q.E.D.

We can assume the condition $\heartsuit$, because $\heartsuit$ is equivalent to that the I1-condition does not hold. Therefore, we can consider only the above three graphs.

Proposition 17.17. *(1) We can make $D_6 + E_6$ and $D_6 + A_5 + A_1$ from the basic graph A_{11} by tie transformations repeated twice. Thus, they belong to $PC = PC(W_{1,0})$. (2) $D_5 + 3A_2 \notin PC(W_{1,0})$.*

Proof. (1) We can make D_{12} from A_{11} easily. From D_{12} we can make them.
(2) To see this we can apply the same method as in the verification of Proposition 17.15 for (24) and (25). Q.E.D.

We complete the case $r = 11$.

Proposition 17.18. (NID12, NID13) *(1) Any graph in the class NID12 does not belong to $PC(W_{1,0})$.*
(2) Any graph in the class NID13 does not belong to $PC(W_{1,0})$.

Proof. (1) Any graph in NID12 contains $D_5 + 3A_2$ as a subgraph. Thus, by Proposition 13.7 (2), they cannot belong to $PC(W_{1,0})$.
(2) For each graph F in NID13 we can find a graph F' in MID12 or NID12 in Proposition 17.13 such that F' is a subgraph of F. Thus, by Proposition 13.7 (2), it cannot belong to $PC(W_{1,0})$, either. Q.E.D.

We have shown Proposition 17.1 under the assumption ((1)).

18
A singular fiber of type II^*, III^*, IV^* in the case $W_{1,0}$

Let G be a Dynkin graph belonging to $PC(W_{1,0})$ with r vertices. Throughout this section we assume that the corresponding elliptic K3 surface $\Phi : Z \to C$ has a singular fiber of type II^*, III^*, or IV^*, which is denoted by F_2.

Proposition 18.1. *(1) G is a subgraph of the Coxeter-Vinberg graph of the unimodular lattice of signature* $(14, 1)$.
(2) G has a component of type E.

Proof. (1) We have the associated embedding $P \oplus Q(G) \hookrightarrow \Lambda_3$ satisfying Looijenga's conditions (L1), (L2). Let $\overline{P}$ denote the sublattice in P of rank 6 with a basis $e_0, \ldots, e_5$ corresponding to C_5 and 5 components of F_1. This $\overline{P}$ is isomorphic to P defined for $J_{3,0}$. It is easy to check that the induced embedding $\overline{P} \oplus Q(G) \hookrightarrow \Lambda_3$ also satisfies the conditions (L1), (L2). Thus, the embedding $Q(G) \hookrightarrow \Lambda_3/\overline{P}$ is full. By Proposition 11.1 the orthogonal complement of $Q(G)$ in $\Lambda_3/\overline{P}$ contains a vector ξ with $\xi^2 = -4$. (Though we can write $\xi = 2\eta$ in $\Lambda_3/\overline{P}$, we do not use this fact.) The orthogonal complement L of $\mathbf{Z}\xi$ in $\Lambda_3/\overline{P}$ is a unimodular lattice of signature $(14, 1)$, and $Q(G)$ is full in L. The claim follows from these facts.
(2) Consider the dual graph G' associated with the set of components of F_2 disjoint from IF. The dual graph of all components of F_2 minus 1 or 2 vertices corresponding to components with multiplicity 1 is G'.

Assume that G' is not of type E. We will deduce a contradiction. One knows immediately that F_2 is of type IV^* and G' is of type D_5.

For $2 \le i \le t$, let $n(F_i)$ denote the number of components disjoint from IF of the singular fiber F_i. $n(F_2) = 5$ and $r = \sum_{i=2}^{t} n(F_i)$. Recall that we can assume $\rho = 7 + r$ without loss of generality. Then, by the equality (2) in the beginning of Section 10, we have $\sum_{i=3}^{t}(m(F_i) - n(F_i) - 1) + a = 0$, since $m(F_1) = 5$ and $m(F_2) = 7$. We have $a = 0$ and $m(F_i) - 1 = n(F_i)$ for $3 \le i \le t$. In particular, the group MW of sections is finite.

Here we recall that we denoted by $\operatorname{Tor} M$ the subgroup of an abelian group M consisting of all elements of finite order.

In our case the section C_6 belongs to $\operatorname{Tor} MW = MW$. We consider $F_i^\# = F_i \cap Z^\#$ for $i = 1, 2$. Recall that they carry the group structure. Since F_1 is of type I_0^*, $\operatorname{Tor} F_1^\# \cong \mathbf{Z}/2 + \mathbf{Z}/2$. One knows that C_6 has order 2 in MW since $MW \to \operatorname{Tor} F_1^\#$ is injective. Thus, $\operatorname{Tor} F_2^\#$ has an element of order 2, since $MW \to \operatorname{Tor} F_2^\#$ is injective. However, $\operatorname{Tor} F_2^\# \cong \mathbf{Z}/3$, since F_2 is of type IV^*. It is a contradiction. Q.E.D.

We consider the following conditions on a Dynkin graph F:

(A) F is a Dynkin graph with components of type A, D or E only.
(B) F is a subgraph of the Coxeter-Vinberg graph of the unimodular lattice of signature $(14, 1)$.
(C) F has a component of type E.

By Proposition 18.1 under our assumption $G \in PC(W_{1,0})$ satisfies these conditions. By the concrete form of the Coxeter-Vinberg graph in Section 11, one knows that under the condition (A) and (C), the condition (B) is equivalent to that F is a subgraph of $E_8 + E_6$ or $2E_7$.

By Theorem 16.4 we need not consider the case where the I1-condition holds. Therefor we can assume $r = 13, 12$ or 11. The first case is $r = 13$.

Proposition 18.2. *Let F be a Dynkin graph with 13 vertices satisfying above (A), (B) and (C). Then, F is a member of class MPE13 or class NIE13 below.*

MPE13: (7 items.)

(1) $E_8 + A_4 + A_1$	(2) $E_8 + D_5$	(3) $E_7 + A_6$
(4) $E_7 + D_6$	(5) $E_7 + D_5 + A_1$	(6) $2E_6 + A_1$
(7) $E_6 + D_7$		

NIE13: (11 items.)

(8) $E_8 + 2A_2 + A_1$	(9) $E_6 + A_4 + A_2 + A_1$	(10) $E_6 + D_5 + A_2$
(11) $E_8 + A_5$	(12) $E_7 + A_5 + A_1$	(13) $E_7 + A_3 + A_2 + A_1$
(14) $E_7 + A_4 + A_2$	(15) $E_7 + E_6$	(16) $E_6 + A_7$
(17) $E_6 + A_6 + A_1$	(18) $E_6 + A_4 + A_3$	

Proposition 18.3. (MPE13) *Every graph in the class MPE13 can be made from one of the essential basic Dynkin graphs by tie transformations repeated twice. They belong to $PC(W_{1,0})$.*

Proof. The following shows an example of the basic graph at the start:

(1) $\leftarrow E_8 + B_1 + G_2$	(2) $\leftarrow B_9 + G_2$	(3) $\leftarrow E_7 + B_3 + G_1$
(4) $\leftarrow E_7 + B_3 + G_1$	(5) $\leftarrow E_8 + B_1 + G_2$	(6) $\leftarrow E_8 + B_1 + G_2$
(7) $\leftarrow B_9 + G_2$		

Q.E.D.

We consider the 11 graphs in NIE13 after the case $r = 12$.

Let us proceed to the case of $r = 12$. Thanks to Proposition 13.5 (1), we can exclude graphs isomorphic to a subgraph of a graph in MPE13 from our argument.

Proposition 18.4. *Let F be a Dynkin graph with 12 vertices satisfying above (A), (B) and (C). Assume that any member of above MPE13 does not contain F as a subgraph. Then, F is one of the four members of the following class MIE12:*

MIE12:	$\diamond 1 \diamond$ $E_8 + 2A_2$	$\diamond 2 \diamond$ $E_7 + 2A_2 + A_1$
(4 items.)	$\diamond 3 \diamond$ $E_6 + A_3 + A_2 + A_1$	$\diamond 4 \diamond$ $E_6 + A_4 + A_2$

Proposition 18.5. (MIE12)
Any graph in the class MIE12 does not belong to $PC(W_{1,0})$.

Proof. For these graphs we can apply the p-adic method for $p = 3$. By arguments similar to the case (8.5) $D_5 + 2A_4$ in the class MID13, we can show that any one of them is not in PC. (See Lemma 17.10.) Q.E.D.

Proposition 18.6. (NIE13)
Any graph in the class NIE13 does not belong to $PC(W_{1,0})$.

Proof. For any graph F in NIE13, we can find a graph F' in MIE12 such that F contains F' as a subgraph. By Proposition 13.5 (2) F never belongs to $PC(W_{1,0})$.

Q.E.D.

Finally, we consider the case $r = 11$. However, in this case, every graph satisfying (A), (B) and (C) is isomorphic to a subgraph of a graph in MPE13. By Proposition 13.5 (1) it can be made from a basic graph by our transformations.

We have shown Proposition 17.1 under the assumption ((2)) in the beginning of Section 17.

19
Combinations of graphs of type A in the case $W_{1,0}$

In this section we assume the condition ((3)) in the beginning of Section 17. As in the previous sections G denotes a Dynkin graph in $PC(W_{1,0})$ with r vertices. The corresponding elliptic K3 surface $\Phi : Z \to C$ has t singular fibers $F_1, \ldots, F_t$ and one of them, say F_1, is of type I_0^* and the others are of type I, II, III or IV. The union $IF = F_1 \cup C_5 \cup C_6$ is the curve at infinity. By $C_0, \ldots, C_4$ we denote the components of IF. We assume that C_0 has multiplicity 2, C_5 intersects C_4, and C_6 intersects C_3. Let A_{n_i} be the dual graph of the set of components disjoint from C_5 of a singular fiber F_i for $2 \leq i \leq t$. (A_0 stands for an empty graph $\emptyset$.)

By ρ we denote the Picard number of Z and by a we denote the rank of the Mordell-Weil group MW of Φ.

Lemma 19.1. *(1) If $\rho = r + 7$, then $a = 0$ or 1. If $\rho = r + 7$ and $a = 1$, then every singular fiber F_i with $2 \leq i \leq t$ has only one component intersecting IF. If $\rho = r + 7$ and $a = 0$, then $G = 2A_5 + G'$ for some Dynkin graph G'.*
(2) $r \leq 13$. If $r = 13$, then $a = 1$.
(3) The number of components of G is at most $18 - r$.

Proof. (1) For $2 \leq i \leq t$ by $n(F_i)$ we denote the number of components of F_i disjoint from IF. By definition $r = \sum_{i=2}^{t} n(F_i)$. By the equality (2) in Section 10, we have $1 - a = \sum_{i=2}^{t}(m(F_i) - n(F_i) - 1)$. Obviously, $m(F_i) \geq n(F_i) + 1$ by definition, and $a \leq 1$.

If $a = 1$, then $m(F_i) = n(F_i) + 1$ for all i with $2 \leq i \leq t$.

Assume $a = 0$. There is a unique singular fiber, say F_2, different from F_1 such that C_5 and C_6 hit different components of F_2. Let S_i be the subgroup of Pic(Z) generated by the classes of components of F_i disjoint from C_5. We can write $[C_6] = m[F] + [F + C_5] + \omega_3 + \chi$, where $\omega_3 \in S_1^*$, $\chi \in S_2^*$ and $m \in \mathbf{Z}$. We have $m = 1$, since $[C_6] \cdot [C_5] = 0$.

Under the isomorphism $S_1^* \cong Q(-D_4)^*$ ω_3 corresponds to the fundamental weight associated with the vertex of the Dynkin graph D_4 with one edge corresponding to the component C_3 of F_1. In particular, $\omega_3^2 = -1$.

On the other hand, under $S_2^* \cong Q(-A_{n_2})^*$ χ corresponds to the fundamental weight associated with the vertex of A_{n_2} corresponding to the component of F_2 hit by C_6.

However, by injectivity of $MW = \mathrm{Tor}\, MW \to F_2^{\#}$, the image of χ in the quotient $Q(-A_{n_2})^*/Q(-A_{n_2})$ has order 2. Thus, n_2 is odd, and χ corresponds to the central vertex of A_{n_2}. In particular, $\chi^2 = -(n_2+1)/4$. We have $-2 = [C_6]^2 = 2-1-(n_2+1)/4$. Thus, $n_2 = 11$. G contains the graph A_{11} minus the central vertex, i.e., $2A_5$.

(2) Since $20 \geq \rho \geq 7+r$, the first claim is obvious.

Assume $r = 13$. Then, $\rho = 20 = 7+r$. We have $a \geq \nu(I_0^*) \geq 1$ by the inequality (4) in Section 10 when J is not constant.

Thus, we can assume moreover that J is constant. Then, $t_1 = 0$ in the equality (3) in Section 10, and we have $2(t-1) = 4+a \leq 5$.

On the other hand, every component of G is of type A_1 or A_2 under our assumption, since every singular fiber except F_1 is of type II, III or IV. Therefore, we have $7 \leq$ the number of components of $G \leq t-1 \leq 2$, which is a contradiction.

(3) We can assume without loss of generality that $\rho = r+7$. If $a = 0$, then the inequality obviously holds by (1). Thus, we assume, moreover, $a = 1$. We apply the equality (3) in Section 10. First, obviously, $t - t_1 - 1 = \nu(II) + \nu(III) + \nu(IV)$ under our assumption. Second, by (1) t_1 is the sum of $\nu(I_1)$ and the number of components of G. The claim follows from the equality (3) in Section 10. Q.E.D.

Note that by Lemma 19.1 (1) $G = \sum A_{n_i}$ if $\rho = r+7$ and $a = 1$.

Let P be the lattice associated with the singularity $W_{1,0}$. $\operatorname{rank} P = 7$. Recall that $e_0, \ldots, e_6$ denote the basis of P which has a one-to-one correspondence with the components C_i's of IF. The surface Z defines an embedding $P \oplus Q(G) \hookrightarrow \Lambda_3$ satisfying Looijenga's conditions (L1) and (L2). The induced embedding $Q(G) \hookrightarrow \Lambda_3/P$ is full and has no obstruction component of type A_{11}, i.e., if G has a component G_0 of type A_{11}, then $[P(Q(G_0), \Lambda_3/P) : Q(G_0)] < 12$. Recall here that we have denoted the primitive hull of a submodule M in L by $P(M, L) = \{x \in L \mid \text{For some non-zero integer } m,\ mx \in M\}$.

We regard $P \oplus Q(G)$ as a submodule of Λ_3 through the induced embedding.

Proposition 19.2. *$PC(W_{1,0}) \subset PC(J_{3,0})$.*

Proof. Let $\overline{P}$ be the sublattice of rank 6 in P generated by $e_0, e_1, \ldots, e_5$. This $\overline{P}$ is isomorphic to P defined in the case of $J_{3,0}$.

If $G \in PC(W_{1,0})$, then we have an embedding $P \oplus Q(G) \hookrightarrow \Lambda_3$ satisfying the conditions (L1), (L2). It is easy to check that the induced embedding $\overline{P} \oplus Q(G) \hookrightarrow \Lambda_3$ also satisfies (L1) and (L2). It implies $G \in PC(J_{3,0})$. Q.E.D.

Lemma 19.3. *Assume that $\rho = r+7$ and $a = 1$.*
(1) $Q(G)$ is primitive in Λ_3.
(2) For every element $\overline{\alpha} \in P(P \oplus Q(G), \Lambda_3)/(P \oplus Q(G))$ with order 2, there is a subset $M \subset N(2) = \{i \mid 2 \leq i \leq t,\ n_i + 1 \equiv 0 \pmod 2\}$ satisfying $\sum_{i \in M}(n_i+1) = 12$ and $\overline{\alpha}$ can be written $\overline{\alpha} = \overline{\omega} + \sum_{i \in M} \overline{\chi}_i$, where $\overline{\omega} \in P^*/P$, and $\overline{\chi}_i \in Q(A_{n_i})^*/Q(A_{n_i})$ $(i \in M)$ have order 2.

Proof. (1) The proof is same as in Lemma 17.7 (1).
(2) We can write $\overline{\alpha} = \overline{\omega} + \sum_{i=2}^{t} \overline{\chi}_i$, where $\overline{\omega} \in P^*/P \cong \mathbf{Z}/12$, and $\overline{\chi}_i \in Q(A_{n_i})^*/Q(A_{n_i}) \cong \mathbf{Z}/(n_i+1)$ for $2 \leq i \leq t$.

Set $M = \{i \mid 2 \leq i \leq t, \overline{\chi}_i \neq 0\}$. If $\overline{\chi}_i \neq 0$ it has order 2 by assumption, and n_i is odd.

If $\overline{\omega} = 0$, we have a contradiction by above (1). Thus, $\overline{\omega} \neq 0$. By assumption, $\overline{\omega}$ has order 2. It can be checked that the element of order 2 in P^*/P is $\omega_0 \bmod P$ where $\omega_0 = (e_1+e_2)/2 + e_0 + e_3 \in P^*$; thus, it is contained in $(P^* \cap \overline{P}^*) + P/P \cong (P^* \cap \overline{P}^*)/\overline{P}$. Namely, we can regard $\overline{\alpha}$ as an element in $(\overline{P} \oplus Q(G))^*/(\overline{P} \oplus Q(G))$. Thus, we have a section C' representing the class $\overline{\alpha}$. Since for every point $a \in C$ the homomorphism from $\widetilde{S}/S$ to the group $F_a^{\#}$ of the fiber over a is injective, $C' \cdot C_5 = 0$. Let S_i be the same group as in the proof of Lemma 19.1 (1). We can write

$$[C'] = [F] + [F + C_5] + \omega + \sum_{i \in M} \chi_i,$$

where F denotes a general fiber, $\chi_i \in S_i^*$ is the fundamental weight associated with the central vertex of the Dynkin graph A_{n_i}. In particular, $\chi_i^2 = -(n_i+1)/4$. The element $\omega \in S_1^*$ corresponds to $\omega_0 \in P_0'^* \cong Q(D_4)$ under $S_1^* \tilde{\to} P_0'^*$. We have $-2 = [C']^2 = ([F]+[F+C_5])^2 + \omega^2 + \sum_{i \in M}(n_i+1)/4$. Therefore, $\sum_{i \in M}(n_i+1) = 12$, since $([F]+[F+C_5])^2 = 2$ and $\omega^2 = -1$. Q.E.D.

Lemma 19.4. *We consider a Dynkin graph $G = \sum_{i \in I} A_{k_i} \in PC(W_{1,0})$ under the condition ((3)). Assume, moreover, that $\rho = r + 7$ and $a = 1$. Set $S = P \oplus Q(G)$ and $\widetilde{S} = P(S, \Lambda_3)$.*
(1) For any prime p with $p \geq 5$ $(\widetilde{S}/S)_p = 0$.
(2) For $p = 2, 3$, $l((\widetilde{S}/S)_p) \leq 1$.

Proof. The proof is same as that of Lemma 17.7 (3), (4).

Lemma 19.5. *Under our assumption ((3)) any Dynkin graph $G \in PC(W_{1,0})$ with $r = 13$ is a member of one of the following 4 classes MPA13, MOA13, MIA13, NIA13. Moreover, the surface Z corresponding to G satisfies $\rho = r + 7$ and $a = 1$:*

MPA13: (2 items.) *(1)* A_{13} *(2)* $A_{12} + A_1$
MOA13: (2 items.) *(3)* $A_{11} + A_2$ *(8)* $A_{11} + 2A_1$
MIA13: (2 items.) *(6)* $A_8 + A_5$ *(17)* $2A_6 + A_1$
NIA13: (19 items.)

(5) $A_9 + A_4$	*(9)* $A_{10} + A_2 + A_1$	*(10)* $A_9 + A_3 + A_1$
(11) $A_9 + 2A_2$	*(12)* $A_8 + A_4 + A_1$	*(18)* $A_6 + A_5 + A_2$
(20) $2A_5 + A_3$	*(21)* $A_5 + 2A_4$	*(23)* $A_9 + A_2 + 2A_1$
(26) $A_7 + A_4 + 2A_1$	*(27)* $A_7 + A_3 + A_2 + A_1$	*(30)* $A_6 + A_4 + A_2 + A_1$
(32) $A_6 + A_3 + 2A_2$	*(34)* $A_5 + A_4 + A_3 + A_1$	*(43)* $A_7 + 2A_2 + 2A_1$
(47) $2A_5 + 3A_1$	*(49)* $A_5 + 2A_3 + 2A_1$	*(53)* $2A_4 + 2A_2 + A_1$
(57) $3A_3 + 2A_2$		

Proof. Under ((3)) every component of G is of type A. By Lemma 19.1 (3) items to be checked are graphs corresponding to the division of 13 into a sum of 5 non-negative integers. By Proposition 19.2 we can assume moreover that the graph belongs to $PC(J_{3,0})$. Recall that there are 25 kinds of such graphs. Twenty-four graphs of 25 formed the class [[1]] in Section 13 and we can make them from $E_8 + F_4$ by two tie transformations. The last one is (57) 3+3+3+2+2, $3A_3 + 2A_2$.

The claim on ρ and a is obvious by Lemma 19.1 (2). Q.E.D.

Remark. To quote them later we have divided 25 graphs into 4 classes. We have appended the numbering in Section 13 to them.

In the remainder of this section we do not necessarily assume that G belongs to $PC = PC(W_{1,0})$.

Lemma 19.6. (MPA13) *If (1) $G = A_{13}$ or (2) $G = A_{12} + A_1$ we can make G from the essential basic Dynkin graph A_{11} by two tie transformations. Thus, $G \in PC$.*

For the class MOA13, MIA13 and NIA13 the corresponding graph G is not a member of PC.

Lemma 19.7. (MOA13) *[1] Two members (3) $A_{11} + A_2$, (8) $A_{11} + 2A_1$ of MOA13 can be made from the basic graph A_{11} by two tie transformations. However, in these cases the component A_{11} remains as an obstruction component.*
[2] $A_{11} + A_2 \notin PC$. $A_{11} + 2A_1 \notin PC$.

Proof. [1] is obvious. We will show [2]. Consider (3) $G = A_{11}+A_2$. Set $S = P \oplus Q(G)$. We assume that there is an embedding $S \hookrightarrow \Lambda_3$ satisfying Looijenga's (L1) and (L2). By $\widetilde{S}$ we denote the primitive hull of S in Λ_3. We set $I = \widetilde{S}/S$. T denotes the orthogonal complement of S in Λ_3. D is the discriminant of T.

We have $S^*/S \cong (\mathbf{Z}/4)^2 \oplus (\mathbf{Z}/3)^3$. The first $\mathbf{Z}/4$-component and the third component isomorphic to $\mathbf{Z}/3$ correspond to P, the second and the fourth correspond to A_{11}, and the last fifth $\mathbf{Z}/3$-component corresponds to A_2. We have $D = 2^4 \cdot 3^3/m^2$ for $m = [\widetilde{S} : S]$.

Assume that $m = [\widetilde{S} : S]$ is odd. $(T^*/T)_2 \cong (S^*/S)_2$. Since the discriminant quadratic form on $(S^*/S)_2 \cong (\mathbf{Z}/4)^2$ can be written $-a^2/4 + b^2/4$, $D \equiv -2^4 \bmod \mathbf{Z}_2^{*2}$. Thus, we have $-3^3 \equiv m^2 \pmod 8$, which is a contradiction. One knows that m is even.

Assume that m is not a multiple of 4. Then, by Lemma 19.3 (2) I_2 is generated by $(2,2) \in (S^*/S)_2$. One can check that the finite quadratic form on $(\widetilde{S}/\widetilde{S})_2 \cong I_2^{\perp}/I_2 \cong \mathbf{Z}/2 + \mathbf{Z}/2$ can be written ab for $(a,b) \in \mathbf{Z}/2 + \mathbf{Z}/2$. Thus, one has $D \equiv -2^2 \bmod \mathbf{Z}_2^{*2}$ and $-3^3 \equiv m'^2 \pmod 8$ for $m' = m/2$, which is a contradiction. Thus, m is a multiple of 4.

Next, we consider $p = 3$. $3 - 2l(I_3) = l((S^*/S)_3) - 2l(I_3) \leq l((\widetilde{S}^*/\widetilde{S})_3) \leq \operatorname{rank} T = 2$ and we have $l(I_3) \geq 1$. Let $\overline{\alpha} \in I_3$ be non-zero element. The discriminant quadratic form on $(S^*/S)_3$ can be written $q_3 \equiv -2x^2/3 + 2(y_1^2 + y_2^2)/3$. Thus, $\overline{\alpha}$ is equal to either $\overline{\alpha}_1 = (\pm 1, \pm 1, 0)$, or $\overline{\alpha}_2 = (\pm 1, 0, \pm 1)$. If $\overline{\alpha} = \overline{\alpha}_2$, $\widetilde{S}$ contains an element $\alpha = \chi + \omega$ where $\chi \in P^*$, $\omega \in Q(A_2)^*$ and $\omega^2 = 2/3$. The image of α under $\Lambda_3 \to \Lambda_3/P$ defines a short root in the primitive hull of $Q(G)$. It contradicts fullness. Thus, $\overline{\alpha} = \overline{\alpha}_1$. Note that the third component of $\overline{\alpha}_1 = (\pm 1, \pm 1, 0)$ is 0. It implies that $(P(P \oplus Q(A_{11}), \Lambda_3)/(P \oplus Q(A_{11})))_3 \neq 0$.

In conclusion, one has $[P(P \oplus Q(A_{11}), \Lambda_3) : P \oplus Q(A_{11})] \geq 12$. It implies that A_{11} is an obstruction component with respect to $Q(G) \hookrightarrow \Lambda_3/P$. Thus, $A_{11} + A_2 \notin PC$.

Consider (8) $G = A_{11} + 2A_1$ next. We define $S = P \oplus Q(G)$, $\widetilde{S}$, T, D, and I as above.

In this case, we have $S^*/S \cong (\mathbf{Z}/4)^2 \oplus (\mathbf{Z}/2)^2 \oplus (\mathbf{Z}/3)^2$ The first $\mathbf{Z}/4$-component and the fifth $\mathbf{Z}/3$-component correspond to P, the second and the sixth to A_{11}, and the middle third, fourth $\mathbf{Z}/2$-component correspond to 2 of A_1.

Since $l((S^*/S)_2) = 4 > 2$, $I_2 \neq 0$. Assume that I_2 has order 2. Then, $(2,2,0,0) \in (S^*/S)_2$ is the generator of I_2 and $l(I_2^{\perp}) \geq 4$. One has $3 \leq l(I_2^{\perp}/I_2) = l((\widetilde{S}^*/\widetilde{S})_2) = l((T^*/T)_2) \leq \operatorname{rank} T = 2$, a contradiction. Thus, I_2 is cyclic of order 4. Let $\overline{\alpha} =$

$(a_1, b, c_1, c_2) \in (S^*/S)_2$ be the generator. Note that $2\overline{\alpha} = (2,2,0,0)$. Since the discriminant quadratic form q_2 on $(S^*/S)_2$ can be written $-a^2/4 + b^2/4 + (c_1^2 + c_2^2)/2$, solving $q_2(\overline{\alpha}) \equiv 0$, one has $\overline{\alpha} = (\pm 1, \pm 1, 0, 0)$.

On the other hand, by considering $T \otimes \mathbf{Z}_3$ one knows $I_3 \neq 0$.

In conclusion, $[P(P \oplus Q(A_{11}), \Lambda_3) : P \oplus Q(A_{11})] \geq 12$, and A_{11} is an obstruction component. Thus, $A_{11} + 2A_1 \notin PC$. Q.E.D.

Lemma 19.8. (M1A13) *(6) $A_8 + A_5 \notin PC$. (17) $2A_6 + A_1 \notin PC$.*

Proof. Consider (6) $G = A_8 + A_5$. For this case we have $l((S^*/S)_3) \geq 3$. Thus, $(\widetilde{S}/S)_3 \neq 0$. For every non-zero element in the 3-Sylow subgroup $(S^*/S)_3$ at which the discriminant quadratic form takes 0, one can construct an extra root with length $\sqrt{2/3}$ or $\sqrt{2}$ in the primitive hull of $Q(G)$ in Λ_3/P. Thus, the corresponding graph $G \notin PC$.

For (17) $G = 2A_6 + A_1$ we can conclude $(\widetilde{S}/S)_7 \neq 0$ by calculating the discriminant of $T \otimes \mathbf{Z}_7$. It contradicts Lemma 19.4 (1). Thus, $2A_6 + A_1 \notin PC$. Q.E.D.

The class NIA13 is treated at the last part of this section.

Let us proceed to the case $r = 12$. We can assume $\rho = r + 7 = 19$. By Lemma 19.1 (1) $a = 0$ or 1

Lemma 19.9. *Under our assumption ((3)) and $r = 12$, $G \in PC$ can be witten in the form $G = \sum_{i=1}^{6} A_{k_i}$ for some non-negative integers $k_1, \ldots, k_6$ with $k_1 \geq k_2 \geq \ldots \geq k_6$ and $\sum k_i = 12$. (A_0 stands for the empty graph.)*

Proof. Obvious by Lemma 19.1 (3).

We have considered division of 12 into a sum of 6 non-negative integers in Section 13. There was 58 divisions. We here divide them into the 5 classes NPA12, MPA12, E12, MIA12, NIA12 below. We use the same numbering as in Section 13.

NPA12: (19 items.)

[1] 12	[3] 10+2	[4] 9+3	[5] 8+4
[6] 7+5	[7] 6+6	[8] 10+1+1	[9] 9+2+1
[10] 8+3+1	[12] 7+4+1	[14] 6+5+1	[15] 6+4+2
[18] 5+4+3	[20] 9+1+1+1	[24] 6+4+1+1	[25] 6+3+2+1
[31] 4+4+3+1	[36] 7+2+1+1+1	[42] 4+4+2+1+1	

MPA12: (10 items.)

[2] 11+1	[11] 8+2+2	[13] 7+3+2	[17] 5+5+2
[21] 8+2+1+1	[22] 7+3+1+1	[27] 5+5+1+1	[30] 5+3+2+2
[46] 3+3+3+2+1	[51] 5+2+2+1+1+1		

E12: (13 items.)

[16] 6+3+3	[33] 4+3+3+2	[34] 3+3+3+3	[35] 8+1+1+1+1
[37] 6+3+1+1+1	[41] 5+2+2+2+1	[45] 4+2+2+2+2	[48] 7+1+1+1+1+1
[49] 6+2+1+1+1+1	[52] 4+4+1+1+1+1	[53] 4+3+2+1+1+1	[57] 3+2+2+2+2+1
[58] 2+2+2+2+2+2			

MIA12: (6 items.)

[19] 4+4+4	[29] 5+3+3+1	[39] 5+4+1+1+1	[43] 4+3+3+1+1
[50] 5+3+1+1+1+1	[55] 3+3+3+1+1+1		

NIA12: (10 items.)

[23] 7+2+2+1	[26] 6+2+2+2	[28] 5+4+2+1	[32] 4+4+2+2
[38] 6+2+2+1+1	[40] 5+3+2+1+1	[44] 4+3+2+2+1	[47] 3+3+2+2+2
[54] 4+2+2+2+1+1	[56] 3+3+2+2+1+1		

Lemma 19.10. (NPA12) *Any graph in the class NPA12 is a subgraph of a graph which can be made from one of the essential basic Dynkin graph by tie transformations repeated twice. Therefore, it belongs to PC.*

- ▷ *A subgraph of* A_{13} *[1], [3]–[7]*
- ▷ *A subgraph of* $A_{12} + A_1$ *[8], [9], [10], [12], [14]*
- ▷ *A subgraph of* $E_7 + A_6$ *[15], [25]*
- ▷ *A subgraph of* $D_9 + A_4$ *[18]*
- ▷ *A subgraph of* $D_{12} + A_1$ *[20]*
- ▷ *A subgraph of* $E_8 + A_4 + A_1$ *[24], [31], [42]*
- ▷ *A subgraph of* $D_{10} + A_2 + A_1$ *[36]*

Lemma 19.11. (MPA12) *Every graph in MPA12 can be made from a basic graph by two transformations as below. Below every arrow except the left one at the bottom three lines indicates a tie transformation.*

[2] $A_{11} + A_1$	$\longleftarrow$	A_{12}	$\longleftarrow$	A_{11}
[11] $A_8 + 2A_2$	$\longleftarrow$	$A_8 + A_2 + A_1$	$\longleftarrow$	$E_8 + B_1 + G_2$
[13] $A_7 + A_3 + A_2$	$\longleftarrow$	$A_7 + A_3 + A_1$	$\longleftarrow$	$E_7 + B_3 + G_1$
[17] $2A_5 + A_2$	$\longleftarrow$	$2A_5 + A_1$	$\longleftarrow$	A_{11}
[21] $A_8 + A_2 + 2A_1$	$\longleftarrow$	$A_8 + A_2 + A_1$	$\longleftarrow$	$E_8 + B_1 + G_2$
[22] $A_7 + A_3 + 2A_1$	$\longleftarrow$	$A_7 + A_3 + A_1$	$\longleftarrow$	$E_7 + B_3 + G_1$
[27] $2A_5 + 2A_1$	$\longleftarrow$	$2A_5 + A_1$	$\longleftarrow$	A_{11}
[30] $A_5 + A_3 + 2A_2$	$\overset{elementary}{\longleftarrow}$	$E_7 + B_3 + G_2$	$\longleftarrow$	$E_7 + B_3 + G_1$
[46] $3A_3 + A_2 + A_1$	$\overset{elementary}{\longleftarrow}$	$E_7 + B_3 + G_2$	$\longleftarrow$	$E_7 + B_3 + G_1$
[51] $A_5 + 2A_2 + 3A_1$	$\overset{elementary}{\longleftarrow}$	$E_8 + B_2 + G_2$	$\longleftarrow$	$E_8 + B_1 + G_2$

Note that in the case [2] $A_{11} + A_1$*, the component* A_{11} *is not an obstruction.*

Lemma 19.12. *If* $r = 12$*,* $\rho = 19$ *and* $a = 0$*, then* G *is either [17]* $2A_5 + A_2$ *or [27]* $2A_5 + 2A_1$.

Proof. By Lemma 19.1 (1) it is obvious.

Since [17] and [27] belong to MPA12, we can assume $a = 1$ in the following. We can apply Lemma 19.3 and Lemma 19.4.

The 13 graphs in the class E12 coincide with the 13 items treated in paragraph **A** just after Proposition 13.5 in Section 13. Therefore, they do not belong to $PC(J_{3,0})$.

Lemma 19.13. (E12) *Every graph in 13 members of the class E12 does not belong to PC.*

Proof. Obvious by Proposition 19.2.

In what follows we define $S = P \oplus Q(G)$, $\widetilde{S}$, $I = \widetilde{S}/S$, T and D corresponding to the graph G under consideration as above.

Lemma 19.14. (MIA12) *[50] $A_5 + A_3 + 4A_1 \notin PC$. [55] $3A_3 + 3A_1 \notin PC$.*

Proof. For [50] and [55] the division of 12 consists of 6 odd numbers. This implies $l((S^*/S)_2) \geq 1 + 6 = 7$. On the other hand, by Lemma 19.4 (2) $l(I_2) \leq 1$. Thus, if $G \in PC$, then we have $5 = 7 - 2 \leq l((S^*/S)_2) - 2l(I_2) \leq l((\widetilde{S}^*/\widetilde{S})_2) = l((T^*/T)_2) \leq \operatorname{rank} T = 3$, a contradiction. Q.E.D.

Lemma 19.15. (MIA12) *[19] $3A_4 \notin PC$.*

Proof. We can conclude $(\widetilde{S}/S)_p \neq 0$ for $p = 5$ in the case [19] by calculating D in 2 ways. Thus, by Lemma 19.4 (1) $G \notin PC$. Q.E.D.

Lemma 19.16. (MIA12) *[29] $A_5 + 2A_3 + A_1 \notin PC$. [39] $A_5 + A_4 + 3A_1 \notin PC$. [43] $A_4 + 2A_3 + 2A_1 \notin PC$.*

Proof. In the cases [29], [39], [43] we have $l((S^*/S)_2) \geq 4$, Thus, $I_2 = (\widetilde{S}/S)_2 \neq 0$. By Lemma 19.3 (2) we can easily find an element $\overline{\alpha} \in I_2$ of order 2. On the other hand, for every $\overline{\beta} \in (S^*/S)_2$ $2\overline{\beta} \neq \overline{\alpha}$; thus, I_2 is generated by $\overline{\alpha}$. Computing the discriminant quadratic form of $\widetilde{S}$ and computing the discriminant D in two different ways, one can deduce a contradiction. Thus, the corresponding graph $G \notin PC$. Q.E.D.

Ten graphs of the class NIA12 are remaining. We treat them after the case $r = 11$. In the following we proceed to the last case $r = 11$.

Proposition 19.17. *For every division $(k_1, k_2, \ldots, k_7)$ of $11 = \sum_{i=1}^{7} k_i$ into a sum of 7 non-negative integers $k_1 \geq k_2 \geq \cdots \geq k_7 \geq 0$, consider the Dynkin graph $G = \sum A_{k_i}$, the root lattice $Q = Q(G)$ of type G, the discriminant $d(G) = d(Q)$ of Q, and the Hasse invariant $\epsilon_p(G) = \epsilon_p(Q)$ of Q, where p is a prime number. The I1-condition for p fails, i.e.,*

$$3d(G) \notin \mathbf{Q}_p^{*2} \text{ or } \epsilon_p(G) = (-1, 3)_p \qquad \heartsuit$$

is satisfied if, and only if, G and p are one in the following list:

$p = 3$, $A_5 + A_4 + 2A_1$, $A_5 + A_3 + 3A_1$, $A_4 + 3A_2 + A_1$, $A_3 + 3A_2 + 2A_1$. *(4 items.)*
$p = 2$, $A_5 + A_3 + 3A_1$, $A_3 + 3A_2 + 2A_1$. *(2 items.)*

Proof. Set $d = d(G)$ and $\epsilon_p = \epsilon_p(G)$.

If $\heartsuit$ is satisfied for $p \geq 5$, then by Lemma 7.5 (3) $p = 5$ and $G = 2A_4 + G'$ where $G' = A_3$, $A_2 + A_1$ or $3A_1$. However, these 3 candidates do not satisfy $\heartsuit$ for $p = 5$.

Assume that $\heartsuit$ holds for $p = 3$. By G' we denote the maximal sum of components with $3 \not| d' = d(G')$. Since $3d \in \mathbf{Q}_3^{*2}$, the total number of components of type A_{11}, A_5, or A_2 is odd. It is easy to check that $G = A_{11}$, $A_5 + 2A_2 + A_1$, $5A_2 + A_1$ does not satisfy $\heartsuit$. If $G = A_5 + G'$, then $\heartsuit \iff d' \equiv 2 \pmod 3$. If $G = 3A_2 + G'$, then $\heartsuit \iff d' \equiv 1 \pmod 3$. If $G = A_2 + G'$, then $3d \in \mathbf{Q}_3^{*2} \iff d' \equiv 1 \pmod 3$, $\epsilon_3 = 1 \iff d' \equiv 2 \pmod 3$. Thus, we get the result.

Assume that for $p = 2$ $\heartsuit$ holds. By Lemma 7.7 if $3d \in \mathbf{Q}_2^{*2}$, then $G = A_{11}$, $A_5 + A_3 + 3A_1$, $3A_3 + A_2$, $A_3 + 3A_2 + 2A_1$, or $A_3 + A_2 + 6A_1$. Computing ϵ_2 one gets the result. Q.E.D.

By Theorem 16.4, we can consider only the 4 graphs in the above proposition.

Lemma 19.18. *We can make $A_5+A_4+2A_1$ and $A_5+A_3+3A_1$ from the basic graph A_{11} by tie transformations repeated twice. Therefore, they are members in PC.*

Proof. Obviously, A_5+A_4 and $A_5+A_3+A_1$ are subgraphs of A_{11}. Thus, we can make the above two graphs from A_{11} by tie transformations repeated twice. In particular, they are members in PC. Q.E.D.

Lemma 19.19. $A_4+3A_2+A_1 \notin PC$. $A_3+3A_2+2A_1 \notin PC$.

Proof. Let us consider $G = A_4+3A_2+A_1$. By using 3-adic integers $\mathbf{Z}_3$, we can show $(\widetilde{S}/S)_3 \neq 0$ for every embedding $S = P \oplus Q(G) \hookrightarrow \Lambda_3$. Any element $\overline{\alpha} \in (\widetilde{S}/S)_3$ satisfies $q(\overline{\alpha}) \equiv 0 \bmod 2\mathbf{Z}$, where q is the discriminant quadratic form of S. However, for every non-zero element $\overline{\alpha} \in (S^*/S)_3$ with $q(\overline{\alpha}) \equiv 0$ we can construct an extra root not in $Q(G)$ but in the primitive hull of $Q(G)$ in Λ_3/P. It contradicts the fullness. Thus, $A_4+3A_2+A_1 \notin PC$.

By the same method we can show $A_3+3A_2+2A_1 \notin PC$. Q.E.D.

Lemma 19.20. (NIA12, NIA13) *(1) Any graph in NIA12 does not belong to PC. (2) Any graph in NIA13 does not belong to PC*

Proof. (1) Every graph in ten members of NIA12 contains $A_4+3A_2+2A_1$ or $A_3+3A_2+2A_1$ as a subgraph. By Proposition 13.5 (2) it never belong to PC.
(2) For every graph G in the 19 graphs of the class NIA13, we can find a graph G' in NIA12 such that G' is a subgraph of G. Thus, by Proposition 13.5 (2) they never belong to PC, either. Q.E.D.

In this section we have shown Proposition 17.1 under the assumption ((3)) in the beginning of Section 17. We complete the verification of Theorem 0.5.

20
The basic graphs for $S_{1,0}$

We proceed to the case $S_{1,0}$ in this section. Only in this case $S_{1,0}$ our method developed so far does not work very well. To overcome this difficulty sometimes we have to consider naive method such as case-by-case checking. We will spend 8 sections to manipulate $S_{1,0}$.

The objective of this section is to show following Theorem 20.1. In this section by $\Lambda = \Lambda_1$ we denote an even unimodular lattice with signature (17, 1). Note that every embedding $P \hookrightarrow \Lambda_N$ is primitive (Proposition 4.5 (1)).

Theorem 20.1. *Let P be the lattice associated with the case $S_{1,0}$.*
(1) the following ⟨i⟩, and ⟨ii⟩ are equivalent:
  ⟨i⟩ The Dynkin graph of the root system of the quotient $\Lambda/\phi(P)$ coincides with G for some embedding $\phi : P \hookrightarrow \Lambda$.
  ⟨ii⟩ $G = A_9+BC_1$, E_8+BC_1, B_8+A_1, E_7+BC_2, or E_6+B_3.
(2) Let ϕ, $\phi' : P \hookrightarrow \Lambda$ be two embeddings. If the Dynkin graphs of $\Lambda/\phi(P)$ and $\Lambda/\phi'(P)$ coincide, then there are lattice isomorphisms $a : P \to P$ and $b : \Lambda \to \Lambda$ with $\phi' = b\phi a$.

For $S_{1,0}$ the quasi-lattice Λ_2/P does not satisfy the equivalent conditions in Proposition 8.4. Thus, we cannot draw the Coxeter-Vinberg graph in finite steps, and we cannot deduce the above theorem from the graph. Moreover, the lattice P has no nice decomposition as for $Q_{2,0}$. Thus, in the following we have to apply more naive method to show the above theorem.

Now, let $\phi : P \hookrightarrow \Lambda$ be an embedding. Let F be the orthogonal complement of $\phi(P)$. We would like to determine the root system $R(F^*)$ of the dual quasi-lattice $F^* \cong \Lambda/\phi(P)$ of F.

Let $b : \Lambda \to \Lambda$ be an orthogonal transformation. Obviously, the root system associated with ϕ and the root system associated with $b\phi$ are isomorphic. Thus, in what follows we can exchange ϕ for $b\phi$ with a suitable b repeatedly.

Lemma 20.2. *(1) $F^*/F \cong \mathbf{Z}/5 \oplus \mathbf{Z}/2 \oplus \mathbf{Z}/2$. The discriminant form of F can be written*

$$q_F(\overline{\alpha}) \equiv -\frac{2}{5}a^2 + \frac{1}{2}(b_1^2 + b_2^2) \bmod 2\mathbf{Z}$$

for $\overline{\alpha} = (a, b_1, b_2) \in \mathbf{Z}/5 \oplus \mathbf{Z}/2 \oplus \mathbf{Z}/2$.
(2) F^* contains no vector α with $\alpha^2 = 2/3$.

Proof. (1) It follows from Proposition 4.2 and Lemma 5.5 (3).
(2) q_F does not take the value $2/3 \bmod 2\mathbf{Z}$. Q.E.D.

Lemma 20.3. (Dynkin [7]) *(1) If two sublattices M_1 and M_2 in the root lattice $Q(D_{16})$ of type D_{16} are isomorphic to the root lattice of type D_4, then $M_2 = w(M_1)$ for some element w in the Weyl group $W(D_{16})$ of $Q(D_{16})$.*
(2) (resp. (3)) If two sublattices M_1 and M_2 in the root lattice $Q(E_8)$ of type E_8 are isomorphic to the root lattice of type D_4 (resp. A_2), then $M_2 = w(M_1)$ for some element w in the Weyl group $W(E_8)$ of $Q(E_8)$.

Lemma 20.4. *Let L be a lattice. The reflection $s_\alpha : L \to L$ associated with a long root $\alpha \in R(L)$ induces the identity on the discriminant group L^*/L.*

Now, we have the orthogonal decomposition $P = P' \oplus H_0$. We have 2 cases.

I. $C(\phi(H_0), \Lambda) \cong \Gamma_{16}$.
II. $C(\phi(H_0), \Lambda) \cong Q(2E_8)$.

Case I.

We identify $C(\phi(H_0), \Lambda)$ with Γ_{16}. $R = \{\alpha \in \Gamma_{16} \mid \alpha^2 = 2\}$ is a root system of type D_{16}, and the sublattice Q generated by R is a root lattice of type D_{16} with $[\Gamma_{16} : Q] = 2$.

Let $\alpha_1, \ldots, \alpha_{16}$ be the root basis of Q. We adopt the same numbering as in Example 2.3. Set

$$\omega_1 = \sum_{i=1}^{14} \alpha_i + \frac{1}{2}(\alpha_{15} + \alpha_{16}),$$

$$\omega_{15} = \frac{1}{2}\left(\sum_{i=1}^{14} i\alpha_i + 8\alpha_{15} + 7\alpha_{16}\right), \quad \omega_{16} = \frac{1}{2}\left(\sum_{i=1}^{14} i\alpha_i + 7\alpha_{15} + 8\alpha_{16}\right).$$

$\omega_i \cdot \alpha_j = \delta_{ij}$ $(1 \le j \le 16,\ i = 1,\ 15,\ 16)$. Thus, $\omega_i \in Q^*$. We denote $\overline{x} = x \bmod Q \in Q^*/Q$ for $x \in Q^*$. Then, we have $\{0, \overline{\omega}_1, \overline{\omega}_{15}, \overline{\omega}_{16}\} = Q^*/Q \cong \mathbf{Z}/2 + \mathbf{Z}/2$. Thus, $\overline{\omega}_1 = \overline{\omega}_{15} + \overline{\omega}_{16}$. Moreover, $\omega_1^2 = 1$ and $\omega_{15}^2 = \omega_{16}^2 = 4$. Exchanging the number 15 and 16 if necessary, we can assume

$$Q^*/Q \supset \Gamma_{16}/Q = \{0, \overline{\omega}_{16}\}.$$

In what follows we assume this equality.

Consider the sublattice $P_0' = \sum_{i=0}^{3} \mathbf{Z}e_i$ in P'. $P_0' \cong Q(D_4)$. Since P_0' is generated by long roots, $\phi(P_0') \subset Q$. Set $\alpha_0 = -(\alpha_1 + 2\sum_{i=2}^{14} \alpha_i + \alpha_{15} + \alpha_{16})$. This α_0 satisfies $\alpha_0^2 = +2$, $\alpha_0 \cdot \alpha_i = 0$ $(1 \le i \le 16,\ i \ne 2)$, and $\alpha_0 \cdot \alpha_2 = -1$.

Set $L = \mathbf{Z}\alpha_{13} + \mathbf{Z}\alpha_{14} + \mathbf{Z}\alpha_{15} + \mathbf{Z}\alpha_{16}$, $M = \sum_{i=0}^{11} \mathbf{Z}\alpha_i$. M is the orthogonal complement of L in Q. By Lemma 20.3 (1) for some roots $\gamma_1, \ldots, \gamma_k \in R$

$$\phi(P_0') = w(L), \qquad w = s_{\gamma_1} s_{\gamma_2} \cdots s_{\gamma_k}.$$

By Lemma 20.4 w maps $\{0, \overline{\omega}_{16}\}$ to itself; thus, w can be extended to an orthogonal transformation b on Γ_{16}. Considering $b^{-1}\phi$ instead of ϕ, we can assume $\phi(P_0') = L$ from the beginning. Set $S = C(\phi(P_0'), \Gamma_{16})$. $d(S) = d(P_0') = 4$. Then, we have $S \cap Q = M$. On the other hand, $d(M) = 4 = d(S)$; thus, one has $S = M$.

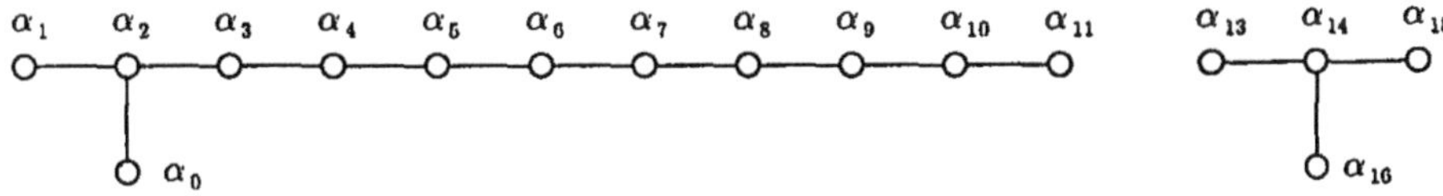

α_1 α_2 α_3 α_4 α_5 α_6 α_7 α_8 α_9 α_{10} α_{11} α_0 α_{13} α_{14} α_{15} α_{16}

M $\qquad$ $L = \varphi(P_0')$

Here, set moreover

$$\chi_0 = \frac{1}{2}\left\{6\alpha_0 + 5\alpha_1 + \sum_{j=2}^{11}(12-j)\alpha_j\right\}, \quad \chi_1 = \frac{1}{2}\left\{5\alpha_0 + 6\alpha_1 + \sum_{j=2}^{11}(12-j)\alpha_j\right\}$$

$$\chi_{11} = \frac{1}{2}(\alpha_0 + \alpha_1) + \sum_{j=2}^{11} \alpha_j, \qquad \tau_{13} = \alpha_{13} + \alpha_{14} + \frac{1}{2}(\alpha_{15} + \alpha_{16}),$$

$$\tau_{15} = \frac{1}{2}\alpha_{13} + \alpha_{14} + \alpha_{15} + \frac{1}{2}\alpha_{16}, \quad \tau_{16} = \frac{1}{2}\alpha_{13} + \alpha_{14} + \frac{1}{2}\alpha_{15} + \alpha_{16}.$$

We can check $\alpha_{12} = -\chi_{11} - \tau_{13}$, $\omega_{16} = -\chi_0 + \tau_{16}$, and

$$\chi_i \cdot \alpha_j = \delta_{ij} \quad (0 \le j \le 16,\ j \ne 12,\ i = 0,\ 1,\ 11)$$
$$\tau_i \cdot \alpha_j = \delta_{ij} \quad (0 \le j \le 16,\ j \ne 12,\ i = 13,\ 15,\ 16).$$

We denote $\overline{x} = x \bmod M \oplus L \in (M \oplus L)^*/(M \oplus L) \cong (M^*/M) \oplus (L^*/L) \cong (\mathbf{Z}/2)^4$ for $x \in (M \oplus L)^*$. We have

$$\begin{aligned}(M \oplus L)^*/(M \oplus L) \supset \Gamma_{16}/(M \oplus L) &= \{0, \overline{\alpha}_{12}, \overline{\omega}_{16}, \overline{\alpha_{12}+\omega_{16}}\} \\ &= \{0, \overline{\chi_{11}+\tau_{13}}, \overline{\chi_0+\tau_{16}}, \overline{\chi_1+\tau_{15}}\}.\end{aligned}$$

Recall that all root bases of L are conjugate under the action of the Weyl group. Thus, composing the action of $W(L)$, we can assume $\{\phi(e_1), \phi(e_2), \phi(e_3), \phi(e_0)\} = \{\alpha_{13}, \alpha_{14}, \alpha_{15}, \alpha_{16}\}$. Since $\phi(e_0)$ and α_{14} correspond to the central vertex of the Dynkin graph D_4, we have

$$\phi(e_0) = \alpha_{14}, \quad \{\phi(e_1), \phi(e_2), \phi(e_3)\} = \{\alpha_{13}, \alpha_{15}, \alpha_{16}\}.$$

Recall here that $P' = P'_0 + \mathbf{Z}f + \mathbf{Z}e'$. (See Section 4.) To describe $\phi : P' \hookrightarrow \Gamma_{16}$ we still need to determine the image $\phi(f)$ and $\phi(e')$.

We write the vector $\phi(f) \in \Gamma_{16} \subset M^* \oplus L^*$ in the form $\phi(f) = \xi + \eta$ with $\xi \in M^*$ and $\eta \in L^*$. Since $\eta \cdot \phi(e_i) = \phi(f) \cdot \phi(e_i) = f \cdot e_i = 0$ (if $i = 0, 1, 2$), $= -1$ (if $i = 3$), η coincides with either $-\tau_{13}$, $-\tau_{15}$, or $-\tau_{16}$. In particular, $\eta^2 = \tau_{13}^2 = \tau_{15}^2 = \tau_{16}^2 = 1$. Thus, $\xi^2 = f^2 - \eta^2 = 3$.

We have 3 subcases.

Subcase I-1. $\phi(e_3) = \alpha_{13}$.

In this case, $\eta = -\tau_{13}$. We can write $\xi = \chi_{11} + a$ with $a \in M$, since $\overline{\phi(f)} = \overline{\xi + \eta} \in \Gamma_{16}/(M \oplus L)$. Here we introduce a convenient model of $M \cong Q(D_{12})$ for calculation.

A model of M.

Let $N = \sum_{i=1}^{12} \mathbf{Z}v_i$ be the lattice defined by $v_i^2 = 1$ $(1 \leq i \leq 12)$ and $v_i \cdot v_j = 0$ $(i \neq j)$. We can define a lattice-embedding $M \hookrightarrow N$ by $\alpha_0 = -v_1 - v_2$, $\alpha_i = v_i - v_{i+1}$ $(1 \leq i \leq 11)$. Then, we have $M = \{\sum_{i=1}^{12} x_i v_i \mid \sum_{i=1}^{12} x_i$ is an even integer.$\}$. The closure of a Weyl chamber C has the following expression:

$$\begin{aligned}\overline{C} &= \{x \in M \otimes \mathbf{R} \mid (x, \alpha_i) \leq 0 \text{ for } 1 \leq i \leq 11\} \\ &= \left\{x = \sum_{i=1}^{12} x_i v_i \in N \otimes \mathbf{R} \mid x_1 \leq x_2 \leq \cdots \leq x_{12},\ x_1 + x_2 \geq 0\right\}.\end{aligned}$$

Adopting this model of M, we have $\chi_{11} = -v_{12}$.

For some element w in the Weyl group of M $w(\xi) \in \overline{C}$. Since the action of w can be extended to Γ_{16}, and since $w(\chi_{11}) - \chi_{11} \in M$, we can assume from the beginning that $\xi \in \overline{C}$. Set $\xi = \sum_{i=1}^{12} x_i v_i$. $a = \xi - \chi_{11} = \sum_{i=1}^{11} x_i v_i + (x_{12}+1)v_{12}$. Therefore, $x_i \in \mathbf{Z}$, $\sum_{i=1}^{12} x_i$ is odd, $3 = \xi^2 = \sum_{i=1}^{12} x_i^2$, and $x_1 \leq \cdots \leq x_{12}$, $x_1 + x_2 \geq 0$.

By these conditions one knows $x_1 = x_2 = \cdots = x_9 = 0$, $x_{10} = x_{11} = x_{12} = 1$. One has $\xi = v_{10} + v_{11} + v_{12} = -3(\alpha_0 + \alpha_1)/2 - 3(\alpha_2 + \alpha_3 + \cdots + \alpha_9) - 2\alpha_{10} - \alpha_{11}$. For $0 \leq i \leq 11$ $i \neq 9$, $\xi \cdot \alpha_i = 0$. $\xi \cdot \alpha_9 = -1$.

Next, we consider $\phi(e')$. Set $\phi(e') = \sum_{i=1}^{12} y_i v_i$. $\phi(e') \in M$ since $\phi(e')$ is orthogonal to $L = \phi(P'_0)$. Thus, $y_i \in \mathbf{Z}$ $(0 \leq i \leq 12)$, and $\sum_{i=1}^{12} y_i$ is even. $2 = e'^2 = \sum_{i=1}^{12} y_i^2$. Now, the reflections s_{α_i} $(0 \leq i \leq 11,\ i \neq 9)$ keep ξ fixed. By the action of the Weyl group generated by them we can assume moreover; $y_1 \leq y_2 \leq \cdots \leq y_9$, $y_{10} \leq y_{11} \leq y_{12}$,

$y_1 + y_2 \geq 0$. The solution satisfying the conditions is unique, and $y_1 = \cdots = y_8 = 1$, $y_9 = 1$, $y_{10} = -1$, $y_{11} = y_{12} = 0$. One has $\phi(e') = v_9 - v_{10} = \alpha_9$.

We can compute F here. $F = C(\phi(P), \Lambda) = C(\phi(P'), \Gamma_{16}) = C(\mathbf{Z}(2\xi)+\mathbf{Z}\alpha_9, M) = \sum_{i=1}^{7} \mathbf{Z}\alpha_i + \mathbf{Z}(\alpha_8 - \alpha_{10}) + \mathbf{Z}(-\alpha_{11})$. Setting $\beta = \alpha_8 - \alpha_{10}$, $\beta^2 = +4$. F is the lattice defined by the following dual graph:

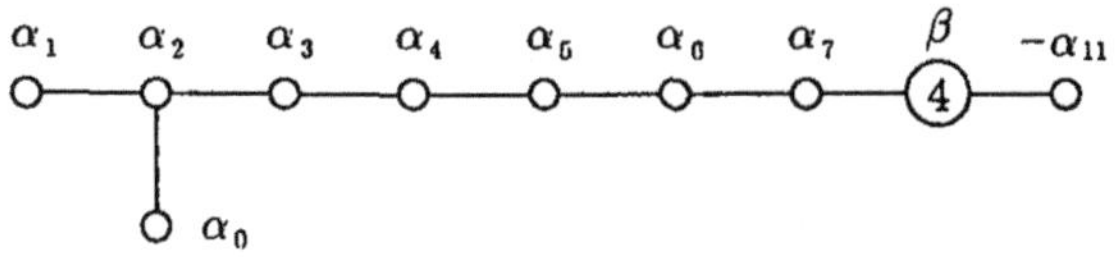

α_1 α_2 α_3 α_4 α_5 α_6 α_7 β $-\alpha_{11}$

4

α_0

Next, we would like to determine the root system of F^*. Several notations are necessary.

$$F_1 = \sum_{i=0}^{7} \mathbf{Z}\alpha_i \cong Q(D_8), \qquad F_2 = C(F_1, F),$$

$$\delta = \alpha_0 + \alpha_1 + 2(\alpha_2 + \cdots + \alpha_7 + \beta) - \alpha_{11}, \qquad \epsilon = \alpha_{11}.$$

We can check that $F_2 = \mathbf{Z}\delta + \mathbf{Z}\epsilon$, $\delta^2 = 10$, $\epsilon^2 = 2$, and $\delta \cdot \epsilon = 0$. $F_1 \oplus F_2 \subset F = (F_1 \oplus F_2) + \mathbf{Z}\beta \subset F^* \subset F_1^* \oplus F_2^*$. Set moreover

$$\mu_0 = \frac{1}{2}\left\{4\alpha_0 + 3\alpha_1 + \sum_{i=2}^{7}(8-i)\alpha_i\right\}, \quad \mu_1 = \frac{1}{2}\left\{3\alpha_0 + 4\alpha_1 + \sum_{i=2}^{7}(8-i)\alpha_i\right\},$$

$$\lambda = \frac{1}{5}\delta, \quad \nu_1 = \frac{1}{2}\delta, \quad \nu_2 = \frac{1}{2}\epsilon.$$

One knows μ_0, $\mu_1 \in F_1^*$, $\mu_i \cdot \alpha_j = \delta_{ij}$ $(i = 0, 1;\ 0 \leq j \leq 7)$ and λ, ν_1, $\nu_2 \in F_2^*$. We denote $\overline{x} = x \bmod (F_1 \oplus F_2) \in (F_1 \oplus F_2)^*/(F_1 \oplus F_2)$ for $x \in (F_1 \oplus F_2)^*$. We have

$$\begin{aligned}(F_1 \oplus F_2)^*/(F_1 \oplus F_2) &= \langle\overline{\lambda}\rangle \oplus \langle\overline{\mu}_0, \overline{\mu}_1\rangle \oplus \langle\overline{\nu}_1\rangle \oplus \langle\overline{\nu}_2\rangle \\ &\cong \mathbf{Z}/5 \oplus (\mathbf{Z}/2 + \mathbf{Z}/2) \oplus \mathbf{Z}/2 \oplus \mathbf{Z}/2.\end{aligned}$$

Here, $\langle\ \rangle$ denotes the group generated by the elements between the angles. The discriminant quadratic form can be written

$$q(x\overline{\lambda} + a_0\overline{\mu}_0 + a_1\overline{\mu}_1 + b_1\overline{\nu}_1 + b_2\overline{\nu}_2) \equiv \frac{2}{5}x^2 + a_0a_1 + \frac{1}{2}(b_1^2 + b_2^2) \bmod 2\mathbf{Z}.$$

Let I be the group of order 2 generated by $\overline{\beta} = \overline{\mu}_0 + \overline{\mu}_1 + \overline{\nu}_1 + \overline{\nu}_2$, and $I^{\perp}$ be the orthogonal complement of I with respect to the discriminant bilinear form. We have

$$F/(F_1 \oplus F_2) = I, \qquad F^*/(F_1 \oplus F_2) = I^{\perp}.$$

Now, we can define the expected minimum square length $m(\overline{x})$ for each element $\overline{x} \in (F_1 \oplus F_2)^*/(F_1 \oplus F_2)$, since $F_1 \oplus F_2$ is a positive definite lattice.

Lemma 20.5. *(1) Let F_1' be the overlattice of F_1 with $F_1'/F_1 = \{\overline{0}, \overline{\mu}_0+\overline{\mu}_1\} \subset F_1^*/F_1$. Then, $F_1' \cong Q(B_8)$ and $R(F^*) = R(F_1') \oplus \hat{R}(F_2^*)$.*
(2) $\hat{R}(F_2^*) = \{x \in F_2^* \mid x^2 = 2 \text{ or } 1\}$ is a root system of type A_1 consisting of 2 vectors $\{\epsilon, -\epsilon\}$.

Proof. We show (2) first. Pick a root $\alpha = x_1\delta/10 + x_2\epsilon/2 \in \hat{R}(F_2^*)$. $x_1, x_2 \in \mathbf{Z}$. If $\alpha^2 = 1$, then $10 = x_1^2+5x_2^2$. This equation has no solution. If $\alpha^2 = 2$, then $20 = x_1^2+5x_2^2$; thus, $x_1 = 0$, $x_2 = \pm 2$.

Next, we show (1). By (2) one knows $R(F^*) \supset \hat{R}(F_2^*)$. Let $\alpha \in R(F_1')$ be an arbitrary root. If $\alpha^2 = 1$, then by definition, $\overline{\alpha} = \overline{\mu}_0+\overline{\mu}_1$. Thus, $b(\overline{\alpha}, \overline{\beta}) \equiv (\overline{\mu}_0+\overline{\mu}_1)^2 \equiv 0 \bmod \mathbf{Z}$. It implies $\overline{\alpha} \in I^{\perp}$; thus, $\alpha \in F^*$. If $\alpha^2 = 2$, then by the definition of F_1', we have $\alpha \in F_1 \subset F \subset F^*$. Thus, one knows $R(F^*) \supset R(F_1') \oplus \hat{R}(F_2^*)$.

We show the opposite inclusion relation. Let $\alpha \in R(F^*)$. By Lemma 20.2 (2) $\alpha^2 \neq 2/3$. We can write $\overline{\alpha} = \overline{\mu}' + \overline{\nu}'$ with $\overline{\mu}' \in F_1^*/F_1$ and $\overline{\nu}' \in F_2^*/F_2$.

Case 1. $\alpha^2 = 1/2$.

If $\overline{\mu}' \neq 0$, then $1/2 = \alpha^2 \geq m(\overline{\alpha}) = m(\overline{\mu}') + m(\overline{\nu}') \geq m(\overline{\mu}') \geq 1$, a contradiction. Thus, $\overline{\mu}' = 0$, and we can write $\alpha = \alpha_1 + \alpha_2$ with $\alpha_1 \in F_1$ and $\alpha_2 \in F_2^*$. If $\alpha_1 \neq 0$, then $\alpha_1^2 \geq 2$; thus $1/2 = \alpha^2 = \alpha_1^2 + \alpha_2^2 \geq 2$, a contradiction. One knows $\alpha \in F_2^*$. Since $q(\overline{\alpha}) \equiv 1/2$, $\overline{\alpha} = \overline{\nu}_1$ or $\overline{\nu}_2$. However, $\overline{\alpha} \in I^{\perp}$ while $\overline{\nu}_1 \notin I^{\perp}$ and $\overline{\nu}_2 \notin I^{\perp}$, a contradiction. This case never takes place.

Case 2. $\alpha^2 = 1$.

First, assume $\overline{\mu}' = 0$. Then, we can write $\alpha = \alpha_1 + \alpha_2$ with $\alpha_1 \in F_1$, $\alpha_2 \in F_2^*$. If $\alpha_1 \neq 0$, then $1 = \alpha^2 = \alpha_1^2 + \alpha_2^2 \geq \alpha_1^2 \geq 2$, a contradiction. Thus, $\alpha \in \hat{R}(F_2^*)$. Second, assume $\overline{\mu}' \neq 0$. We have $1 = \alpha^2 \geq m(\overline{\alpha}) = m(\overline{\mu}') + m(\overline{\nu}') \geq m(\overline{\mu}') \geq 1$. Thus, $\overline{\nu}' = 0$, and $m(\overline{\mu}') = 1$. One knows $\overline{\alpha} = \overline{\mu}' = \overline{\mu}_0 + \overline{\mu}_1$ and $\alpha \in R(F_1')$.

Case 3. $\alpha^2 = 2$.

If $\overline{\mu}' = 0$, then $\overline{\nu}' = 0$ since $\overline{\alpha} = \overline{\nu}'$ satisfies $q(\overline{\alpha}) \equiv 0$. $\alpha \in F_1 \oplus F_2$. Either $\alpha \in F_1$ or $\alpha \in F_2$ holds, since both F_1 and F_2 are even lattices. If $\overline{\mu}' = \overline{\mu}_1 + \overline{\mu}_2$, then $\overline{\alpha} = \overline{\mu}_1+\overline{\mu}_2+\overline{\nu}_1+\overline{\nu}_2$, since $q(\overline{\alpha}) \equiv 0$. $2 = \alpha^2 \geq m(\overline{\alpha}) = m(\overline{\mu}_1+\overline{\mu}_2)+m(\overline{\nu}_1)+m(\overline{\nu}_2) = 1+5/2+1/2 = 4$, which is a contradiction. If $\overline{\mu}' = \overline{\mu}_i$ ($i = 1$ or 2), then $\overline{\nu}' = 0$ and $\overline{\alpha} = \overline{\mu}_i$, since $q(\overline{\alpha}) \equiv 0$. On the other hand, $\overline{\mu}_i \notin I^{\perp}$, since $b(\overline{\mu}_i, \overline{\mu}_1+\overline{\mu}_2+\overline{\nu}_1+\overline{\nu}_2) \equiv 1/2 \bmod \mathbf{Z}$. It contradicts $\overline{\alpha} \in I^{\perp}$. Since $q(\overline{\alpha}) \equiv 0$, no other case occurs. Q.E.D.

Proposition 20.6. *In Subcase I-1 $R(F^*)$ is of type $B_8 + A_1$.*

Subcase I-2. $\phi(e_3) = \alpha_{16}$.

Writing $\phi(f) = \xi + \eta$ ($\xi \in M^*$, $\eta \in N^*$), one has $\eta = -\tau_{16}$ and $\xi = -\chi_0 + a$ for some $a \in M$. We use the above model $M \hookrightarrow N$. Since $\chi_0 = -\sum_{i=1}^{12} v_i/2$, we can write $\xi = \sum_{i=1}^{12}(a_i + 1/2)v_i$ with $a_i \in \mathbf{Z}$, where $\sum_{i=1}^{12} a_i$ is an even integer. One has $3 = \xi^2 = \sum_{i=1}^{12}(a_i + 1/2)^2$. Moreover, we can assume $\xi \in \overline{C}$, i.e., $a_1 \leq a_2 \leq \cdots \leq a_{12}$, $a_1 + a_2 + 1 \geq 0$. The solution is only $a_1 = \cdots = a_{12} = 0$, and we have $\xi = -\chi_0 = \sum_{i=1}^{12} v_i/2$. $\xi \cdot \alpha_i = 0$ $(1 \leq i \leq 11)$, $\xi \cdot \alpha_0 = -1$.

Next, we consider $\phi(e')$. $\phi(e') \in M$, since $\phi(e')$ is orthogonal to $\phi(P_0') = L$. We can write $\phi(e') = \sum_{i=1}^{12} y_i v_i$ with $y_i \in \mathbf{Z}$ and $\sum_{i=1}^{12} y_i$ is an even integer. $2 = e'^2 = \sum_{i=1}^{12} y_i^2$. Moreover, we can assume $y_1 \leq y_2 \leq \cdots \leq y_{12}$ because of the action of the Weyl group generated by s_{α_i} $(1 \leq i \leq 11)$, the reflections fixing ξ. In addition $-1 = e' \cdot f =$

$\sum_{i=1}^{12} y_i/2$. The solution satisfying the conditions is unique and $\phi(e') = -v_1 - v_2 = \alpha_0$. We can calculate F. $F = C(\phi(P'), \Gamma_{16}) = C(\mathbf{Z}\chi_0 + \mathbf{Z}\alpha_0, M) = \mathbf{Z}\alpha_1 + \sum_{i=3}^{11} \mathbf{Z}\alpha_i \cong Q(A_9 + A_1)$.

Next, we would like to compute the root system $R(F^*)$ of F^*. Set $F_1 = \mathbf{Z}\alpha_1$ and $F_2 = \sum_{i=3}^{11} \mathbf{Z}\alpha_i$. $F_1 \cong Q(A_1)$. $F_2 \cong Q(A_9)$.

Lemma 20.7. *(1) $R(F^*) = R(F_1^*) \oplus R(F_2)$.*
(2) $R(F_1^*)$ is of type BC_1. $R(F_2)$ is of type A_9.

Proof. (2) is trivial. Also the inclusion relation $R(F^*) \supset R(F_1^*) \oplus R(F_2)$ is trivial. We show the opposite inclusion relation. Let $\alpha \in R(F^*)$ be an arbitrary root. We write $\alpha = \alpha_1 + \alpha_2$ with $\alpha_i \in F_i^*$. Let $\omega \in F_2^*$ be the fundamental weight corresponding to one end of the Dynkin graph A_9. A number k with $0 \leq k \leq 9$ and $\overline{\alpha}_2 = k\overline{\omega} \in F_2^*/F_2$ is uniquely determined. One has $m(k\overline{\omega}) = k(10-k)/10$.

Case 1. $\alpha^2 = 1/2$.

If $\overline{\alpha}_2 \neq 0$, then $1/2 = \alpha^2 \geq m(\overline{\alpha}_2) \geq 9/10$, a contradiction. Thus, $\alpha_2 \in F_2$. If $\alpha_2 \neq 0$, then $1/2 = \alpha^2 = \alpha_1^2 + \alpha_2^2 \geq \alpha_2^2 \geq 2$, a contradiction. Thus, $\alpha \in F_1^*$.

Case 2. $\alpha^2 = 1$.

If $\overline{\alpha}_2 = k\overline{\omega}$, $2 \leq k \leq 8$, then $1 = \alpha^2 \geq m(k\overline{\omega}) \geq 2 \times 8/10 = 8/5$, a contradiction. If either $\overline{\alpha}_2 = \overline{\omega}$ or $\overline{\alpha}_2 = 9\overline{\omega}$, then $1 = \alpha^2 \geq m(\overline{\alpha}_1) + m(\overline{\alpha}_2) \geq m(\overline{\alpha}_2) = 9/10$. On the other hand, $m(\overline{\alpha}_1) + m(\overline{\alpha}_2) = m(\overline{\alpha}) \equiv 1 \bmod 2\mathbf{Z}$. Thus, one knows $1 = m(\overline{\alpha}_1) + m(\overline{\alpha}_2)$ and $1/10 = m(\overline{\alpha}_1) \equiv q_{F_1}(\overline{\alpha}_1) \bmod 2\mathbf{Z}$. However, q_{F_1} takes only values $0 \bmod 2\mathbf{Z}$ and $1/2 \bmod 2\mathbf{Z}$. Thus, we have a contradiction. One knows $\overline{\alpha}_2 = 0$ and $\alpha_2 \in F_2$. If $\alpha_2 \neq 0$, then $1 = \alpha_1^2 + \alpha_2^2 \geq \alpha_2^2 \geq 2$, a contradiction. Thus, $\alpha_2 = 0$ and $\alpha = \alpha_1 \in F_1$.

Case 3. $\alpha^2 = 2$.

We have $\alpha \in F_1 \oplus F_2$, since $F^*/F \cong F_1^*/F_1 \oplus F_2^*/F_2$ contains no isotropic element. One knows either $\alpha \in F_1$ or $\alpha \in F_2$, since F_1 and F_2 are even lattices. Q.E.D.

Subcase I-3. $\phi(e_3) = \alpha_{15}$.

Let b be the orthogonal transformation of $M \oplus L$ such that it exchanges α_0 and α_1, exchanges α_{15} and α_{16}, and fixes α_2, α_3, $\ldots$, α_{11}, α_{13}, and α_{15}. It is extended to $M^* \oplus L^*$. We denote the extension by the same character b. One has $b(L) = L$, $b(\chi_{11} + \tau_{13}) = \chi_{11} + \tau_{13}$, $b(\chi_0 + \tau_{16}) = \chi_1 + \tau_{15}$, and $b(\chi_1 + \tau_{15}) = \chi_0 + \tau_{16}$. Thus, b sends Γ_{16} to Γ_{16} itself. Exchanging ϕ for $b\phi$, Subcase I-3 is reduced to above subcase I-2, since $b\phi(e_3) = \alpha_{16}$.

Proposition 20.8. *In Subcase I-2 and I-3 $R(F^*)$ is of type $A_9 + BC_1$.*

Case II.

In this case, we have the decomposition $C(\phi(H_0), \Lambda) = Q_1 \oplus Q_2$, $Q_1 \cong Q_2 \cong Q(E_8)$. Note that if $x = x_1 + x_2$, $x_i \in Q_i$, and $2 = x^2 = x_1^2 + x_2^2$, then $x_1 = 0$ or $x_2 = 0$. Thus, we have a number $i = 1$, 2, with $\phi(P_0') \subset Q_i$. Exchanging the number if necessary, we assume $\phi(P_0') \subset Q_1$ in the following. Moreover, we have a number $i = 1$, 2 with $\phi(e') \in Q_i$.

We have 3 subcases.

II-1 $\phi(P') \subset Q_1$.

II-2 $\phi(P'_0 + \mathbf{Z}e') \subset Q_1$, $\phi(P') \not\subset Q_1$.

II-3 $\phi(P'_0 + \mathbf{Z}e') \not\subset Q_1$.

By Lemma 20.3 We can assume $L = \phi(P'_0)$. Set $M = C(L, Q_1)$. $M \cong Q(D_4)$. Set $\alpha_i = \phi(e_i)$ $(i = 0, 1, 2, 3)$. We associate a root basis β_0, β_1, β_2, β_3 with the vertices of the Dynkin graph D_4 as in the following figure:

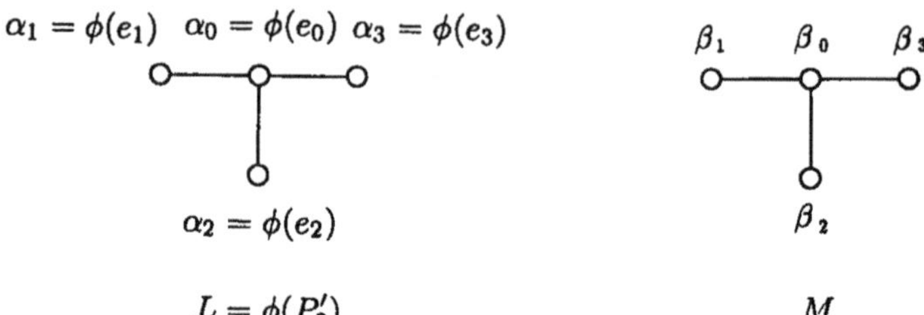

$L = \phi(P'_0)$ $\qquad\qquad$ M

Let $\omega_0, \omega_1, \omega_2, \omega_3$ (resp. $\chi_0, \chi_1, \chi_2, \chi_3$) denote the dual basis of $\alpha_0, \ldots, \alpha_3$ (resp. $\beta_0, \ldots, \beta_3$). Exchanging the numbers of $\beta_1, \beta_2, \beta_3$ if necessary, we can assume

$$(M \oplus L)^*/(M \oplus L) \supset Q_1/(M \oplus L) = \{0, \overline{\omega}_1 + \overline{\chi}_1, \overline{\omega}_2 + \overline{\chi}_2, \overline{\omega}_3 + \overline{\chi}_3\}.$$

We here denote $\overline{x} = x \bmod (M \oplus L)$ for $x \in (M \oplus L)^*$. For some $a \in M$ and $f_2 \in Q_2$ we can write $\phi(f) = -\omega_3 - \chi_3 + a + f_2$. We here introduce a model of M.

A model of M.

Let $N = \sum_{i=1}^4 \mathbf{Z}v_i$ be the lattice of rank 4 defined by the conditions $v_i^2 = 1$ $(1 \le i \le 4)$, and $v_i \cdot v_j = 0$ $(i \ne j)$. We define a lattice-embedding $M \hookrightarrow N$ by $\beta_1 = -v_1 - v_2$, $\beta_2 = v_1 - v_2$, $\beta_0 = v_2 - v_3$, $\beta_3 = v_3 - v_4$. $M = \{\sum x_i v_i \in N \mid \sum x_i \text{ is an even integer.}\}$. $\chi_3 = -v_4$. A Weyl chamber C of M can be written

$$\begin{aligned}\overline{C} &= \{t \in M \otimes \mathbf{R} \mid (t, \beta_i) \le 0 \ \ (0 \le i \le 3)\} \\ &= \{\textstyle\sum x_i v_i \in N \otimes \mathbf{R} \mid x_1 + x_2 \ge 0, \ x_1 \le x_2 \le x_3 \le x_4\}\end{aligned}$$

In this model we can write $-\chi_3 + a = \sum_{i=1}^4 a_i v_i$ with $a_i \in \mathbf{Z}$, and $\sum a_i$ is odd. Moreover,

$$a_1 + a_2 \ge 0, \quad a_1 \le a_2 \le a_3 \le a_4,$$

since we can assume $-\chi_3 + a \in \overline{C}$.

Subcase II-1 $\phi(P') \subset Q_1$.

In this case, $f_2 = 0$ and $4 = f^2 = \omega_3^2 + (-\chi_3 + a)^2$. Since $\omega_3^2 = 1$, $(-\chi_3 + a)^2 = \sum_{i=1}^4 a_i^2 = 3$. We have a unique solution $a_1 = 0$, $a_2 = a_3 = a_4 = 1$. Thus, $-\chi_3 + a = v_2 + v_3 + v_4 = -3(\beta_1 + \beta_2)/2 - 2\beta_0 - \beta_3 = -\chi_1 - \chi_2$. Next, we consider $\phi(e')$. Since $\phi(e') \in Q_1$ and since $\phi(e')$ is orthogonal to L, $\phi(e') \in M$. We can write $\phi(e') = \sum_{i=1}^4 y_i v_i$ where $y_i \in \mathbf{Z}$ and $\sum y_i$ is even. Now, since s_{β_0} and s_{β_3} fix $-\chi_3 + a$, we can assume moreover $y_2 \le y_3 \le y_4$ by considering the action of them. Furthermore, $-1 = f \cdot e' = y_2 + y_3 + y_4$ and $2 = e'^2 = \sum_{i=1}^4 y_i^2$. We have just 2 solutions satisfying the conditions; $\phi(e') = -v_1 - v_2 = \beta_1$ and $\phi(e') = v_1 - v_2 = \beta_2$.

In the latter case $\phi(e') = \beta_2$ we proceed like the following. Let b be the automorphism of the lattice $M \oplus L$ keeping α_0, α_3, β_0, and β_3 fixed, however exchanging α_1 and α_2, and exchanging β_1 and β_2. b has a natural extension to $M^* \oplus L^*$. We denote the extension by the same character b. This satisfies $b(\phi(f)) = \phi(f)$, $b(L) = L$, $b(\omega_3 + \chi_3) = \omega_3 + \chi_3$, $b(\omega_1 + \chi_1) = \omega_2 + \chi_2$, and $b(\omega_2 + \chi_2) = \omega_1 + \chi_1$; thus, Q_1 is sent Q_1 itself. By considering $b\phi$ instead of ϕ, the latter case is reduced to the former case, since $b(\phi(e')) = b(\beta_2) = \beta_1$. Therefore, we assume $\phi(e') = \beta_1$ below.

We have $F = C(\phi(P'), Q_1 \oplus Q_2) = C(\phi(P'), Q_1) \oplus Q_2 = C(\mathbf{Z}(\chi_1+\chi_2)+\mathbf{Z}\beta_1, M) \oplus Q_2 = \mathbf{Z}\xi \oplus \mathbf{Z}\beta_3 \oplus Q_2$, where $\xi = 2\beta_0 + \beta_1 + \beta_2 + \beta_3$ and $Q_2 \cong Q(E_8)$. Note that $\xi^2 = 10$ and $\xi \cdot \beta_3 = 0$.

We calculate $R(F^*)$ using this expression. Set $F_1 = \mathbf{Z}\xi + \mathbf{Z}\beta_3$ and $F_2 = Q_2$.

Lemma 20.9. *(1) $R(F^*) = R(F_1^*) \oplus R(F_2)$.*
(2) $R(F_1^*)$ is of type BC_1. $R(F_2)$ is of type E_8.

Proof. The inclusion relation $R(F^*) \supset R(F_1^*) \oplus R(F_2)$ and the latter half of (2) is obvious.

Let $\alpha \in R(F^*)$ be a root. We write it in the form $\alpha = \alpha_1 + \alpha_2$ with $\alpha_i \in F_i^*$. Note that $F_2^* = F_2$. Thus, if $\alpha_2 \neq 0$, then $\alpha_2^2 \geq 2$.

Case 1. $\alpha^2 = 1/2$.

Since $\alpha^2 \geq \alpha_2^2$, we have $\alpha_2 = 0$ and $\alpha = \alpha_1 \in F_1^*$. We can write $\alpha = x_1\xi/10 + x_2\beta_3/2$ with $x_1, x_2 \in \mathbf{Z}$. Under the assumption one has $5 = x_1^2 + 5x_2^2$. Thus, $x_1 = 0$, $x_2 = \pm 1$. One has $\alpha = \pm\beta_3/2$.

Case 2. $\alpha^2 = 1$.

By the same reason as in Case 1, one has $\alpha = \alpha_1 \in F_1^*$. Setting $\alpha = x_1\xi/10 + x_2\beta_3/2$ with $x_1, x_2 \in \mathbf{Z}$, one has $10 = x_1^2 + 5x_2^2$. This equation has no solution. This case never takes place.

Case 3. $\alpha^2 = 2$.

First, assume $\alpha_2 \neq 0$. Since $2 = \alpha^2 = \alpha_1^2 + \alpha_2^2 \geq \alpha_2^2 \geq 2$, one has $\alpha_1^2 = 0$, $\alpha_1 = 0$ and $\alpha = \alpha_2 \in F_2$. Second, if $\alpha_2 = 0$, then $\alpha = \alpha_1 \in F_1$. Setting $\alpha = x_1\xi/10 + x_2\beta_3/2$ with $x_1, x_2 \in \mathbf{Z}$, one has $20 = x_1^2 + 5x_2^2$. One knows $x_1 = 0$, $x_2 = \pm 2$. Thus, $\alpha = \pm\beta_2$.

Q.E.D.

Proposition 20.10. *In Subcase II-1 $R(F^*)$ is of type $E_8 + BC_1$.*

Subcase II-2 $\phi(P_0' + \mathbf{Z}e') \subset Q_1$, $\phi(P') \not\subset Q_1$.

We can write $\phi(f) = f_1 + f_2$ with $f_i \in Q_i$. By assumption $f_2 \neq 0$. On the other hand, $f_1 = -\omega_3 - \chi_3 + a \neq 0$. Thus, $2 = f_2^2 = f_1^2 = \omega_3^2 + (-\chi_3 + a)^2$, since $4 = f^2 = f_1^2 + f_2^2$. Since $\omega_3^2 = 1$, $1 = (-\chi_3 + a)^2 = \sum_{i=1}^4 a_i^2$. The solution is unique and $a_1 = a_2 = a_3 = 0$, $a_4 = 1$. $f_1 = -\omega_3 - \chi_3$.

Next, we consider $\phi(e')$. Since $\phi(e') \in M$, we can write $\phi(e') = \sum_{i=1}^4 y_i v_i$ with $y_i \in \mathbf{Z}$ and $\sum_{i=1}^4 y_i$ is even. $2 = e'^2 = \sum_{i=1}^4 y_i^2$. $-1 = e' \cdot f = y_4$. By the action of the Weyl group generated by the reflections s_{β_0}, s_{β_1} and s_{β_2} fixing f_1, we can assume moreover

$$y_1 + y_2 \geq 0, \quad y_1 \leq y_2 \leq y_3.$$

The solution is unique and $y_1 = y_2 = 0$, $y_3 = 1$, $y_4 = -1$. Thus, $e' = \beta_3$.

We can compute F here. Set $\beta_4 = -(2\beta_0 + \beta_1 + \beta_2 + \beta_3)$, $M_1 = \mathbf{Z}\beta_3$, and $M_2 = \mathbf{Z}\beta_1 \oplus \mathbf{Z}\beta_2 \oplus \mathbf{Z}\beta_4$. M_1, $M_2 \subset M$. $M_2 = C(M_1, M)$. $\chi_3 = \beta_0 + \beta_1/2 + \beta_2/2 + \beta_3 = \beta_3/2 - \beta_4/2$. Note that corresponding to the inclusion relations $\phi(f) \in Q_1 \oplus Q_2 \subset L^* \oplus M^* \oplus Q_2 \subset L^* \oplus M_1^* \oplus M_2^* \oplus Q_2$, we can write $\phi(f) = -\omega_3 - \beta_3/2 + \beta_4/2 + f_2$ with $-\omega_3 \in L^*$, $-\beta_3/2 \in M_1^*$, $\beta_3/2 \in M_2^*$, and $f_2 \in Q_2$. $C(\phi(P_0' + \mathbf{Z}e'), Q_1 \oplus Q_2) = M_2 \oplus Q_2$. Thus, $F = C(\phi(P'), Q_1 \oplus Q_2) = \{x \in M_2 \oplus Q_2 \mid (x, \phi(f)) = 0\} = \{x \in M_2 \oplus Q_2 \mid (x, \beta_4/2 + f_2) = 0\} = C(\mathbf{Z}(\beta_4 + 2f_2), M_2 \oplus Q_2) = C(\mathbf{Z}(\beta_4 + 2f_2), \mathbf{Z}\beta_4 \oplus Q_2) \oplus \mathbf{Z}\beta_1 \oplus \mathbf{Z}\beta_2$.

We choose a root basis of Q_2 as in the right figure. Set $\gamma_0 = -(3\gamma_1 + 2\gamma_2 + 4\gamma_3 + 6\gamma_4 + 5\gamma_5 + 4\gamma_6 + 3\gamma_7 + 2\gamma_8)$. $\gamma_0^2 = 2$. $\gamma_0 \cdot \gamma_i = 0$ $(1 \le i \le 7)$. $\gamma_0 \cdot \gamma_8 = -1$.

We can assume $f_2 = \gamma_0$, since every long root in Q_2 is conjugate with respect to the Weyl group. Then, for a vector $x = a\beta_4 + \sum_{i=1}^{8} b_i\gamma_i$

$$(x, \beta_4 + 2f_2) = 0 \iff 2a - 2b_8 = 0 \iff x \in \sum_{i=1}^{7} \mathbf{Z}\gamma_i + \mathbf{Z}(\gamma_8 + \beta_4).$$

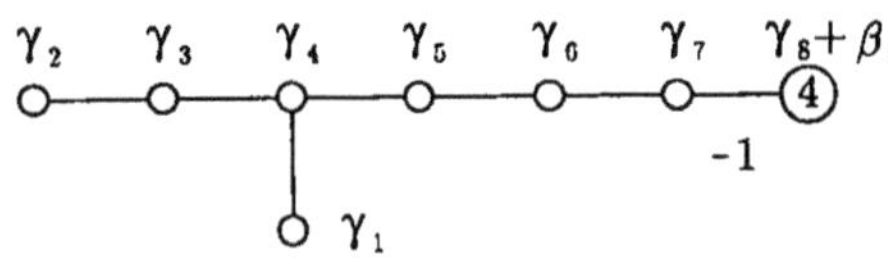

Therefore, $F_2 = C(\mathbf{Z}(\beta_4 + 2f_2), \mathbf{Z}\beta_4 + Q_2)$ is the lattice of rank 8 defined by the left dual graph. Set $F_1 = \mathbf{Z}\beta_1 \oplus \mathbf{Z}\beta_2 \cong Q(2A_1)$. $F = F_1 \oplus F_2$. We would like to compute $R(F^*)$ using this decomposition. Some preparation is necessary. Set $F_2 \supset F_3 = \sum_{i=1}^{7} \mathbf{Z}\gamma_i \cong Q(E_7)$. Set $\xi = 3\gamma_1 + 2\gamma_2 + 4\gamma_3 + 6\gamma_4 + 5\gamma_5 + 4\gamma_6 + 3\gamma_7 + 2(\gamma_8 + \beta_4) \in F_2$. $\xi^2 = 10$. $\xi \cdot (\gamma_8 + \beta_4) = 5$. $\xi \cdot \gamma_i = 0$ $(1 \le i \le 7)$.

Setting $F_4 = \mathbf{Z}\xi$, F_2 is an overlattice of $F_3 \oplus F_4$.

Lemma 20.11. *(1) $F_2^*/F_2 \cong \mathbf{Z}/5$. The discriminant quadratic form is given by $q_{F_2}(x) \equiv 2x^2/5 \bmod 2\mathbf{Z}$ for $x \in \mathbf{Z}/5$.*
(2) $R(F_2^*) = R(F_2)$.
(3) $R(F_2) = R(F_3)$.
(4) $R(F_3)$ is of type E_7.

Proof. (1) Note that $F^*/F \cong \mathbf{Z}/5 \oplus \mathbf{Z}/2 \oplus \mathbf{Z}/2$, $F_1^*/F_1 \cong \mathbf{Z}/2 \oplus \mathbf{Z}/2$ and $F^*/F \cong F_1^*/F_1 \oplus F_2^*/F_2$. The discriminant form of F_2 is equal to the restriction of that of F to the 5-Sylow subgroup.
(2) It follows from the fact that q_{F_2} does not take values $1/2 \bmod 2\mathbf{Z}$, $2/3 \bmod 2\mathbf{Z}$ and $1 \bmod 2\mathbf{Z}$.
(3) Let $\alpha \in R(F_2)$ be a root. $\alpha^2 = 2$. Writing $\alpha = \sum_{i=1}^{7} x_i\gamma_i + y(\gamma_8 + \beta_4)$, set $\alpha' = \sum_{i=1}^{7} x_i\gamma_i + y\gamma_8$. We have $2 = \alpha^2 = \alpha'^2 + 2y^2$. If $y \neq 0$, then $y = \pm 1$, $\alpha' = 0$ and $\alpha = 0$, a contradiction. Thus, $y = 0$ and $\alpha \in R(F_3)$. The opposite inclusion relation is obvious.
(4) is trivial. Q.E.D.

Lemma 20.12. *$R(F^*) = R(F_1^*) \oplus R(F_3)$. $R(F_1^*)$ is of type BC_2, while $R(F_3)$ is of type E_7.*

Proof. Let $\alpha \in R(F^*)$ be a root. We can write it in the form $\alpha = \alpha_1 + \alpha_2$ with $\alpha_i \in F_i$ $i = 1, 2$. By Lemma 20.11 it suffices to show that $\alpha_1 \neq 0$ implies $\alpha_2 = 0$. In the following we assume $\alpha_1 \neq 0$.

Case 1. $\alpha^2 = 1/2$.

Since we can write $\alpha_1^2 = s/2$ with a positive integer s, $1/2 = \alpha^2 = \alpha_1^2 + \alpha_2^2 \geq \alpha_1^2 \geq 1/2$. Thus, $\alpha_2 = 0$.

Case 2. $\alpha^2 = 1$.

If $\alpha_1^2 \geq 1$, then by the same argument as in Case 1, we can conclude $\alpha_2 = 0$. Thus, it suffices to deduce a contradiction under the assumption $0 < \alpha_1^2 < 1$. Since we can write $\alpha_1^2 = s/2$ with $s \in \mathbf{Z}$, we have $\alpha_1^2 = 1/2$ under the assumption. Thus, $\alpha_2^2 = 1/2$. However, by Lemma 20.11 (2) such an α_2 does not exist, which is a contradiction.

Case 3. $\alpha^2 = 2$.

If $\alpha_1^2 \geq 2$, then by the same argument as above we have $\alpha_2 = 0$. Therefore, in the following, we assume $0 < \alpha_1^2 < 2$.

$2 = \alpha_1^2 + \alpha_2^2$. Set $\alpha_1^2 = s/2$ and $\alpha_2^2 = t/5$. $s, t \in \mathbf{Z}$. $20 = 5s + 2t$. Thus, s is even and t is a multiple of 5. In conclusion, $\alpha_1^2 = \alpha_2^2 = 1$. However, by Lemma 20.11 (2) such an α_2 does not exist, which is a contradiction. Q.E.D.

Proposition 20.13. *In Subcase II-2 $R(F^*)$ is of type $E_7 + BC_2$.*

Subcase II-3 $\phi(P_0' + \mathbf{Z}e') \not\subset Q_1$.

By assumption $\phi(e') \in Q_2$. Writing $\phi(f) = f_1 + f_2$ with $f_i \in Q_i$, we have $-1 = \phi(f) \cdot \phi(e') = f_2 \cdot \phi(e')$. Thus, $f_2 \neq 0$. Thus, one knows $f_1 = -\omega_3 - \chi_3$, $f_2^2 = 2$ by similar arguments to that in II-2. Let $\gamma_1, \ldots, \gamma_8$ be the same root basis of Q_2 as in II-2. Let $\gamma_0 \in Q_2$ be the same root as above. Note that f_2 and $\phi(e')$ form a root basis of type A_2. By Lemma 30.3 (3) we can assume $\{f_2, \phi(e')\} = \{\gamma_0, \gamma_8\}$. We have two cases.

(a) $f_2 = \gamma_8$, $\phi(e') = \gamma_0$.

(b) $f_2 = \gamma_0$, $\phi(e') = \gamma_8$.

We consider the case (b). Set $\gamma_9 = \gamma_0 + \gamma_8$. $\gamma_9^2 = 2$. $s_{\gamma_9}(\gamma_0) = \gamma_0 - (\gamma_0, \gamma_9)\gamma_9 = -\gamma_8$. $s_{\gamma_9}(\gamma_8) = \gamma_8 - (\gamma_8, \gamma_9)\gamma_9 = -\gamma_0$. Now, there exists an element $w_0 \in W(Q_2)$ of the Weyl group of type E_8 such that $w_0(x) = -x$ for any $x \in Q_2$. If we replace ϕ by $w_0 s_{\gamma_9}\phi$, then (a) holds. Thus, in the following, we assume (a).

Set $K = \sum_{i=1}^{7} \mathbf{Z}\gamma_i \cong Q(E_7)$. $K = C(\mathbf{Z}\gamma_0, Q_2)$. Thus, $C(\phi(P_0' + \mathbf{Z}e'), Q_1 \oplus Q_2) = M \oplus K$. We have $F = C(\phi(P'), Q_1 \oplus Q_2) = \{x \in M \oplus K \mid (x, \phi(f)) = 0\} = \{x \in M \oplus L \mid (x, -\chi_3 + \gamma_8) = 0\} = \sum_{i=0}^{2} \mathbf{Z}\beta_i + \sum_{i=1}^{6} \mathbf{Z}\gamma_i + \mathbf{Z}(\beta_3 - \gamma_7)$. F is the lattice of rank 10 defined by the following dual graph:

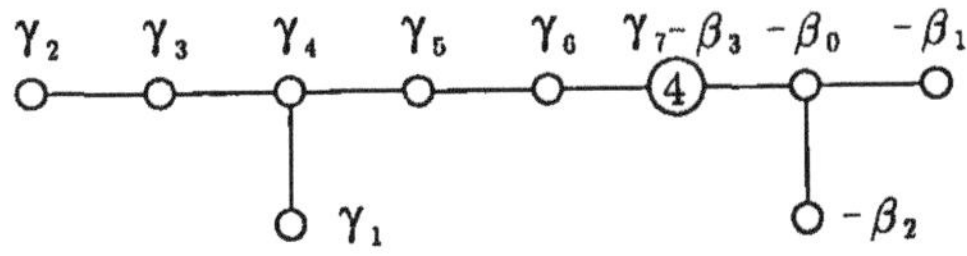

Next, we compute $R(F^*)$ using the above expression. Corresponding to the inclusion relations $-\chi_3+\gamma_8 \in M^* \oplus Q_2 \subset M^* \oplus K^* \oplus (\mathbf{Z}\gamma_0)^*$, we can write $\gamma_8 = -\xi_7 - \gamma_0/2$ with $\xi_7 = 3\gamma_1/2+\gamma_2+2\gamma_3+3\gamma_4+5\gamma_5/2+2\gamma_6+3\gamma_7/2 \in K^*$. $\xi_7^2 = 3/2$. Set $\lambda = 2\chi_3+2\xi_7$. $\lambda \in M \oplus K$. $\lambda^2 = 10$. λ is primitive in $M \oplus K$ and $F = C(\mathbf{Z}\lambda, M \oplus K)$.

Lemma 20.14. *(1)* $F \oplus \mathbf{Z}\lambda \subset M \oplus K \subset M^* \oplus K^* \subset F^* \oplus (\mathbf{Z}\lambda)^*$.
(2) $R(F^*) \subset M^* \oplus K^*$

Proof. (1) is obvious.
(2) $I = (M \oplus K)/(F \oplus \mathbf{Z}\lambda) \subset (F \oplus \mathbf{Z}\lambda)^*/(F \oplus \mathbf{Z}\lambda) = D$ is an isotropic subgroup. Let m be the order of I. Since $|d(F \oplus \mathbf{Z}\lambda)| = m^2|d(M \oplus K)|$, one knows $m = 5$. Thus, I is contained in the 5-Sylow subgroup of D.

On the other hand, for any root $\alpha \in F^*$ $q_F(\overline{\alpha}) \equiv 1/2$, 1 or $0 \bmod 2\mathbf{Z}$, where $\overline{\alpha} = \alpha \bmod F \in F^*/F$. By Lemma 20.2 (1) $\overline{\alpha}$ is contained in the 2-Sylow subgroup of F^*/F. Moreover, since F^*/F is a direct summand of D, $\overline{\alpha}$ is contained in the 2-Sylow subgroup of D. One knows that $\overline{\alpha}$ and I are orthogonal with respect to the discriminant bilinear form. It implies $\alpha \in M^* \oplus K^*$. Q.E.D.

Lemma 20.15. *(1)* M^*, K^* *and* $M^* \oplus K^*$ *are root modules.*
(2) $R(M^* \oplus K^*) = R(M^*) \oplus R(K^*)$.

Proof. (1) It is obvious since $M \cong Q(D_4)$, $K \cong Q(E_7)$.
(2) It is obvious that the left-hand side contains the right-hand side.

Let $\alpha \in R(M^* \oplus K^*)$ be a root. We can write it in the form $\alpha = \alpha_1 + \alpha_2$ with $\alpha_1 \in M^*$, $\alpha_2 \in K^*$. We denote $\overline{\alpha}_1 = \alpha_1 \bmod M \in M^*/M$, $\overline{\alpha}_2 = \alpha_2 \bmod K \in K^*/K$.

Case 1. $\alpha^2 = 1/2$ or 1.

If $\overline{\alpha}_2 \neq 1$, then $1 \geq \alpha^2 \geq \alpha_2^2 \geq m(\overline{\alpha}_2) = 3/2$, a contradiction. Thus, $\alpha_2 \in K$. If $\alpha_2 \neq 0$, then $1 \geq \alpha^2 \geq \alpha_2^2 \geq 2$, a contradiction. Thus, $\alpha = \alpha_1 \in R(M^*)$.

Case 2. $\alpha^2 = 2$.

Since the discriminant group of $M \oplus K$ has no isotropic element, $\alpha_1 \in M$ and $\alpha_2 \in K$. Since M and K are positive definite even lattices, $\alpha_1 = 0$ or $\alpha_2 = 0$. In the former case $\alpha = \alpha_2 \in R(K^*)$. In the latter case $\alpha = \alpha_1 \in R(M^*)$. Q.E.D.

Proposition 20.16. *In Subcase II-3* $R(F^*)$ *is of type* $E_6 + B_3$.

Proof. By Lemma 20.14 $R(F^*) = R(M^* \oplus K^*) \cap F^*$. By Lemma 20.15 $R(M^* \oplus K^*) \cap F^* = (R(M^*) \cap F^*) \oplus (Q(K^*) \cap F^*)$. On the other hand, $R(M^*) \cap F^* = \{\alpha \in R(M^*) \mid (\alpha, \chi_3) = 0\} = \{\alpha = \sum_{i=0}^{3} c_i\beta_i \in R(M^*) \mid c_3 = 0\}$. This is a root system of type B_3 whose root basis is β_0, β_1, $(\beta_1 - \beta_2)/2$. $R(K^*) \cap F^* = R(K) \cap F^* = \{\alpha \in R(K) \mid (\alpha, \xi_7) = 0\} = \{\alpha = \sum_{i=1}^{7} c_i\gamma_i \in R(K) \mid c_7 = 0\}$ is of type E_6. Q.E.D.

We complete the verification of Theorem 20.1.

Remark. We got 5 basic graphs in case $S_{1,0}$. Four of them except $A_9 + BC_1$ have only 9 vertices. However, rank $F^* = 10$. Thus, in the corresponding cases to 4 of them the root system never spans $F^* \cong \Lambda/\phi(P)$. The equivalent conditions in Proposition 8.4 are never satisfied and we can never draw the Coxeter-Vinberg graph of Λ_2/P.

Now, readers might notice that we can regard elements $\delta \in \Lambda_N/P$ with $\delta^2 = 2/5$ or $1/10$ as roots in addition to original roots defined in Section 5. Indeed, if $\delta^2 = 2/5$, then $\delta^\vee = 2\delta/\delta^2 = 5\delta \in F$. If $\delta^2 = 1/10$, then $\delta^\vee = 20\delta \in F$. In both cases $(x, \delta^\vee) \in \mathbf{Z}$ for any $x \in \Lambda_N/P$ and $s_\delta(x) = x - (x, \delta^\vee)\delta$ is well-defined. One may ask what takes place under this increment of roots. The answer is as follows: In the cases corresponding to 4 basic graphs except $E_6 + B_3$, $F^* \cong \Lambda/\phi(P)$ is spanned by roots if we regard also elements δ with $\delta^2 = 2/5$ or $1/10$ as roots. However, in the case $E_6 + B_3$ this is not the case. In conclusion, even under the above increment of roots we cannot draw the Coxeter-Vinberg graph of Λ_2/P. We can show that no further increment of roots is possible. (More precisely, no larger reflection group acts on Λ_2/P.) This is the reason the verification of Theorem 20.1 becomes so long.

21
The second reduction for $S_{1,0}$ 1

We continue the study of the case $S_{1,0}$. The next step is to show that if there is a full embedding $Q(G) \hookrightarrow \Lambda_2/P$, then a Dynkin graph G can be made from one of the basic graphs by one elementary or tie transformation. If this can be shown, the second reduction, i.e., the reduction of the problem form Λ_2/P to Λ_1/P becomes possible.

However, it turns out that this claim is not true. We will introduce sub-basic graphs to manipulate the cases where this reduction is impossible.

Because of the absence of the Coxeter-Vinberg graph the argument becomes slightly long. We divide the case into 3 subcases.

((1)) G has a component of type BC.
((2)) G has a component of type B_k with $k \geq 4$.
((3)) Otherwise.

Note that $Q(BC_1)^{\oplus k} \cong Q(BC_k)$ and $Q(B_1)^{\oplus k} \cong Q(B_k)$. Because of these properties, a component of type BC or B plays a special role in ((1)) or ((2)). In this section we consider the subcase ((1)).

Proposition 21.1. *The Dynkin graph G of a positive definite submodule $L \subset \Lambda_N/P$ satisfies the following in the case $S_{1,0}$:*
(1) It contains no vertex ◎ corresponding to a root with length $\sqrt{2/3}$.
(2) It contains no subgraph •——• of type F_2.
(3) If it contains a component of type B, then it contains no component of type BC.

Proof. Recall that $F^* \cong \Lambda_N/P$ and the discriminant quadratic form on F^*/F is given by

$$q_F(a, b_1, b_2) \equiv -\frac{2}{5}a^2 + \frac{1}{2}(b_1^2 + b_2^2) \bmod 2\mathbf{Z}$$

for $(a, b_1, b_2) \in \mathbf{Z}/5 \oplus \mathbf{Z}/2 \oplus \mathbf{Z}/2 \cong F^*/F$.
(1) Easy. (See Lemma 20.2 (2).)

(2) Assume that G contains F_2. Then, we have two roots α_1, $\alpha_2 \in L$ with $\alpha_1^2 = \alpha_2^2 = 1$ and $\alpha_1 \cdot \alpha_2 = -1/2$. Also, $\alpha_3 = \alpha_1 + \alpha_2$ is a root with length 1. We have $q_F(\overline{\alpha}_i) \equiv \alpha_i^2 = 1 \bmod 2\mathbf{Z}$ for $i = 1, 2, 3$, where we denote $\overline{x} = x \bmod F \in F^*/F$ for a vector $x \in \Lambda_N/P \cong F^*$. Here, $(0, 1, 1) \in \mathbf{Z}/5 \oplus \mathbf{Z}/2 \oplus \mathbf{Z}/2$ is the unique element with value $1 \bmod 2\mathbf{Z}$ under q_F. Thus, $\overline{\alpha}_1 = \overline{\alpha}_2 = \overline{\alpha}_3 = (0, 1, 1)$. We have $0 \neq \overline{\alpha}_3 = 2\overline{\alpha}_1 = 0$, since $(0, 1, 1)$ has order 2. It is a contradiction.
(3) Assume that G contains both a component of type B and a component of type BC. Then, we have a root α, $\beta \in L$ with $\alpha^2 = 1$, $\beta^2 = 1/2$ and $\alpha \cdot \beta = 0$. Set $\gamma = \alpha + \beta$. We have $q_F(\overline{\gamma}) \equiv \gamma^2 = 3/2 \bmod 2\mathbf{Z}$. However, q_F never takes value $3/2 \bmod 2\mathbf{Z}$, a contradiction. Q.E.D.

Lemma 21.2. *The automorphism of the discriminant group $P^*/P \cong \mathbf{Z}/5 \oplus \mathbf{Z}/2 \oplus \mathbf{Z}/2$ of P given by $(a, b_1, b_2) \in \mathbf{Z}/5 \oplus \mathbf{Z}/2 \oplus \mathbf{Z}/2 \mapsto (a, b_2, b_1)$ is induced by an automorphism of the lattice P.*

Proof. $P = P' \oplus H_0$. P' has a basis e_0, e_1, e_2, e_3, f, e'. (Section 4.) We can define an automorphism of P' exchanging e_1 and e_2 and keeping e_0, e_3, f and e' fixed. It has an extension to P keeping every vector in H_0 fixed. Referring to Proposition 4.2 one can check that it induces the desired permutation on P^*/P. Q.E.D.

Lemma 21.3. *(1) $P \oplus Q(A_1)$ has just 2 proper over lattices.*
(2) These 2 overlattices are isomorphic.
(3) For each P_C of these 2 over lattices the root system of the quotient P_C/P is of type BC_1.

Proof. Let α denote the generator of $Q(A_1)$. $\alpha^2 = 2$. By $w_0, \ldots, w_4$, z we denote the basis of P^* introduced in Section 4 before Proposition 4.2. $P^*/P \oplus Q(A_1)^*/Q(A_1)$ has only 2 isotropic elements $5\overline{w}_1 + \overline{\alpha/2}$, $5\overline{w}_2 + \overline{\alpha/2}$. Thus, (1) and (3) follow. The claim (2) follows from Lemma 21.2. Q.E.D.

By P_C we denote the even overlattice of $P \oplus Q(A_1)$ with index 2.

Proposition 21.4. *We consider a fixed embedding $P \hookrightarrow \Lambda_2$. Let $\alpha \in \Lambda_2/P$ be a vector with $\alpha^2 = 1/2$.*
(1) The inverse image of $\mathbf{Z}\alpha$ by the canonical surjective homomorphism $\Lambda_2 \to \Lambda_2/P$ is isomorphic to P_C.
(2) We identify the inverse image of $\mathbf{Z}\alpha$ with P_C. Let $\pi_C : \Lambda_2/P \to (\Lambda_2/P)/\mathbf{Z}\alpha \cong \Lambda_2/P_C$ denote the canonical surjective homomorphism. The restriction of π_C to the orthogonal complement of $\mathbf{Z}\alpha$ in Λ_2/P is bijective.

Proof. (1) F denotes the orthogonal complement of P in Λ_2. Recall that we can regard naturally $\Lambda_2/P \supset F = \{x \in \Lambda_2/P \mid x^2 \in 2\mathbf{Z}\}$. (Proposition 5.6 (2).) Thus, $\alpha \notin F$. $\beta = 2\alpha$ satisfies $\beta^2 = 2$ and $\beta \in F$. We can regard β as a vector in Λ_2. Then, the inverse image of $\mathbf{Z}\alpha$ contains $P \oplus \mathbf{Z}\beta$. If the image of $a \in \Lambda_2$ is α, then $2a - \beta \in P$. Thus, the inverse image of $\mathbf{Z}\alpha$ has index 2 over $P \oplus \mathbf{Z}\beta$ and, thus, isomorphic to P_C.
(2) Let $\Gamma = C(\mathbf{Z}\beta, \Lambda_2)$, and $F_C = C(P_C, \Lambda_2)$. $P \oplus F_C \subset \Gamma$. $|d(\Gamma)| = 2$. $|d(P \oplus F_C)| = 20 \times 10 = 200$. Thus, $[\Gamma : P \oplus F_C] = \sqrt{200/2} = 10$. Note that $\Gamma/P = C(\mathbf{Z}\alpha, \Lambda_2/P)$.

Since $\mathrm{Ker}\,\pi_C = \mathbf{Q}\alpha \cap \Lambda_2/P$, $\mathrm{Ker}\,\pi_C \cap \Gamma/P = 0$. The restriction of π_C to Γ/P is injective. In particular, $[\pi_C(\Gamma/P) : \pi_C(P \oplus F_C)] = 10$.

On the other hand, we have natural identifications $\pi_C(P \oplus F_C) = F_C \subset \Lambda_2/P_C$ and $\Lambda_2/P_C \cong F_C^*$. Then, $[\Lambda_2/P_C : F_C] = [F_C^* : F_C] = |d(F_C)| = 10$. Thus, $\pi_C(\Gamma/P) = \Lambda_2/P_C$. Q.E.D.

Corollary 21.5. *Λ_2/P_C is a regular root module.*

Proof. By Proposition 21.4 (2) Λ_2/P_C is isomorphic to a submodule of Λ_2/P. Q.E.D.

Lemma 21.6. *For any two embeddings ι, $\iota' : P_C \hookrightarrow \Lambda_2$ there is an orthogonal transformation ψ of Λ_2 with $\iota' = \psi\iota$.*

Proof. Since the discriminant group of P_C contains no isotropic element, every embedding into an even lattice is primitive. Thus, the lemma follows from Lemma 12.4 (2). Q.E.D.

Proposition 21.7. *Assume that a positive definite full submodule $L \subset \Lambda_2/P$ has the Dynkin graph in the form $BC_k + G'$ with $k \geq 1$. Then, $k = 1$ or 2 and there exists a positive definite full submodule $L_C \subset \Lambda_2/P_C$ satisfying the following conditions (1) and (2):*

(1) The Dynkin graph of L_C coincides with $BC_{k-1} + G'$. (BC_0 denotes the empty graph.)

(2) L_C has an isotropic vector in a nice position in Λ_2/P_C if, and only if, L has an isotropic vector in a nice position in Λ_2/P.

Proof. By Proposition 21.1 (2) $k \leq 2$. Let R_1 be the component of type BC_k of the root system of L. Let $\alpha \in R_1$ be a root with $\alpha^2 = 1/2$. We use the same notation Γ/P, π_C etc. as in Proposition 21.4. Set $L_C = \pi_C(L \cap \Gamma/P)$. $L \cap \Gamma/P$ is a positive definite full submodule of Γ/P, and so is L_C by Proposition 21.4. Since all roots in R_1 orthogonal to α form a root system of type BC_{k-1}, (1) holds.

Let $u_C \in \Lambda_2/P_C$ be an isotropic vector in a nice position with respect to L_C. Since a root basis of type BC_{k-1} does not contain a long root, u_C is orthogonal to the component of type BC_{k-1} in the root system of L_C. Let $u \in \Gamma/P$ be the isotropic vector with $\pi_C(u) = u_C$. u is orthogonal to α and R_1, and is in a nice position with respect to L.

Conversely, let $u \in \Lambda_2/P$ be an isotropic vector in a nice position with respect to L. Since a root basis of type BC_k contains no long root, $\alpha \cdot u = 0$. Thus, $u \in \Gamma/P$. $\pi(u)$ is in a nice position with respect to L_C. Q.E.D.

Next, we would like to draw the Coxeter-Vinberg graph of Λ_2/P_C. Set $F_C = C(P_C, \Lambda_2)$ for a fixed embedding $P_C \hookrightarrow \Lambda_2$. By Lemma 21.6 the isomorphism class of F_C and that of Λ_2/P_C do not depend on the choice of the embedding.

Lemma 21.8. *$F_C \cong Q(A_9) \oplus H$. $\Lambda_2/P_C \cong Q(A_9)^* \oplus H$.*

Proof. The latter half follows from the former half. On the other hand, we can check that the discriminant quadratic form of P_C coincides with (-1) times the discriminant quadratic form of $Q(A_9) \oplus H$. Thus, by Lemma 5.5 (4) we have a primitive embedding $P_C \hookrightarrow \Lambda_2$ the orthogonal complement of whose image is isomorphic to $Q(A_9) \oplus H$. By Lemma 21.6 one gets the former half. Q.E.D.

Now, let K be the odd unimodular lattice with signature (11, 1). It has a basis $v_0, v_1, \ldots, v_{11}$ with $v_0^2 = -1$, $v_i^2 = +1$ $(1 \le i \le 11)$, and $v_i \cdot v_j = 0$ $(i \ne j)$. Set $w = v_0 + v_1 + \cdots + v_{11} \in K$. $w^2 = 10$. Set $M = C(\mathbf{Z}w, K)$. One can check $M \cong Q(A_9) \oplus H$. (See the beginning of Section 16.)

Let $p : K \otimes \mathbf{Q} \to M \otimes \mathbf{Q}$ denote the orthogonal projection. Set $r = p(v_0) = (11v_0 + v_1 + \cdots + v_{11})/10$. $r^2 = -11/10$. One has an isomorphism of quasi-lattices

$$\Lambda_2/P_C \cong M + \mathbf{Z}r.$$

We apply Vinberg's algorithm with the controlling vector r. As the root basis orthogonal to r, we take

$$e_i = -v_i + v_{i+1} \quad (1 \le i \le 10).$$

In addition to roots α with $\alpha^2 = 2$ or $1/2$, we consider roots δ with $\delta^2 = 2/5$, which can be called a *dummy* root. Indeed, since $M^*/M \cong \mathbf{Z}/5 \oplus \mathbf{Z}/2$, $\delta^\vee = 2\delta/\delta^2 = 5\delta \in M$. Thus, for every $x \in M^*$ the reflection $s_\delta(x) = x - (x, \delta^\vee)\delta$ is well-defined. To denote a vertex in the graph corresponding to a short root with length $\sqrt{2/5}$ we use the following symbol: ⊙.

By the algorithm we get successively

$$\begin{aligned}
e_{11} &= v_0 + v_1 + v_2 - v_{11} \\
e_{12} &= \{3v_0 + v_1 + \cdots + v_7 - (v_8 + \cdots + v_{11})\}/2 \\
e_{13} &= \{7v_0 + 2(v_1 + \cdots + v_8) - 3(v_9 + v_{10} + v_{11})\}/5 \\
e_{14} &= \{8v_0 + 3(v_1 + \cdots + v_6) - 2(v_7 + v_8 + v_9 + v_{10})\}/5 \\
e_{15} &= \{13v_0 + 3(v_1 + \cdots + v_9) - 7(v_{10} + v_{11})\}/5.
\end{aligned}$$

The calculation so far is not so complicated. However, it becomes harder and harder gradually. To get the next fundamental root e_{16} we take a short cut using the symmetry of the graph (Vinberg [23]).

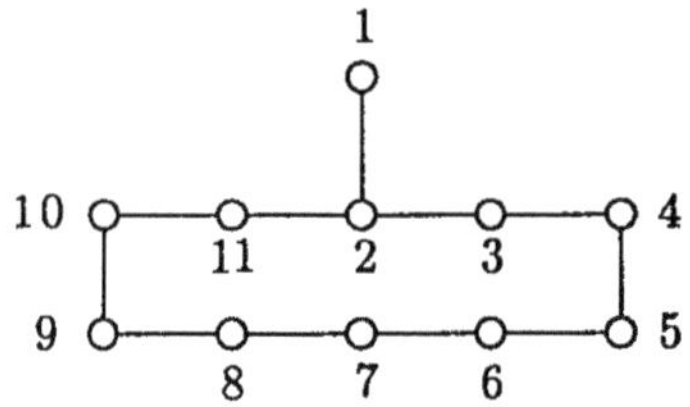

Set $\Delta^* = \{e_i \mid 1 \le i \le 11\}$. The right figure is the graph drawn from Δ^*. It has $\mathbf{Z}/2$-symmetry defined by $\psi : \Delta^* \to \Delta^*$ with $\psi(e_i) = e_i$ for $i = 1, 2$, $\psi(e_j) = e_{14-j}$ for $3 \le j \le 11$. Since Δ^* generates M, ψ induces an orthogonal transformation ψ of M and it extends to M^*. Consider a ψ-invariant vector $g = e_1 + 2\sum_{i=2}^{11} e_i = 2v_0 + v_1 + v_2$. $g^2 < 0$. $\psi(g) = g$. Thus, ψ preserves 2 connected components $\Sigma_\pm$ of the negative cone in $M \otimes \mathbf{R}$.

Let C^* be the polyhedron in $M \otimes \mathbf{R}$ bounded by hyperplanes orthogonal to a member of Δ^*. Let C be the fundamental polyhedron of the Weyl group which will be constructed by Vinberg's algorithm. We consider polyhedrons $\hat{C}^* = C^*/\mathbf{R}_+$ and $\hat{C} = C/\mathbf{R}_+$ in $(M \otimes \mathbf{R} - \{0\})/\mathbf{R}_+$. Obviously, $\hat{C}^* \supset \hat{C}$. $\hat{C}^*$ has at least one vertex in $\hat{\Sigma} = \Sigma_+/\mathbf{R}_+$, for example the point corresponding to r. By the proof of Lemma 8.7 one knows that every vertex p of $\hat{C}^*$ in $\hat{\Sigma}$ is a vertex of $\hat{C}$. Thus, p has a neighborhood $U(p)$ with $U(p)\cap\hat{C}^* = U(p)\cap\hat{C}$. We can choose $U(p)$ such that $\psi(U(p)) = U(\psi(p))$. We have $U(\psi(p))\cap\psi(\hat{C}) = \psi(U(p))\cap\psi(\hat{C}) = \psi(U(p)\cap\hat{C}) = \psi(U(p)\cap\hat{C}^*) = \psi(U(p))\cap\psi(\hat{C}^*) = U(\psi(p))\cap\hat{C}^* = U(\psi(p))\cap\hat{C}$. This implies that $\psi(\hat{C})\cap\hat{C}$ has an interior point. Thus, $\psi(C) = C$. One knows that $\psi(e_i)$ $(1 \le i \le 15)$ are also fundamental roots. In particular,

$$e_{16} = \psi(e_{15}) = \{17v_0 + 7(v_1 + \cdots + v_5) - 3(v_6 + \cdots + v_{11})\}/5$$

is a fundamental root.

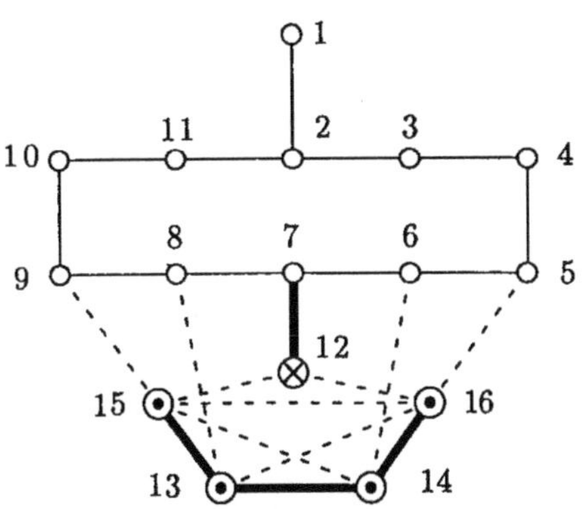

The left figure is the graph of the set S of these 16 roots. It is easy to see that this satisfies the condition $\langle a\rangle$ in Proposition 8.5. (We need to take an extended Dynkin graph ⊙━━⊙ of type $A_1(5)$ into consideration.) Consider the condition $\langle b\rangle$ in Proposition 8.5. Also a graph with a dotted edge and with only 2 vertices one of which is ⊙ is to be taken into account as a Lannér subgraph. Every Lannér subgraph is a graph with only 2 vertices and a dotted edge. It contains 9 such Lannér subgraphs. We apply Lemma 8.6 to them. Let Γ be one of 7 Lannér subgraphs excluding the one with vertex 9 and 15 and the one with vertex 5 and 16. In this case, the graph Ξ defined in Lemma 8.6 has 9 vertices. Since $9 + 2 = \operatorname{rank} M$, $\langle b\rangle$ is satisfied.

Consider the remaining cases. Let Γ be the graph with vertex 9 and 15, or the graph with vertex 5 and 16. By the symmetry of the graph both cases can be treated similarly. Therefore, in particular, we consider the graph Γ with vertex 5 and 16. In this case, the graph Ξ in Lemma 8.6 has vertices 1, 2, 3, 7, 8, 9, 10, 11 and is a Dynkin graph of type D_8. Note that it has only 8 vertices.

We introduce some notations here. Let $\overline{C}_1$ be the closed polyhedron in $M \otimes \mathbf{R}$ bounded by hyperplanes orthogonal to e_i with $1 \le i \le 16$ such that $\overline{C}_1 \cap \Sigma_+ \ne \emptyset$. For a subset $T \subset M \otimes \mathbf{R}$ by $T^\perp$ we denote the collections of vectors in $M \otimes \mathbf{R}$ orthogonal to T. $T^\perp$ is a $\mathbf{R}$-linear subspace. Set $V = (S(\Gamma) \cup S(\Xi))^\perp$. $\dim V = 1$. The condition in Lemma 8.6 is equivalent to $\overline{C}_1 \cap V = \{0\}$. Assume the contrary. Then, $V \cap \overline{C}_1$ is a 1-dimensional facet of $\overline{C}_1$. Let Υ be the subgraph with the vertex 5 plus the vertices of Ξ. Υ is a Dynkin graph of type $D_8 + A_1$. Thus, $S(\Upsilon)^\perp \cap \overline{C}_1$ is a 2-dimensional facet of $\overline{C}_1$ containing $V \cap \overline{C}_1$. The graph Ω_1 with vertex 14 plus the vertices of Υ is a Dynkin graph of type $D_8 + A_1 + A_1(5)$. The graph Ω_2 with vertex 6 plus the vertices of Υ is a Dynkin graph of type D_{10}. $S(\Upsilon)^\perp \cap \overline{C}_1$ contains 3 of 1-dimensional facets $V \cap \overline{C}_1$, $S(\Omega_1)^\perp \cap \overline{C}_1$, $S(\Omega_2)^\perp \cap \overline{C}_1$, which is a contradiction. Thus, by Lemma 8.6 one knows $\langle b\rangle$ is satisfied. The above is the Coxeter-Vinberg graph for Λ_2/P_C. (See Vinberg [20] section 4.2.)

Corollary 21.9. *If the Dynkin graph of a positive definite full submodule $L \subset \Lambda_2/P$ has a component of type BC_1, then there is an isotropic vector $u \in \Lambda_2/P$ in a nice position with respect to L.*

Remark. In the case where the Dynkin graph in Corollary 21.9 has a component of type BC_2 there does not necessarily exist an isotropic vector in a nice position. However, the influence of this condition does not appear in Theorem 0.6 and Theorem 24.13. (Theorem 24.13 is a version of Theorem 0.6 with Ik-condition.) This is because Lemma 21.10 below holds.

Lemma 21.10. *Let $G' + BC_2$ be a Dynkin graph containing a component of type BC_2. Let G be a Dynkin graph obtained from $G' + BC_2$ by one tie or elementary transformation. If G contains no vertex corresponding to a short root, then G can be obtained even from $G' + BC_1$ by the same transformation.*

22
The second reduction for $S_{1,0}$ 2

In this section we consider the case where a positive definite full submodule in Λ_2/P (P is the lattice in the case $S_{1,0}$.) has a Dynkin graph with a component of type B_k with $k \geq 4$.

Lemma 22.1. *Any automorphism on the discriminant group Q^*/Q of the root lattice $Q = Q(D_4)$ of type D_4 is induced by an automorphism of the lattice Q.*

Proof. Q has the action of the symmetric group S_3 of degree 3 induced by the symmetry of the Dynkin graph D_4. This S_3 acts Q^*/Q faithfully. On the other hand, the automorphism group of the group Q^*/Q is isomorphic to S_3. Thus, we get the lemma. Q.E.D.

Lemma 22.2. *(1) There are 3 even overlattices with index 2 over $P \oplus Q(D_4)$, and they are isomorphic.*
(2) For any one P_B of such 3 overlattices the root system of the quotient P_B/P is of type B_4.

Proof. Let $\overline{z}$, $\overline{g}_1$, $\overline{g}_2$ be the generators of $P^*/P \cong P'^*/P'$ as in Proposition 4.2. Let ω_3, ω_4 be fundamental weights of $Q = Q(D_4)$. (See Example 3.3.) We have only 3 isotropic elements $\overline{g}_1 + \overline{g}_2 + \overline{\omega}_3$, $\overline{g}_1 + \overline{g}_2 + \overline{\omega}_4$, $\overline{g}_1 + \overline{g}_2 + \overline{\omega}_3 + \overline{\omega}_4$ in $P^*/P \oplus Q^*/Q$. (We denote $\overline{x} = x \bmod Q$ for $x \in Q^*$.) Thus, we get the former half of (1). The latter half follows from Lemma 22.1. (2) is obvious. Q.E.D.

By P_B we denote an even overlattice with index 2 over $P \oplus Q(D_4)$. P_B has signature (11, 1).

Lemma 22.3. *(1) $P_B^*/P_B \cong \mathbf{Z}/5 \oplus \mathbf{Z}/2 \oplus \mathbf{Z}/2$. The discriminant quadratic form q can be written*

$$q(\overline{\alpha}) \equiv \frac{2}{5}a^2 + \frac{1}{2}(b_1^2 + b_2^2) \bmod 2\mathbf{Z}$$

for $\overline{\alpha} = (a, b_1, b_2) \in \mathbf{Z}/5 \oplus \mathbf{Z}/2 \oplus \mathbf{Z}/2$.
(2) For any two embeddings $\iota, \iota' : P_B \hookrightarrow \Lambda_2$ we have an orthogonal transformation $\psi : \Lambda_2 \to \Lambda_2$ with $\iota' = \psi\iota$.

Proof. (1) follows from Lemma 4.4.
(2) Since the discriminant group of P_B has no isotropic element, every embedding of P_B into an even lattice is always primitive. By Lemma 12.4 (2) we have (2). Q.E.D.

Proposition 22.4. *We fix an embedding $P \hookrightarrow \Lambda_2$ and a full embedding $Q(B_4) \hookrightarrow \Lambda_2/P$.*
(1) The inverse image of $Q(B_4)$ by the canonical surjective homomorphism $\Lambda_2 \to \Lambda_2/P$ is isomorphic to P_B.
(2) Let F_B denote the orthogonal complement of P_B for a fixed embedding $P_B \hookrightarrow \Lambda_2$. $F_B^* \cong \Lambda_2/P_B$. $F_B^*/F_B \cong \mathbf{Z}/5 \oplus \mathbf{Z}/2 \oplus \mathbf{Z}/2$. The discriminant quadratic form on F_B^*/F_B is given by

$$q(\overline{\alpha}) \equiv -\frac{2}{5}a^2 - \frac{1}{2}(b_1^2 + b_2^2) \bmod 2\mathbf{Z}$$

for $\overline{\alpha} = (a, b_1, b_2) \in \mathbf{Z}/5 \oplus \mathbf{Z}/2 \oplus \mathbf{Z}/2$.
(3) $\Lambda_2/P_B \cong F_B^*$ is a regular root module.
(4) Let $\overline{x}_1 \in F_B^*/F_B$ be the unique element with $q(\overline{x}_1) \equiv 1 \bmod 2\mathbf{Z}$. Let L_B be the inverse image of the subgroup of order 2 generated by $\overline{x}_1$ by the canonical surjective homomorphism $F_B^* \to F_B^*/F_B$. L_B is an odd lattice with discriminant -5. $F_B \subset L_B \subset L_B^* \subset F_B^*$.
(5) Let Ξ be the orthogonal complement of $Q(B_4)$ in Λ_2/P. By $\pi_B : \Lambda_2/P \to (\Lambda_2/P)/Q(B_4) \cong \Lambda_2/P_B \cong F_B^*$ we denote the composition of the canonical homomorphisms. The restriction of π_B to Ξ gives an isomorphism of quasi-lattices between Ξ and L_B^*.

Proof. Recall that the orthogonal complement F of P in Λ_2 has natural identification $F = \{x \in \Lambda_2/P \mid x^2 \in \mathbf{Z}\}$. (Proposition 5.6 (2).)
(1) Thus, $F \cap Q(B_4) \cong Q(D_4)$. Thus, the inverse image of $Q(B_4)$ contains $P \oplus Q(D_4)$ and the index over it is equal to $2 = [Q(B_4) : Q(D_4)]$. Thus, (1) follows from Lemma 22.2.
(2) It follows from Lemma 5.5 (3) and Lemma 22.3.
(3), (4) It follows from above (2).
(5) It is obvious that $\pi_B|\Xi$ is injective. Let Γ be the orthogonal complement of $Q(D_4) = Q(B_4) \cap F$ in Λ_2 with respect to the embedding $Q(D_4) \subset F \subset \Lambda_2$. $\Xi = \Gamma/P$. Γ is an overlattice of $P \oplus F_B$. Set $I = \Gamma/(P \oplus F_B)$. I is regarded as a subgroup of $P^*/P \oplus F_B^*/F_B$. The discriminant form on $\Gamma^*/\Gamma \cong I^\perp/I$ coincides with (-1) times that on $Q(D_4)^*/Q(D_4)$ by Lemma 5.5 (3). (In our present case it is equal to that on $Q(D_4)^*/Q(D_4)$ itself.) On the other hand, Proposition 4.2 and Lemma 22.4 give the discriminant forms on P^*/P and F_B^*/F_B respectively.

Let $\overline{w}$, $\overline{h}_1$, $\overline{h}_2$ be the generator of the respective components of $F_B^*/F_B \cong \mathbf{Z}/5 \oplus \mathbf{Z}/2 \oplus \mathbf{Z}/2$. It is not difficult to see that I is the group generated by $\overline{z} \pm \overline{w}$ and $\overline{g}_1 + \overline{g}_2 + \overline{h}_1 + \overline{h}_2$. It implies that $\pi_B(\Xi)/F_B$ coincides with the group generated by $\overline{w}$ and $\overline{h}_1 + \overline{h}_2$, which is equal to L_B^*/F_B. Thus, $\pi_B(\Xi) = L_B^*$. Q.E.D.

Proposition 22.5. *The following two conditions are equivalent. Let $k \geq 4$ be an integer:*
(1) There exists a positive definite full submodule $M \subset \Lambda_2/P$ whose Dynkin graph has the form $B_k + G'$.
(2) There exists a positive definite full submodule $M_B \subset L_B^*$ whose Dynkin graph has the form $B_{k-4} + G'$. (B_0 stands for the empty graph.)

Proof. (1)$\Rightarrow$(2) By assumption $Q(B_4) \subset Q(B_k) \subset M \subset \Lambda_2/P$. $Q(B_4)$ is full in Λ_2/P. Let Ξ be the orthogonal complement of $Q(B_4)$ in Λ_2/P. Let R be the root system of M and $R = \oplus_{i=1}^m R_i$ be the irreducible decomposition. We assume that R_1 is of type B_k. If $i \geq 2$, $R_i \subset \Xi$. $R_1 \cap \Xi$ is a root system of type B_{k-4}. $M \cap \Xi$ is a positive definite full submodule of Ξ whose Dynkin graph is $B_{k-4} + G'$. $M_B = \pi_B(M \cap \Xi)$ satisfies (2).
(2)$\Rightarrow$(1) We fix an embedding $P \hookrightarrow \Lambda_2$ and a full embedding $Q(B_4) \hookrightarrow \Lambda_2/P$. By ρ we denote the inverse mapping of $\pi_B : \Xi \xrightarrow{\sim} L_B^*$. Let R be the root system of M_B and $R = \oplus_{i=1}^k R_i$ be the irreducible decomposition. We assume that R_1 is of type B_{k-4}.

Now, let M be the primitive hull of $\rho(M_B) \oplus Q(B_4)$. Let $\widetilde{R}$ be the root system of M and $\widetilde{R} = \oplus_{i=1}^l \widetilde{R}_i$ be the irreducible decomposition. We assume $Q(B_4) \subset \widetilde{R}_1$. By the fullness of M_B $\widetilde{R} \cap \Xi = \rho(R)$. Here, $\widetilde{R} \cap \Xi = (\widetilde{R}_1 \cap \Xi) \oplus \bigoplus_{i=2}^l \widetilde{R}_i$ and $\rho(R) = \rho(R_1) \oplus \bigoplus_{i=2}^k \widetilde{R}_i$.

Since the component containing a short root with length 1 is unique (Proposition 5.4 (3).) if it exists, $\rho(R_1) = \widetilde{R}_1 \cap \Xi$ and $\bigoplus_{i=2}^l \widetilde{R}_i = \bigoplus_{i=2}^k \rho(R_i)$.

Case 1. $k \geq 5$.

Since $\widetilde{R}_1$ contains the direct sum of $R(B_4)$ and $\widetilde{R}_1 \cap \Xi$, $\widetilde{R}_1$ is of type B. Assume that it is of type B_s. Then, $\widetilde{R}_1 \cap \Xi$ is of type B_{s-4}. Thus, $s = k$. The Dynkin graph of M is $B_k + G'$.

Case 2. $k = 4$.

In this case, $\widetilde{R}_1 \cap \Xi = \emptyset$. If rank $\widetilde{R}_1 >$ rank $R(B_4) = 4$, then $\widetilde{R}_1$ is of type B. If it is of type B, then $\widetilde{R}_1 \cap \Xi \neq \emptyset$. We have a contradiction. Thus, rank $\widetilde{R}_1 = 4$. If $\widetilde{R}_1$ is not of type B_4, then it is of type F_4. However, it contradicts 21.1 (2). Thus, $\widetilde{R}_1$ is of type B_4 and M has the Dynkin graph $B_4 + G'$. Q.E.D.

Next, we would like to draw the Coxeter-Vinberg graph of L_B^*.

Proposition 22.6. *(1) $F_B \cong (Q(D_4) \oplus H \oplus \mathbf{Z}\lambda \oplus \mathbf{Z}\xi) + \mathbf{Z}(\omega + \lambda/2 + \xi/2)$. Here, $\lambda^2 = 10$, $\xi^2 = 2$. ω stands for a fundamental weight of $Q(D_4)$ with $\omega^2 = 1$.*
(2) $L_B \cong Q(B_4) \oplus H \oplus (\mathbf{Z}\xi_1 + \mathbf{Z}\xi_2)$. Here, $\xi_1^2 = 3$, $\xi_2^2 = 2$, and $\xi_1 \cdot \xi_2 = 1$.
(3) $L_B^* \cong Q(B_4) \oplus H \oplus (\mathbf{Z}w_1 + \mathbf{Z}w_2)$. Here, $w_1^2 = 2/5$, $w_2^2 = 3/5$ and $w_1 \cdot w_2 = -1/5$.

Proof. (1) The signature of F_B is (7, 1) and it is equal to the signature of the right-hand side of (1). On the other hand, we can calculate the discriminant quadratic form of the right-hand side of (1) by Lemma 4.4. One sees that it is equal to that of F_B. Thus, by Lemma 5.5 (4) and by Lemma 12.4 (2) one gets (1).
(2) Under the identification in (1) one sees easily $\omega \in F_B^*$. Since $\omega^2 = 1$, $L_B = F_B + \mathbf{Z}\omega = Q(B_4) \oplus H \oplus [(\mathbf{Z}\lambda \oplus \mathbf{Z}\xi) + \mathbf{Z}(\lambda/2 + \xi/2)]$. $\xi_1 = \lambda/2 + \xi/2$ and $\xi_2 = \xi$ satisfy the conditions.
(3) Set $w_1 = 2\xi_1/5 - \xi_2/5$ and $w_2 = -\xi_1/5 + 3\xi_2/5$. $\mathbf{Z}w_1 + \mathbf{Z}w_2 = (\mathbf{Z}\xi_1 + \mathbf{Z}\xi_2)^*$. They satisfy the conditions. Q.E.D.

Remark. $3w_1 + w_2 = \xi_1$. $w_1 + 2w_2 = \xi_2$.

Set $M = \sum_{i=0}^{5} \mathbf{Z}v_i$ where $v_0^2 = -1$, $v_i^2 = +1$ $(1 \le i \le 5)$ and $v_i \cdot v_j = 0$ $(i \ne j)$. One sees $M \cong Q(B_4) \oplus H$. Next, set $N = \mathbf{Z}\xi_1 + \mathbf{Z}\xi_2$ with $\xi_1^2 = 3$, $\xi_2^2 = 2$ and $\xi_1 \cdot \xi_2 = 1$. One has $N^* = \mathbf{Z}w_1 + \mathbf{Z}w_2$ where $w_1 = 2\xi_1/5 - \xi_2/5$, $w_2 = -\xi_1/5 + 3\xi_2/5$. One can check $w_1^2 = 2/5$, $w_2^2 = 3/5$ and $w_1 \cdot w_2 = -1/5$.

Lemma 22.7. *(1) $N = \{x \in N^* \mid x \cdot y \in \mathbf{Z}$ for all $y \in N^*\} = \{x \in N^* \mid x^2 \in \mathbf{Z}\}$.*
(2) If $\alpha \ne 0$ for $\alpha \in N$, then $\alpha^2 \ge 2$.
(3) If $\alpha^2 = 2$ for $\alpha \in N$, then $\alpha = \pm\xi_2$.

Proof. (1) The former half is obvious by definition. Now, for $x \in N^*$ and $y \in N$, $(x+y)^2 - x^2 = 2x \cdot y + y^2 \in \mathbf{Z}$. Thus, we can define a mapping $\overline{q} : N^*/N \to \mathbf{Q}/\mathbf{Z}$ by $q(x \bmod N) \equiv x^2 \bmod \mathbf{Z}$. Since $\#(N^*/N) = |d(N)| = 5$, $N^*/N \cong \mathbf{Z}/5$. Since $q(\pm w_1 \bmod N) \equiv 2/5$ and $q(\pm 2w_1 \bmod N) \equiv -2/5$, we have $q^{-1}(0 \bmod \mathbf{Z}) = \{0\}$. This implies the latter half.
(2), (3) Set $\alpha = z_1\xi_1 + z_2\xi_2$. We have $\alpha^2 = 3z_1^2 + 2z_1z_2 + 2z_2^2 = z_1^2 + 2(z_1^2 + z_1z_2 + z_2^2)$. One has (2) and (3), since $z_1^2 + z_1z_2 + z_2^2$ is positive definite. Q.E.D.

Lemma 22.8. *The following holds for $\alpha \in N^*$:*
(1) If $\alpha^2 = 2/5$, then $\alpha = \pm w_1$.
(2) $\alpha^2 \ne 1$.
(3) If $\alpha^2 = 2$, then $\alpha = \pm(w_1 + 2w_2) = \pm\xi_2$ and $\alpha \cdot w_1 = 0$.

Proof. Set $\alpha = y_1w_1 + y_2w_2$. $\alpha^2 = (2y_1^2 - 2y_1y_2 + 3y_2^2)/5 = \{2(y_1^2 - y_1y_2 + y_2^2) + y_2^2\}/5$.
(1) Since $y_1^2 - y_1y_2 + y_2^2$ is positive definite, one knows (1).
(2), (3) They follow from Lemma 22.7. Q.E.D.

For $L_B^* = M \oplus N^*$ we carry out Vinberg's algorithm. We consider also a dummy root with length $\sqrt{2/5}$ denoted by ⦿ in addition to a root with length $\sqrt{2}$ denoted by o and a root with length 1 •. v_0 is the controlling vector. The following is a root basis orthogonal to v_0:

$$e_1 = -v_1 + v_2, \quad e_2 = -v_2 + v_3, \quad e_3 = -v_3 + v_4, \quad e_4 = -v_4 + v_5,$$
$$e_5 = -v_5, \quad e_6 = -\xi_2, \quad e_7 = -w_1.$$

By the algorithm we get successively

$$\begin{aligned} e_8 &= v_0 + v_1 + \xi_2 = v_0 + v_1 + w_1 + 2w_2 \\ e_9 &= v_0 + \xi_1 = v_0 + 3w_1 + w_2 \\ e_{10} &= v_0 + v_1 + v_2 + v_3 \\ e_{11} &= v_0 + v_1 + w_1. \end{aligned}$$

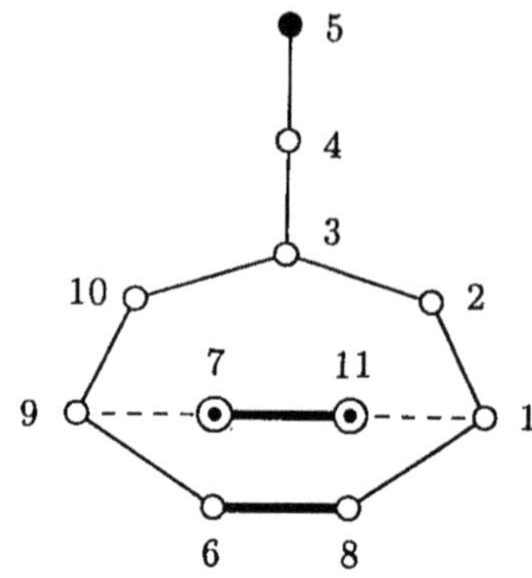

The right figure is the graph associated with the above 11 roots. It is easy to see that this is the Coxeter-Vinberg graph for L_B^*. (Proposition 8.5. Lemma 8.6.)

Corollary 22.9. *Let $G' + B_k$ ($k \geq 4$) be a Dynkin graph. The following conditions are equivalent:*
(1) There exists a positive definite full submodule $M \subset \Lambda_2/P$ whose Dynkin graph coincides with $G' + B_k$.
(2) $G' + B_k$ is a subgraph of one of the following 5 sub-basic graphs:

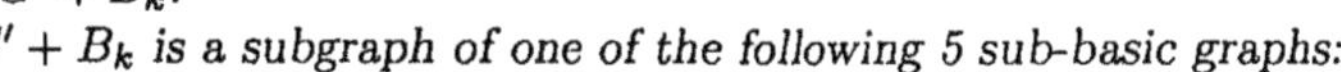

$$B_{10} + A_1, \quad B_9 + A_2, \quad E_7 + B_4, \quad A_6 + B_5, \quad B_6 + A_3 + A_2.$$

Remark. Trying to deduce the existence of a nice isotropic vector from the above Coxeter-Vinberg graph, one sees that it is hard to give simple description. Therefore, we introduce the concept of sub-basic graphs here.

23
The second reduction for $S_{1,0}$ 3

We continue to study the existence of an isotropic vector in a nice position in Λ_2/P in the case of $S_{1,0}$. We would like to show the following in this section:

Proposition 23.1. *Let G be a Dynkin graph. If there is a full embedding $Q(G) \hookrightarrow \Lambda_2/P$, then one of the following three conditions holds:*
(1) G can be made either from one of basic graphs by one transformation.
(2) G is a subgraph of one of the sub-basic graphs.
(3) G has a component of type BC_2 or B_1.

This section and the next section are devoted to the verification of Proposition 23.1. Note that by Lemma 16.3 and Lemma 22.10 the influence of the above case (3) does not appear in our main results on $PC(S_{1,0})$.

First, we consider the relation with the case of $Z_{1,0}$. Recall that the lattice $P = P(2,2,3,4)$ defined in the case $S_{1,0}$ has a decomposition $P = P' \oplus H_0$, and P' has a basis $e_0, \ldots, e_3, f, e'$. Set $P_1' = \sum_{i=0}^{3} \mathbf{Z}e_i + \mathbf{Z}e'$ and $P_1 = P_1' \oplus H_0$. P_1 is the lattice with a basis $e_0, \ldots, e_6$ and it is isomorphic to the lattice $P = P(2,2,2,4)$ defined in the case $Z_{1,0}$. $P_1' \cong Q(D_4 + A_1)$.

Lemma 23.2. *The vector $g_1 = 2e_0 + e_1 + e_2 + 2e_3 + 2f + e' \in P'$ $(= 5z)$ is primitive in P' and orthogonal to P_1'. $g_1^2 = 10$. P' is an overlattice of $P_1' \oplus \mathbf{Z}g_1$.*

F stands for the orthogonal complement of P in Λ_N with respect to a fixed embedding $P \hookrightarrow \Lambda_N$. Set $\Gamma_1 = C(\mathbf{Z}g_1, \Lambda_N)$. $P_1 \subset \Gamma_1$. $C(\mathbf{Z}g_1, \Lambda_N/P_1) = \Gamma_1/P_1$.

Lemma 23.3. *Let $\pi_1 : \Lambda_N/P_1 \to \Lambda_N/P$ be the canonical surjective homomorphism. The restriction of π_1 to Γ_1/P_1 is an injective homomorphism preserving the bilinear forms. The image $\pi_1(\Gamma_1/P_1)$ coincides with the inverse image of the 2-Sylow subgroup of F^*/F by the canonical surjective homomorphism $\Lambda_N/P \cong F^* \to F^*/F$.*

Proof. Injectivity and preserving the bilinear forms follow from the definition easily. We consider the image. Now, Γ_1 is an overlattice of $F \oplus P_1$. Set $m = [\Gamma_1 : F \oplus P_1]$. $m^2|d(\Gamma_1)| = |d(F \oplus P_1)| = |d(F)||d(P_1)| = |d(P)||d(P_1)| = 20 \times 8$. $|d(\Gamma_1)| = |d(\mathbf{Z}g_1)| = 10$. One knows $m = 4$. Thus, $\pi(\Gamma_1/P_1)$ is an overmodule over $F = (F \oplus P)/P$ $(\subset \Lambda_N/P)$ with index 4. On the other hand, $(\Lambda_N/P)/F \cong F^*/F \cong P^*/P \cong \mathbf{Z}/5 \oplus \mathbf{Z}/2 \oplus \mathbf{Z}/2$. The subgroup with order 4 in the right-hand side is unique and it coincides with the 2-Sylow subgroup. Q.E.D.

Remark. The 2-Sylow subgroup of P^*/P contains all special elements in P^*/P. (It contains ones of type B and ones of type BC.) However, it does not contain any element of the third kind with the associated number 9. Thus, for any positive definite submodule $L \subset \Lambda_2/P$ an equality for root systems $R(L) = R(L \cap \pi_1(\Gamma_1/P_1))$ holds. However, we cannot decide whether or not a component A_9 of the root system of L is an obstruction component only by the information on the embedding $L \cap \pi_1(\Gamma_1/P_1) \hookrightarrow \pi_1(\Gamma_1/P_1)$.

Corollary 23.4. *The Dynkin graph of any positive definite full submodule $L \subset \Lambda_2/P(2,2,3,4)$ is a subgraph of the Coxeter-Vinberg graph of $\Lambda_2/P(2,2,2,4)$ defined in the case $Z_{1,0}$. (See Section 9.)*

Next, we consider the relation with the case $W_{1,0}$. We consider the sublattice $P_2' = \sum_{i=1}^{3} \mathbf{Z}e_i + \mathbf{Z}f \subset P'$. Set $P_2 = P_2' \oplus H_0$. P_2 is a sublattice $P = P(2,2,3,4)$ generated by the basis e_0, e_1, e_2, e_3, e_4, e_5, e_7, and isomorphic to $P = P(2,2,3,3)$ defined in the case $W_{1,0}$.

Lemma 23.5. *The vector $g_2 = 2e_0 + e_1 + e_2 + 2e_3 + 2f + 6e' \in P'$ $(= 10w_4)$ is primitive in P' and orthogonal to P_2'. $g_2^2 = 60$. P' is an overlattice of $P_2' \oplus \mathbf{Z}g_2$.*

F stands for the same as above. $\Lambda_N/P \cong F^*$. Let $\bar{t}$, $\bar{h}_1$, $\bar{h}_2$ be the generator of the respective components of $F^*/F \cong \mathbf{Z}/5 \oplus \mathbf{Z}/2 \oplus \mathbf{Z}/2$. The discriminant quadratic form can be written $q_F(a\bar{t} + b_1\bar{h}_1 + b_2\bar{h}_2) \equiv -2a^2/5 + (b_1^2 + b_2^2)/2 \bmod 2\mathbf{Z}$.

Set $\Gamma_2 = C(\mathbf{Z}g_2, \Lambda_N)$. $P_2 \subset \Gamma_2$. $C(\mathbf{Z}g_2, \Lambda_N/P_2) = \Gamma_2/P_2$.

Lemma 23.6. *Let $\pi_2 : \Lambda_N/P_2 \to \Lambda_N/P$ be the canonical surjective homomorphism.*
(1) The restriction of π_2 to Γ_2/P_2 is injective and preserves the bilinear forms.
(2) Let J be a subgroup of order 2 in F^*/F generated by $\overline{h}_1 + \overline{h}_2$. Let K be the inverse image in F^* of J. K is an odd lattice with the discriminant $d(K) = \pm 5$. $\pi_2(\Gamma_2/P_2) = K$.

Proof. (1) it follows easily from the definition. (2) Obviously, K is an odd lattice. $|d(K)| = |d(F)|/[K : F]^2 = 20/2^2 = 5$. Next, we consider $\pi_2(\Gamma_2/P_2)$. Γ_2 is an overlattice of $F \oplus P_2$. Set $l = [\Gamma_2 : F \oplus P_2]$. $l^2|d(\Gamma_2)| = |d(F \oplus P_2)| = |d(F)||d(P_2)| = 20 \times 12$. $|d(\Gamma_2)| = |d(\mathbf{Z}g_2)| = 60$. One sees $l = 2$. Thus, $I = \Gamma_2/(F \oplus P_2) \subset (F \oplus P_2)^*/(F \oplus P_2) = F^*/F \oplus P_2^*/P_2$ is a cyclic subgroup of order 2. Let $\overline{\alpha} \in I$ be the non-zero element. Since Γ_2 is an even lattice, $q_{F\oplus P_2}(\overline{\alpha}) \equiv 0 \bmod 2\mathbf{Z}$. Here, $q_{F\oplus P_2} = q_F \oplus q_{P_2}$. By the above expression of q_F and by the expression of q_{P_2} in Proposition 4.1 ($P_2 \cong P(2,2,3,3)$.), one knows $\overline{\alpha} = \overline{h}_1 + \overline{h}_2 + \overline{q}$, where $\overline{q}$ is the unique element of order 2 in P_2^*/P_2. This implies that the image of Γ_2/P_2 under the canonical homomorphism $\Lambda_N/P_2 \to \Lambda_N/P \cong F^* \to F^*/F$ coincides with J. Q.E.D.

Remark. The subgroup in P^*/P corresponding to $J \subset F^*/F$ contains all special elements of type B, but it does not contain any special element of type BC. Moreover, it does not contain any element of the third kind with the associated number 9.

Thus, for any positive definite full submodule $L \subset \Lambda_N/P$ we have the equality of the reduced root systems; $\hat{R}(L) = \hat{R}(L \cap \pi_2(\Gamma_2/P_2))$. However, we cannot judge how different the root system of L and the reduced root system of L are, and we cannot decide whether or not an A_9-component is an obstruction component, if we see only the embedding $L \cap \pi_2(\Gamma_2/P_2) \hookrightarrow \pi_2(\Gamma_2/P_2)$.

Corollary 23.7. *Let G be the Dynkin graph of a positive definite full submodule $L \subset \Lambda_2/P(2,2,3,4)$. Let G' be the Dynkin graph obtained from G by replacing any component of type BC_1 and any component of type BC_2 by a component of type A_1 and a component of type B_2 respectively. Then, G' is a subgraph of the Coxeter-Vinberg graph of $\Lambda_2/P(2,2,3,3)$ defined in the case $W_{1,0}$. (See Section 16.)*

Now, let G be a Dynkin graph with a full embedding $Q(G) \hookrightarrow \Lambda_2/P$. By r we denote the number of vertices of G. By the results so far we can consider only Dynkin graphs G without a BC-component and without a component of type B_k with $k \geq 4$. Recall Lemma 16.3 here. Thanks to it, we need not consider the case where G has a component B_1. By Proposition 5.4 (3) G has at most one component of type B. Moreover, by Proposition 21.1 we can assume that G has no subgraph in the form •——• or ⓞ. Thus, we can assume the following conditions (1) and (2):

(1) Any component of G is of type A, B, D or E.
(2) If G has a component of type B, then such a component is unique and is of type B_2 or B_3.

Thanks to Corollary 23.4 and Corollary 23.7, we can assume the following:

(3) G is a subgraph of the Coxeter-Vinberg graph of $\Lambda_2/P(2,2,2,4)$ corresponding to the case $Z_{1,0}$ in Section 9 and it is also a subgraph of the Coxeter-Vinberg graph of $\Lambda_2/P(2,2,3,3)$ corresponding to the case $W_{1,0}$ in Section16.

Moreover, we can assume that the I(Q, $\Lambda_2/\widetilde{P}$)-condition fails for $Q = Q(G)$. Thus, in particular, we have the following conditions (4), (5) and (6):
(4) $8 \le r \le 11$.
(5) If $r = 9$, then for $p = 3$, 5, or 7 the following (5_p) holds:

$$(5_p) \qquad \epsilon_p(G) \neq (5,-2)_p(-5, d(G))_p$$

(6) If $r = 8$, then for $p = 2$, 3, 5 or 7 the following (6_p) holds:

$$(6_p) \qquad -5d(G) \in \mathbf{Q}_p^{*2} \text{ and } \epsilon_p(G) \neq (-1,-1)_p(5,2)_p$$

In (5) and (6) we used the abbreviation $d(G) = d(Q(G))$ and $\epsilon_p(G) = \epsilon_p(Q(G))$. (5) and (6) are a part of the I(Q, $\Lambda_2/\widetilde{P}$)-condition. (See Section 7.) By Lemma 7.5 (3), (4) we need not consider a prime number $p \ge 11$.

Proposition 23.8. *In the case $r = 11$ G satisfies the above conditions (1), (2) and (3) if, and only if, G belongs one of the following 2 classes MP11 and NI11 of Dynkin graphs:*

MP11: (2 items.) A_{11}, $A_9 + A_2$.

NI11: (14 items.)

(without a B-component. 5 items.)
$D_{10} + A_1$, $A_7 + A_3 + A_1$, $E_7 + A_4$, $A_6 + A_5$,
$D_6 + A_5$,

(with a B_2-component. 1 item.) $A_9 + B_2$,

(with a B_3-component. 8 items.)
$A_8 + B_3$, $E_8 + B_3$, $D_7 + B_3 + A_1$, $E_7 + B_3 + A_1$,
$A_6 + B_3 + A_2$, $D_6 + B_3 + A_2$, $A_5 + A_3 + B_3$, $A_4 + A_3 + B_3 + A_1$.

Proposition 23.9. *In the case $r = 10$ G satisfies the above conditions (1), (2) and (3) if, and only if, G belongs one of the following 4 classes MP10, NP10, MI10 and NI10 of Dynkin graphs:*

MP10: (14 items.)

(without a B-component. 5 items.)
D_{10}, $E_8 + A_2$, $E_7 + A_3$,
$E_6 + A_4$, $A_5 + D_5$,

(with a B_2-component. 4 items.)
$A_6 + A_2 + B_2$, $A_7 + B_2 + A_1$, $2A_3 + A_2 + B_2$,
$E_7 + B_2 + A_1$,

(with a B_3-component. 5 items.)
$E_7 + B_3$, $A_6 + B_3 + A_1$, $E_6 + B_3 + A_1$,
$A_5 + B_3 + A_2$, $A_3 + B_3 + 2A_2$.

NP10: *(10 items.) (All have no B-component.)*
A_{10}, $A_9 + A_1$, $A_8 + A_2$,
$A_7 + A_3$, $A_7 + A_2 + A_1$, $A_6 + A_4$,
$A_6 + 2A_2$, $2A_5$, $A_5 + A_3 + A_2$,
$2A_4 + A_2$.

MI10: (3 items.)
$3A_3 + A_1$, $A_5 + B_3 + 2A_1$, $A_4 + B_3 + A_2 + A_1$.

NI10: (32 items.)

(*without a B-component. 14 items.*)

$D_9 + A_1$,	$D_8 + 2A_1$,	$A_7 + 3A_1$,
$D_7 + A_2 + A_1$,	$E_7 + A_2 + A_1$,	$A_6 + A_3 + A_1$,
$D_6 + A_4$,	$D_6 + A_3 + A_1$,	$D_6 + 2A_2$,
$A_5 + A_4 + A_1$,	$A_5 + D_4 + A_1$,	$A_5 + A_3 + 2A_1$,
$D_5 + A_4 + A_1$,	$A_4 + A_3 + A_2 + A_1$,	

(*with a B_2-component. 8 items.*)

$A_8 + B_2$,	$E_8 + B_2$,	$D_7 + B_2 + A_1$,
$D_6 + A_2 + B_2$,	$A_5 + A_3 + B_2$,	$D_5 + A_2 + B_2 + A_1$,
$2A_4 + B_2$,	$A_4 + A_3 + B_2 + A_1$,	

(*with a B_3-component. 10 items.*)

$A_7 + B_3$,	$D_7 + B_3$,	$D_6 + B_3 + A_1$,
$D_5 + B_3 + A_2$,	$D_5 + B_3 + 2A_1$,	$A_4 + A_3 + B_3$,
$A_4 + B_3 + 3A_1$,	$D_4 + B_3 + A_2 + A_1$,	$2A_3 + B_3 + A_1$,
$A_3 + B_3 + A_2 + 2A_1$.		

Lemma 23.10. *Let G be a Dynkin graph with 9 vertices satisfying (1) and (2). G satisfies the condition (5_7) if, and only if, G is one of the following 4 graphs:*

$$A_6 + A_3, \quad A_6 + B_3, \quad A_6 + B_2 + A_1, \quad A_6 + 3A_1.$$

Proof. If $7 \not| d(G)$, then $\epsilon_7(G) = 1 = (5, 2)_7(-5, d(G))_7$. Thus, $7 \mid d(G)$ and we can write $G = A_6 + G'$. We have 5 possibilities as G' and ones except $A_2 + A_1$ satisfy the condition. Q.E.D.

Lemma 23.11. *Let G be a Dynkin graph with 9 vertices. We assume that G satisfies (1) and (2). The following 3 conditions are equivalent:*

⟨1⟩ *$5 \mid d(G)$ and G satisfies (5_5).*

⟨2⟩ *We can write $G = A_4 + G'$ with $5 \not| d(G')$.*

⟨3⟩ *G is one of the following 13 graphs:*

$A_5 + A_4$,	$D_5 + A_4$,	$A_4 + D_4 + A_1$,	$A_4 + A_3 + A_2$,
$A_4 + A_3 + B_2$,	$A_4 + A_3 + 2A_1$,	$A_4 + B_3 + A_2$,	$A_4 + B_3 + 2A_1$,
$A_4 + 2A_2 + A_1$,	$A_4 + A_2 + B_2 + A_1$,	$A_4 + A_2 + 3A_1$,	$A_4 + B_2 + 3A_1$,
$A_4 + 5A_1$.			

Proof. ⟨1⟩⇒⟨2⟩ By assumption, G contains A_4 or A_9. $G = A_9$ and $G = 2A_4 + A_1$ do not satisfy the condition.

⟨2⟩⇒⟨1⟩ Set $d' = d(G')$. We abbreviate the lower index 5. $\epsilon = (-1, 5)(5, d') = (5, d')$. $(5, -2)(-5, d(G)) = -(-5, 5d') = -(5, d')$.

⟨2⟩⇔⟨3⟩ is easy. Q.E.D.

Lemma 23.12. *Let G be a Dynkin graph with 9 vertices. We assume that G satisfies (1) and (2). The following 3 conditions are equivalent:*

⟨1⟩ *$5 \not| d(G)$ and G satisfies (5_5).*

⟨2⟩ *G contains no component of type A_4 or A_9 and the total number of components of type A_1, A_2, A_6, A_7, E_6, or E_7 is even.*

⟨3⟩ *G is one of the following 38 graphs:*

(without a B-component. 21 items.)

D_9, A_7+A_2, D_7+2A_1, E_7+A_2,
A_6+3A_1, D_6+A_3, $D_6+A_2+A_1$, E_6+3A_1,
A_5+D_4, A_5+2A_2, A_5+4A_1, D_5+D_4,
D_5+2A_2, D_5+4A_1, $D_4+A_3+2A_1$, $D_4+A_2+3A_1$,
$3A_3$, $A_3+2A_2+2A_1$, A_3+6A_1, $3A_2+3A_1$,
A_2+7A_1,

(with a B_2-component. 11 items.)

D_7+B_2, $A_6+B_2+A_1$, $E_6+B_2+A_1$, $A_5+B_2+2A_1$,
$D_5+B_2+2A_1$, $D_4+A_3+B_2$, $D_4+A_2+B_2+A_1$, $A_3+2A_2+B_2$,
$A_3+B_2+4A_1$, $3A_2+B_2+A_1$, $A_2+B_2+5A_1$,

(with a B_3-component. 6 items.)

D_6+B_3, $D_4+B_3+2A_1$, $2A_3+B_3$, $A_3+B_3+A_2+A_1$,
$B_3+2A_2+2A_1$, B_3+6A_1.

Proof. ⟨1⟩⇒⟨2⟩ Since $\epsilon_5(G)=1$ and $(5,-2)_5=-1$ we have $(-5, d(G))_5 = \left(\frac{d(G)}{5}\right) = +1$, under the assumption. Let G_0 be a connected Dynkin graph with the number of vertices less than or equal to 9 satisfying the above (1) and (2). Then, $\left(\frac{d(G_0)}{5}\right) = -1 \Leftrightarrow G_0 = A_7, E_7, A_6, E_6, A_2$ or A_1. Thus, we have ⟨2⟩. Similarly we have ⟨2⟩⇒⟨1⟩. ⟨2⟩⇔⟨3⟩ is easy. Q.E.D.

Lemma 23.13. *Let G be a Dynkin graph with 9 vertices. We assume that G satisfies (1) and (2). The following 4 conditions are equivalent:*

⟨1⟩ *G satisfies* (5_3).

⟨2⟩ $\epsilon_3(G)=-1$.

⟨3⟩ *G contains a component of type* A_2, A_5, E_6 *or* A_8 *and* $\epsilon_3(G)=-1$.

⟨4⟩ *G is one of the following 25 graphs:*

(without a B-component. 14 items.)

D_7+A_2, E_6+3A_1, A_5+D_4, A_5+4A_1,
D_5+2A_2, $D_5+A_2+2A_1$, $A_4+2A_2+A_1$, $A_4+A_2+3A_1$,
$D_4+A_3+A_2$, $D_4+2A_2+A_1$, $A_3+2A_2+2A_1$, $A_3+A_2+4A_1$,
$3A_2+3A_1$, $2A_2+5A_1$,

(with a B_2-component. 8 items.)

$E_6+B_2+A_1$, $A_5+B_2+2A_1$, $D_5+B_2+A_2$, $A_4+A_2+B_2+A_1$,
$A_3+2A_2+B_2$, $A_3+A_2+B_2+2A_1$, $3A_2+B_2+A_1$, $2A_2+B_2+3A_1$,

(with a B_3-component. 3 items.)

$D_4+B_3+A_2$, $B_3+2A_2+2A_1$, $B_3+A_2+4A_1$.

Proof. Set $d(G)=3^k m$ with $3 \not| m$. $(5,-2)_3(-5,d(G))_3 = (-5, 3^k m)_3 = (-5, 3)_3^k = +1$. Thus, one knows ⟨1⟩⇔⟨2⟩. By Lemma 7.5 (4) ⟨2⟩ implies $3 \mid d(G)$. Thus, we have ⟨2⟩⇒⟨3⟩. The implication ⟨3⟩⇒⟨2⟩ is trivial. Calculating ϵ_3 on each possibility, one has ⟨3⟩⇔⟨4⟩. Q.E.D.

Proposition 23.14. *In the case where a Dynkin graph G has 9 vertices, G satisfies (1), (2) and (5) if, and only if, G belongs to one of the following 4 classes NP9, MI9, NI9, E9:*

NP9: (27 items.)

(without a B-component. 14 items.)

D_9, $A_7 + A_2$, $D_7 + A_2$,
$E_7 + A_2$, $A_6 + A_3$, $D_6 + A_3$,
$A_5 + A_4$, $A_5 + D_4$, $A_5 + 2A_2$,
$D_5 + A_4$, $D_5 + 2A_2$, $A_4 + A_3 + A_2$,
$A_4 + 2A_2 + A_1$, $3A_3$,

(with a B_2-component. 7 items.)

$A_6 + B_2 + A_1$, $E_6 + B_2 + A_1$, $A_5 + B_2 + 2A_1$,
$D_5 + B_2 + 2A_1$, $A_4 + A_2 + B_2 + A_1$, $A_3 + 2A_2 + B_2$,
$A_3 + A_2 + B_2 + 2A_1$,

(with a B_3-component. 6 items.)

$A_6 + B_3$, $D_6 + B_3$, $A_4 + B_3 + A_2$,
$A_4 + B_3 + 2A_1$, $A_3 + B_3 + A_2 + A_1$, $B_3 + 2A_2 + 2A_1$.

MI9: (3 items.)

$A_4 + A_3 + B_2$, $2A_3 + B_3$, $3A_2 + B_2 + A_1$.

NI9: (19 items.)

(without a B-component. 11 items.)

$D_7 + 2A_1$, $A_6 + 3A_1$, $D_6 + A_2 + A_1$,
$A_5 + 4A_1$, $D_5 + A_2 + 2A_1$, $A_4 + D_4 + A_1$,
$A_4 + A_3 + 2A_1$, $A_4 + A_2 + 3A_1$, $D_4 + A_3 + 2A_1$,
$D_4 + 2A_2 + A_1$, $A_3 + 2A_2 + 2A_1$,

(with a B_2-component. 5 items.)

$D_7 + B_2$, $D_5 + A_2 + B_2$, $A_4 + B_2 + 3A_1$,
$D_4 + A_2 + B_2 + A_1$, $2A_2 + B_2 + 3A_1$,

(with a B_3-component. 3 items.)

$D_4 + B_3 + A_2$, $D_4 + B_3 + 2A_1$, $B_3 + A_2 + 4A_1$.

E9: (15 items.)

(without a B-component. 11 items.)

$E_6 + 3A_1$, $D_5 + D_4$, $D_5 + 4A_1$,
$A_5 + 4A_1$, $D_4 + A_3 + A_2$, $D_4 + A_2 + 3A_1$,
$A_3 + A_2 + 4A_1$, $A_3 + 6A_1$, $3A_2 + 3A_1$,
$2A_2 + 5A_1$, $A_2 + 7A_1$,

(with a B_2-component. 3 items.)

$D_4 + A_3 + B_2$, $A_3 + B_2 + 4A_1$, $A_2 + B_2 + 5A_1$,

(with a B_3-component. 1 item.) $B_3 + 6A_1$.

Lemma 23.15. *Let G be a Dynkin graph with 8 vertices. We assume that G satisfies (1) and (2). The following 2 conditions are equivalent:*

⟨1⟩ *$5|d(G)$ and the condition (6) is satisfied.*

⟨2⟩ *G is one of the following 4:*

$$A_4 + D_4, \quad A_4 + 2A_2, \quad A_4 + B_2 + 2A_1, \quad A_4 + 4A_1.$$

Proof. ⟨2⟩⇒⟨1⟩ Obvious. ⟨1⟩⇒⟨2⟩ Under the assumption we can write $G = A_4 + G'$. (G' has 4 vertices.) Set $d' = d(G')$. G' has no component of type A_6. Thus, $7 \not| 5d' = d(G)$ and $\epsilon_7(G) = 1 = (-1, -1)_7(5, 2)_7$.

There are 7 possibilities for G'. Checking the condition (6_p) for $p = 2, 3, 5$, one gets ⟨2⟩. Q.E.D.

Lemma 23.16. *Let G be a Dynkin graph with 8 vertices. We assume that G satisfies (1) and (2). We here assume moreover $5 \not| d(G)$. By p we denote a prime number. The following 2 conditions are equivalent:*
⟨1⟩ $\epsilon_p(G) = -1$.
⟨2⟩ *G and p are one in the following list:*

$A_6 + A_2$	$p = 3, 2;$	$A_6 + B_2$	$p = 7, 2;$
$A_6 + 2A_1$	$p = 7, 2;$	$D_6 + A_2$	$p = 3, 2;$
$A_5 + A_3$	$p = 3, 2;$	$A_5 + B_3$	$p = 3, 2;$
$D_4 + 2A_2$	$p = 3, 2;$	$D_4 + A_2 + B_2$	$p = 3, 2;$
$D_4 + A_2 + 2A_1$	$p = 3, 2;$	$2A_3 + A_2$	$p = 3, 2;$
$A_3 + 2A_2 + A_1$	$p = 3, 2;$	$A_3 + B_3 + A_2$	$p = 3, 2;$
$B_3 + 2A_2 + A_1$	$p = 3, 2;$	$2A_2 + B_2 + 2A_1$	$p = 3, 2;$
$2A_2 + 4A_1$	$p = 3, 2;$	$A_2 + B_2 + 4A_1$	$p = 3, 2;$
$A_2 + 6A_1$	$p = 3, 2.$		

Proof. If $p \geq 11$ or $p = 5$, then $p \not| d(G)$ under the assumption; thus, $\epsilon_p(G) = 1$. By the product formula one has $\epsilon_2(G) = \epsilon_3(G)\epsilon_7(G)$.
⟨2⟩⇒⟨1⟩ When $p \neq 2$, we can calculate ϵ_p very easily. We need not calculate ϵ_2 by the product formula.
⟨1⟩⇒⟨2⟩ If we make the list of G with $\epsilon_7(G) = -1$ and the list of G with $\epsilon_3(G) = -1$, then the case of $\epsilon_2(G) = -1$ follows from the product formula.

Assume $\epsilon_7(G) = -1$. $7 \mid d(G)$. Thus, we can write $G = A_6 + G'$. We have only 3 possibilities as G'. One knows $G = A_6 + B_2$ or $A_6 + 2A_1$.

Assume $\epsilon_3(G) = -1$. $3 \mid d(G)$. G contains A_2, A_5, E_6 or A_8. Calculating ϵ_3 for possible G we get the above result. Q.E.D.

Note that if $5 \not| d(G)$, then $-5d(G) \notin \mathbf{Q}_5^{*2}$; thus, ($6_5$) fails. Therefore, to check the condition (6) in the case $5 \not| d(G)$ it suffices to check $-5d(G) \in \mathbf{Q}_p^{*2}$ or not for each pair (G, p) in the list in Lemma 23.16. We get the following proposition:

Proposition 23.17. *We consider the case where a Dynkin graph G has 8 vertices. Then, G satisfies (1), (2) and (6) if, and only if, G belongs one of the following 3 classes NP8, MI8 and E8:*

NP8: (9 items.)

(without a B-component. 5 items.)

$D_6 + A_2$, $A_4 + D_4$, $A_4 + 2A_2$, $D_4 + 2A_2$, $2A_3 + A_2$,

(with a B_2-component. 3 items.)

$A_4 + B_2 + 2A_1$, $2A_2 + B_2 + 2A_1$, $A_2 + B_2 + 4A_1$,

(with a B_3-component. 1 item.) $A_3 + B_3 + A_2$.

MI8: (4 items.)
$D_4 + A_2 + B_2$, $A_4 + 4A_1$, $D_4 + A_2 + 2A_1$, $2A_2 + 4A_1$.
E8: (1 item.) $A_2 + 6A_1$.

In the above we got 13 classes of Dynkin graphs.

MP11			NI11	
MP10	NP10	MI10	NI10	
	NP9	MI9	NI9	E9
	NP8	MI8		E8

If there is a counter-example to Proposition 23.1, it should be a member of one of the above classes. In the next section we show that for every graph G in the above classes Proposition 23.1 holds.

24
The second reduction for $S_{1,0}$ 4

Proposition 24.1. (MP11, MP10) *Every Dynkin graph belonging to MP11 or MP10 can be made from one of the basic graphs by one tie transformation, if it is not a subgraph of one of the sub-basic graphs.*

Proof. Indeed, by one tie transformation we can make A_{11}, $A_9 + A_2$, D_{11} from $A_9 + BC_1$; we can make $E_8 + A_2$, $E_7 + A_3$, $E_6 + A_4$, $A_5 + D_5$ from $E_8 + BC_1$; and we can make $E_7 + B_3$, $A_6 + B_3 + A_1$, $E_6 + B_3 + A_1$, $A_5 + B_3 + A_2$, $A_5 + B_3 + 2A_2$ from $E_6 + B_3$. The remaining 4 graphs $A_6 + A_2 + B_2$, $A_7 + B_2 + A_1$, $2A_3 + A_2 + B_2$, $E_7 + A_1 + B_2$ are subgraphs of $B_9 + A_2$, $B_{10} + A_1$, $B_6 + A_3 + A_2$, $B_4 + E_7$ respectively. Q.E.D.

Proposition 24.2. (NP10, NP9, NP8)
(1) Every graph G belonging to NP10 is a subgraph of a graph belonging to MP11.
(2) Every graph G belonging to NP9 is a subgraph of a graph belonging to MP10 or NP10.
(3) Every graph G belonging to NP8 is a subgraph of a graph belonging to MP9.

Corollary 24.3. *Every Dynkin graph belonging to NP10, NP9, or NP8 can be made from one of the basic graphs by one tie or elementary transformation, if it is not a subgraph of one of the sub-basic graphs.*

Lemma 24.4. (MI10) *For $G = 3A_3 + A_1$ or $A_5 + B_3 + 2A_1$ there is no full embedding $Q(G) \hookrightarrow \Lambda_2/P$.*

Proof. First, we consider $G = 3A_3 + A_1$. Assume the contrary. We will deduce a contradiction. Let $S = P \oplus Q(G) \hookrightarrow \Lambda_2$ be the induced embedding. Set $I = \widetilde{S}/S$. For the 2-Sylow subgroup $(\widetilde{S}/S)_2 \cong \mathbf{Z}/4 \oplus \mathbf{Z}/4 \oplus \mathbf{Z}/4 \oplus \mathbf{Z}/2 \oplus \mathbf{Z}/2 \oplus \mathbf{Z}/2$. Here, the first direct summands of the right-hand side, and the second and the third are induced by an A_3-component respectively. The fourth and the fifth are by P, and the sixth by the A_1-component. The discriminant quadratic form can be written $q(\overline{\alpha}) \equiv 3(a_1^2 + a_2^2 +$

$a_3^2)/4-(b_1^2+b_2^2)/2+c^2/2 \bmod 2\mathbf{Z}$ for $\overline{\alpha}=(a_1,a_2,a_3,b_1,b_2,c)\in \mathbf{Z}/4^{\oplus 3}\oplus\mathbf{Z}/2^{\oplus 3}$. Since $l((S^*/S)_2)=6$, $l(I_2)\geq 2$. An isotropic element in $(S^*/S)_2$ is one of the following:

$$\begin{array}{ll}
\overline{\alpha}_1=(\{2,2,0\},0,0,0), & \overline{\alpha}_2=(0,0,0,\{1,0\},1),\\
\overline{\alpha}_3=(\{2,0,0\},1,1,0), & \overline{\alpha}_4=(\{\pm1,\pm1,0\},0,0,1),\\
\overline{\alpha}_5=(\{\pm1,\pm1,2\},\{1,0\},0), & \overline{\alpha}_6=(\{\pm1,\pm1,2\},1,1,1),\\
\overline{\beta}_1=(2,2,2,1,1,0), & \overline{\beta}_2=(\{2,2,0\},\{1,0\},1).
\end{array}$$

Recall that $\{x_1,x_2,\ldots,x_k\}$ stands for $x_{\sigma(1)},x_{\sigma(2)},\cdots,x_{\sigma(k)}$ for some permutation σ. If $\overline{\alpha}_1\in I_2$ (resp. $\overline{\alpha}_2\in I_2$, $\overline{\alpha}_3\in I_2$, $\overline{\alpha}_4\in I_2$), then the primitive hull in Λ_2/P of the component $Q(2A_3)$ (resp. $Q(A_1)$, $Q(A_3)$, $Q(2A_3+A_1)$) of $Q(G)$ contains a root system of type D_6 (resp. BC_1, B_3, E_7). It contradicts fullness. If $\overline{\alpha}_5\in I_2$, then $2\overline{\alpha}_5=\overline{\alpha}_1\in I_2$. If $\overline{\alpha}_6\in I_2$, then $2\overline{\alpha}_6=\overline{\alpha}_1\in I_2$. We get a contradiction in both cases. Thus, every non-zero element in I_2 has the form of $\overline{\beta}_1$ or $\overline{\beta}_2$. Since I_2 is not cyclic it contains an element in the form $(*,*,*,*,*,0)$; thus, $\overline{\beta}_1\in I_2$. Also, $\overline{\beta}_2\in I_2$ since I_2 is not cyclic. Thus, $\overline{\beta}_1+\overline{\beta}_2=(\{0,0,2\},\{0,1\},1)\in I_2$, which is not isotropic. We get a contradiction.

We proceed to the next case. Assume that we have a full embedding $Q=Q(A_5+B_3+2A_1)\hookrightarrow\Lambda_2/P$. Since $F\cap Q(A_5+B_3+2A_1)=Q(A_5+A_3+2A_1)$, one has the induced embedding $S=P\oplus Q(A_5+A_3+2A_1)\hookrightarrow\Lambda_2$. Note that because Q contains a component B_3, a certain element in S^*/S belongs $\widetilde{S}/S$. Noting this point, by the same method as above we can deduce a contradiction. Q.E.D.

Lemma 24.5. (MI10) *There is no full embedding $Q(A_4+B_3+A_2+A_1)\hookrightarrow\Lambda_2/P$.*

Proof. Assuming the contrary, we will deduce a contradiction. Consider the induced embedding $S=P\oplus Q(A_4+A_3+A_2+A_1)\hookrightarrow\Lambda_2$. By T we denote the orthogonal complement of S. T is a lattice with signature (1, 1). Let m be the order of $I=\widetilde{S}/S$. $|d(T)|=|d(S)|/m^2=3\times 2^5\times 5^2/m^2$. Thus, $d(T)=-3\times 2^5\times 5^2/m^2$.

On the other hand, $(S^*/S)_5\cong\mathbf{Z}/5\oplus\mathbf{Z}/5$. The first component of the right-hand side is induced by P and the second by the A_4-component. The discriminant quadratic form on $(S^*/S)_5$ can be given by $q_5(\overline{\alpha})\equiv 2x^2/5+4y^2/5 \bmod 2\mathbf{Z}$ for $\overline{\alpha}=(x,y)\in\mathbf{Z}/5\oplus\mathbf{Z}/5$. In particular, if $\overline{\alpha}\neq 0$, then $q_5(\overline{\alpha})\not\equiv 0$. Thus, $I_5=0$ and $5\not| m$. One knows $(S^*/S)_5\cong(\widetilde{S}^*/\widetilde{S})_5\cong(T\otimes\mathbf{Z}_5)^*/(T\otimes\mathbf{Z}_5)$. The discriminant quadratic form of $T\otimes\mathbf{Z}_5$ coincides with $-q_5$. By Lemma 12.2 $T\otimes\mathbf{Z}_5$ is equivalent to the 5-adic lattice defined by the diagonal 2×2 matrix whose diagonal entries are -2×5 and $-2^2\times 5$. Thus, $d(T)\equiv 2^3\times 5^2 \bmod \mathbf{Z}_5^{*2}$.

By the above 2 expressions of $d(T)$ there is an element $\theta\in\mathbf{Z}_5^*$ with $-3=\theta^2$. It implies that $-3\equiv x^2 \pmod 5$ for some integer x, a contradiction. Q.E.D.

Lemma 24.6. (MI9) *There is no full embedding $Q(3A_2+B_2+A_1)\hookrightarrow\Lambda_2/P$.*

Proof. Assume the contrary. We will deduce a contradiction. Consider the induced embedding $S=P\oplus Q(3A_3+3A_1)\hookrightarrow\Lambda_2$. Let $I=\widetilde{S}/S$ and m be the order of I. Let T denote the orthogonal complement of S. T is a lattice with signature (2, 1). We have $d(T)=-5\times 2^5\times 3^3/m^2$. Now, $(S^*/S)_3\cong\mathbf{Z}/3\oplus\mathbf{Z}/3\oplus\mathbf{Z}/3$. Each component of the right-hand side is induced by an A_2-component. The discriminant quadratic form on $(S^*/S)_3$ can be written $q_3(\overline{\alpha})\equiv 2(a_1^2+a_2^2+a_3^2)/3 \bmod 2\mathbf{Z}$.

Assume $I_3\neq 0$. Let $\overline{\alpha}\in I_3$ be a non-zero element. Since $q_3(\overline{\alpha})\equiv 0$, $\overline{\alpha}=(\pm1,\pm1,\pm1)$. This implies that the primitive hull of the component $Q(3A_2)$ of Q

contains a root system of type $\widetilde{E}_6$. It contradicts fullness. One can conclude $I_3 = 0$ and $3 \not| m$. Hence $(S^*/S)_3 \cong (\widetilde{S}^*/\widetilde{S})_3 \cong (T \otimes \mathbf{Z}_3)^*/(T \otimes \mathbf{Z}_3)$. The discriminant quadratic form of $T \otimes \mathbf{Z}_3$ is $-q_3$. By Lemma 12.2 $T \otimes \mathbf{Z}_3$ is isomorphic to the 3-adic lattice defined by the diagonal 3×3 matrix whose diagonal entries are all -2×3. Thus, $d(T) \equiv -2^3 \times 3^3 \bmod \mathbf{Z}_3^{*2}$. Comparing 2 expressions of $d(T)$, one has $5 \in \mathbf{Z}_3^{*2}$. It implies that $5 \equiv x^2 \pmod 3$ for some integer x, a contradiction. Q.E.D.

Lemma 24.7. (MI9, MI8) *For $G = A_4 + A_3 + B_2$, $2A_3 + B_3$, or $D_4 + A_2 + B_2$ there is no full embedding $Q(G) \hookrightarrow \Lambda_2/P$.*

Proof. We will deduce a contradiction, assuming the contrary. We consider $G = A_4 + A_3 + B_2$ first. Let $S = P \oplus Q(A_4 + A_3 + 2A_1) \hookrightarrow \Lambda_2$ be the induced embedding. $T = C(S, \Lambda_2)$ is a lattice with signature (2, 1). Set $I = \widetilde{S}/S$ and let m be the order of I. $(S^*/S)_2 \cong \mathbf{Z}/4 \oplus \mathbf{Z}/2 \oplus \mathbf{Z}/2 \oplus \mathbf{Z}/2 \oplus \mathbf{Z}/2$. The first component of the right-hand side corresponds to A_3, the second and the third to P, and the fourth and the fifth to $2A_1$ (B_2). The discriminant quadratic form on it can be written $q_2(\overline{\alpha}) \equiv 3a^2/4 - (b_1^2 + b_2^2)/2 + (c_1^2 + c_2^2)/2 \bmod 2\mathbf{Z}$ for $\overline{\alpha} = (a, b_1, b_2, c_1, c_2) \in \mathbf{Z}/4 \oplus (\mathbf{Z}/2)^{\oplus 4}$. $3 = \operatorname{rank} T \geq l((\widetilde{S}/S)_2) \geq l((S^*/S)_2) - 2l(I_2) = 5 - 2(I_2)$. Thus, $I_2 \neq 0$. Let $\overline{\alpha} \in I_2$ be a non-zero element. Since $q(\overline{\alpha}) \equiv 0$, $\overline{\alpha}$ coincides with one of the following:

$$\overline{\alpha}_1 = (0, \{1,0\}, \{1,0\}), \qquad \overline{\alpha}_2 = (2,1,1,0,0),$$
$$\overline{\alpha}_3 = (2,0,0,1,1), \qquad \overline{\epsilon} = (0,1,1,1,1).$$

If $\overline{\alpha} = \overline{\alpha}_1$ (resp. $\overline{\alpha} = \overline{\alpha}_2$, $\overline{\alpha} = \overline{\alpha}_3$), then the primitive hull of the B_2-component (resp. A_3-component, (A_3+B_2)-component) of $Q(A_4+A_3+B_2)$ contains a root system of type BC_2 (resp. B_3, B_5), which contradicts fullness. Thus, $\overline{\alpha} = \overline{\epsilon}$. I_2 is the group of order 2 generated by $\overline{\epsilon}$. In particular, $m = 2n$ for some odd integer n, and $d(T) = -5^2 \times 2^4/n^2$. On the other hand, $I_2^{\perp}/I_2$ is the orthogonal direct sum of the group of order 4 generated by $(1,0,0,0)$ mod I_2 and the group of type $(2,2)$ generated by $(0,1,0,1,0)$ mod I_2 and $(0,1,0,0,1)$ mod I_2. Thus, $(\widetilde{S}^*/\widetilde{S})_2 \cong I_2^{\perp}/I_2 \cong \mathbf{Z}/4 \oplus (\mathbf{Z}/2 + \mathbf{Z}/2)$. The discriminant form is given by $q_2'(a, b_1, b_2) \equiv 3a^2/4 + b_1 b_2 \bmod 2\mathbf{Z}$ for an element (a, b_1, b_2) in the right-hand side. The discriminant quadratic form of $T \otimes \mathbf{Z}_2$ is $-q_2'$. By Lemma 12.2 $T \otimes \mathbf{Z}_2$ is isomorphic to the 2-adic lattice defined by the matrix $A = \begin{pmatrix} 0 & 2 & 0 \\ 2 & 0 & 0 \\ 0 & 0 & -3 \times 2^2 \end{pmatrix}$. It implies $d(T) \equiv \det A = 3 \times 2^4 \bmod \mathbf{Z}_2^{*2}$. Comparing 2 expressions of $d(T)$, one knows $-3 \in \mathbf{Z}_2^{*2}$, which is equivalent to $-3 \equiv 1 \pmod 8$, a contradiction.

For the remaining 2 cases $2A_3 + B_3$, $D_4 + A_2 + B_2$ the argument is similar. Q.E.D.

Lemma 24.8. (MI8) *For $G = D_4 + A_2 + 2A_1$, $A_4 + 4A_1$ or $2A_2 + 4A_1$ there is no full embedding $Q(G) \hookrightarrow \Lambda_2/P$.*

Proof. Assume the contrary. In the first case $G = D_4 + A_2 + 2A_1$. We consider the induced embedding $S = P \oplus Q(G) \hookrightarrow \Lambda_2$. $T = C(S, \Lambda_2)$ is a lattice with signature (3, 1). Let $I = \widetilde{S}/S$ and m be the order of I. $d(T) = -5 \times 3 \times 2^6/m^2$.

On the other hand, $(S^*/S)_2 \cong (\mathbf{Z}/2)^{\oplus 4} \oplus (\mathbf{Z}/2 + \mathbf{Z}/2)$. The first component of the right-hand side and the second are induced by P, the third and the fourth correspond to 2 A_1-components, and the fifth and the sixth correspond to the D_4-component. The discriminant quadratic form on it is given by $q(\overline{\alpha}) \equiv -(a_1^2+a_2^2)/2+(b_1^2+b_2^2)/2+c_1^2+c_1c_2+c_2^2 \bmod 2\mathbf{Z}$ for $\overline{\alpha} = (a_1, a_2, b_1, b_2, c_1, c_2)$. Here, $4 = \text{rank}\, T \geq l((S^*/S)_2) - 2l(I_2) = 6-2l(I_2)$. Thus, $I_2 \neq 0$. A non-zero element $\overline{\alpha}$ on I_2 is an isotropic element in $(S^*/S)_2$ and coincides one of the following:

$$\overline{\alpha}_1 = (0,0,0,1,1,\langle 1,0\rangle), \qquad \overline{\alpha}_2 = (\{1,0\},\{1,0\},0,0),$$
$$\overline{\alpha}_3 = (1,1,1,1,0,0), \qquad \overline{\alpha}_4 = (1,1,1,1,0,0).$$

Here, $\langle 1,0\rangle$ stands for one of (1, 0), (0, 1) or (1, 1). If $\overline{\alpha} = \overline{\alpha}_1$ (resp. $\overline{\alpha} = \overline{\alpha}_2$, $\overline{\alpha} = \overline{\alpha}_3$, $\overline{\alpha} = \overline{\alpha}_4$), then the primitive hull in Λ_2/P of the D_4-component (resp. A_1-, (D_4+2A_1)-, $2A_1$-component) of $Q(G)$ contains a root system of type B_4 (resp. BC_1, D_6, B_2), which contradicts fullness.

For the other 2 cases $A_4 + 4A_1$, $2A_2 + 4A_1$ the argument is similar. Q.E.D.

Remark. The argument in the case $D_4 + A_2 + B_2$ and that in the case $D_4 + A_2 + 2A_1$ have common parts. In the both cases we consider an embedding $P \oplus Q(D_4 + A_2 + 2A_1) \hookrightarrow \Lambda_2$. The argument for $D_4 + A_2 + 2A_1$ is slightly longer.

Corollary 24.9. *Every graph G belonging to $MI10$, $MI9$ or $MI8$ has no full embedding $Q(G) \hookrightarrow \Lambda_2/P$.*

Lemma 24.10. (NI11, NI10, NI9)
(1) Every graph belonging to NI9 has a subgraph belonging to MI8.
(2) Every graph belonging to NI10 has a subgraph belonging to MI9 or NI9.
(3) Every graph belonging to NI11 has a subgraph belonging to MI10 or NI10.

Corollary 24.11. *For every graph G belonging to NI9, NI10, NI11 there is no full embedding $Q(G) \hookrightarrow \Lambda_2/P$.*

Proposition 24.12. (E9, E8) *Every graph G belonging to E8 or E9 has no full embedding $Q(G) \hookrightarrow \Lambda_2/P$.*

Proof. G does not satisfy the condition (3) in Section 23. Q.E.D.

By the above Proposition 23.1 has been established. The following theorem follows from Proposition 23.1, Lemma 16.3 and Lemma 21.10:

Theorem 24.13. *We consider the Ik-conditions in the case $S_{1,0}$. (See Section 7.) By r we denote the number of vertices in the Dynkin graph G. $Q = Q(G)$ stands for the root (quasi-)lattice of type G. $r = \text{rank}\, Q$.*

[I] The following two conditions (a) and (b) are equivalent:
(a) $G \in PC(S_{1,0})$ and the I2-condition holds for the root lattice $Q = Q(G)$ of type G.
(b) G contains no vertex corresponding to a short root and can be obtained from one of the 5 basic Dynkin graphs in Theorem 0.6 by elementary transformations repeated twice.

[II] The following two conditions (A) and (B) are also equivalent:
(A) $G \in PC(S_{1,0})$ and the I1-condition holds for the root lattice $Q = Q(G)$ of type G.

(B) G contains no vertex corresponding to a short root and either the following (B-1) or (B-2) holds:

(B-1) G can be obtained from one of the 5 sub-basic Dynkin graphs in Theorem 0.6 by one elementary transformation.

(B-2) G can be obtained from one of the 5 basic Dynkin graphs in Theorem 0.6 by one of the following 3 kinds of procedures:

The procedures:

⟨1⟩ elementary transformations repeated twice

⟨2⟩ an elementary transformation following after a tie transformation

⟨3⟩ a tie transformation following after an elementary transformation.

25
Two singular fibers of type I^* in the case $S_{1,0}$

In the case $S_{1,0}$ to show the following is remaining:

Proposition 25.1. *Assume that $G \in PC(S_{1,0})$ and that $Q = Q(G)$ does not satisfy the I1-condition. Then, with respect to some full embedding $Q(G) \hookrightarrow \Lambda_3/P$ without an obstruction component A_9, there exists a primitive isotropic vector u in Λ_3/P in a nice position, i.e., such that either u is orthogonal to $Q(G)$, or there is a root basis $\Delta \subset Q(G)$ and a long root $\theta \in \Delta$ such that $\alpha \cdot u = 0$ for every $\alpha \in \Delta$ with $\alpha \neq \theta$ and $\theta \cdot u = 1$.*

We can apply the same method developed in Section 17 through 19 for $W_{1,0}$.

Now, if $G \in PC = PC(S_{1,0})$, then we have an elliptic K3 surface $\Phi : Z \to C$ corresponding to a full embedding $Q(G) \hookrightarrow \Lambda_3/P$. By $\Sigma = \{c_1, \ldots, c_t\}$ we denote the set of critical values of Φ. By F_i with $1 \leq i \leq t$ we denote the singular fiber $\Phi^{-1}(c_i)$. The elliptic surface $Z \to C$ has a singular fiber F_1 of type I_0^* in our situation by definition. Moreover, it has two sections $s_0, s_1 : C \to Z$ whose images $C_5 = s_0(C)$ and $C_7 = s_1(C)$ are disjoint. Exchanging the numbering, we can assume the singular fiber F_2 has a component C_6 intersecting C_5 but disjoint from C_7. The union $IF = F_1 \cup C_5 \cup C_6 \cup C_7$ is called the curve at infinity. The lattice P has signature $(7,1)$ and it is defined associated with the dual graph of components of IF. The union $\mathcal{E}$ of smooth rational curves on Z disjoint from IF coincides with the union of components disjoint from IF of singular fibers of $Z \to C$. The dual graph of $\mathcal{E}$ is G by definition.

Recall that $m(F_i)$ denotes the number of irreducible components of F_i. By $n(F_i)$ we denote the number of components disjoint from IF. We have $\sum_{i=1}^{t} n(F_i) = r =$ the number of vertices in G.

Lemma 25.2. *(1) $m(F_1) = 5$, $n(F_1) = 0$.*
(2) $n(F_2) \leq m(F_2) - 2$. If $n(F_2) = m(F_2) - 2$, then F_2 is of type I_2 or III.
(3) For $3 \leq i \leq t$ $n(F_i) \leq m(F_i) - 1$.

Proof. (1) and (3) are trivial. We consider (2). Since C_6 is a component of F_2 disjoint from C_7, we have the inequality. Assume the equality holds. Let D be a component of F_2 intersecting C_6 with $D \neq C_6$. Under the assumption D coincides with the component of F_2 intersecting C_7. This implies that F_2 is never of type I^*, II^*, III^* or IV^*. Obviously, F_2 is not of type I_0, II, or I_1. If F_2 is of type IV or of type I_k with $k \geq 3$, then F_2 has 2 components intersecting C_6 apart from C_6. Thus, in this case, the equality does not hold. Q.E.D.

Recall that by Theorem 2.2 we can assume moreover that the Picard number ρ of Z satisfies $\rho = r + 8$. By a we denote the rank of the Mordell-Weil group MW of Φ.

Proposition 25.3. *Assume $\rho = r + 8$. Exchanging the numbers of $F_3, \ldots, F_t$ if necessary, one of the following (A), (B), (C) holds:*

(A) $a = 1$. F_2 is either of type I_2 or of type III. C_5 and C_7 hit the same component of F_i for $3 \leq i \leq t$.

(B) $a = 0$. F_2 is either of type I_2 or of type III. The type of F_3 is either I_{2b} with $b \geq 1$, I_b^* with $b \geq 0$, III, or III^*. For $i \geq 4$ F_i is of type I_{b_i} with $b_i \geq 1$. C_5 and C_7 hit different components of F_3. For $i \geq 4$ C_5 and C_7 hit the same component of F_i. The dual graph of components of F_3 disjoint from C_5 and C_7 is one in the following table, depending on the type of F_3:

the type of F_3	*the dual graph*
I_{2b} $(b \geq 1)$	$2A_{b-1}$
I_b^* (b even)	D_{b+3} or A_{b+3}
I_b^* (b odd)	D_{b+3}
III	empty graph
III^*	E_6

(C) $a = 0$. F_2 is of type I_b^* or of type III^*. For $i \geq 3$ C_5 and C_7 hit the same component of F_i and F_i is of type I_{b_i} with $b_i \geq 1$. The dual graph of the components of F_2 disjoint from C_5 and C_7 are as in the following table:

the type of F_2	*the dual graph*
I_b^* (b even)	D_{b+2} or $A_{b+1} + A_1$
I_b^* (b odd)	D_{b+2}
III^*	D_5

Proof. Recall the equality (2) in Section 10. By this equality and by Lemma 25.2 (1) we have

$$1 - a = \{m(F_2) - 2 - n(F_2)\} + \sum_{i=3}^{t} \{m(F_i) - 1 - n(F_i)\}.$$

By Lemma 25.2 (2) and (3) the right-hand side is non-negative. Thus, $a = 0$ or 1. If $a = 1$, then the equalities in Lemma 25.2 (2), (3) hold.

Consider the case $a = 0$. We give the group structure with the unit element C_5 to the set MW of sections of Φ. The group homomorphism $\mathrm{Tor}\, MW = MW \to F_i^{\#}$ is injective for every i. Since the intersection point of C_7 and F_1 has order 2 in $F_1^{\#}$, $C_7 \in MW$ has order 2. The intersection point of C_7 and F_i is of order 2 in $F_i^{\#}$ for every i. Since $F_i^{\#}$ has a point of order 2, the type of F_i $(2 \leq i \leq t)$ is either I_b, I_b^*, III, or III^*.

In the case where $m(F_2) = n(F_2)+2$, exchanging the numbers, we can assume that $m(F_3) = n(F_3) + 2$, and $m(F_i) = n(F_i) + 1$ for $4 \le i \le t$. The claim on F_2 follows from Lemma 25.2 (2). The intersection point of C_7 and F_3 has order 2 in $F_3^{\#}$ and it belongs to a different component from the one containing the intersection point of C_5 and F_3. Thus, F_3 is not of type I_b with b odd. For $i \ge 4$ a point of order 2 which is the intersection of C_7 and F_i and the unit element lie on the same component; thus F_i is of type I.

In the case $m(F_2) \ne n(F_2) + 2$, $m(F_2) = n(F_2) + 3$ and $m(F_i) = n(F_i) + 1$ for $3 \le i \le t$. We consider F_2. Since $m(F_2) \ge 3$, F_2 is not of type III. Assume F_2 is of type I_b. The component of F_2 intersecting C_5 is C_6, and C_6 and C_7 are disjoint. The intersection point of C_7 and F_2 is a point of order 2 on $F_2^{\#}$ not belonging to C_6. Thus, b has to be even. In particular, $b \ge 4$. Two components of F_2 different from C_6 intersect C_6 and they are different from the component intersecting C_7. It implies $m(F_2) \ge n(F_2) + 4$, which is a contradiction. For $i \ge 3$ by the same reason as above F_i is of type I.

The restriction on the dual graph follows from considering the position of points of order 2 on the singular fiber. Q.E.D.

We consider the same division of the case into three subcases as in Section 17.

((1)) The surface $Z \to C$ has another singular fiber of type I^* apart from F_1.

((2)) $Z \to C$ has a singular fiber of type II^*, III^* or IV^*.

((3)) $Z \to C$ has no singular fiber of type I^*, II^*, III^* or IV^* apart from F_1.

For each case we apply the same method as in Section 17 through 19, because in the both cases $W_{1,0}$ and $S_{1,0}$ IF contains 2 sections.

The case ((i)) is discussed in Section 24+i. In particular, this section is devoted to ((1)).

In the following by G we denote a Dynkin graph belonging to $PC(W_{1,0})$ with the number of vertices r. We assume the equality $\rho = r + 8$ and that the I1-condition fails for $Q = Q(G)$. By Theorem 2.2 we do not lose any generality, if we assume $\rho = r + 8$.

We assume in this section that the singular fiber F_1 is of type I_0^*, F_2 contains a component C_6 of IF, and F_3 is a singular fiber of $\Phi : Z \to C$ of type I^*, if F_2 is not of type I^*.

We have an embedding $S = P \oplus Q(G) \hookrightarrow \Lambda_3$ satisfying (L1) and (L2). Recall that the lattice P has a basis $e_0, e_1, \ldots, e_7$ whose mutual intersection numbers are described by the dual graph in Section 4 "The case of $S_{1,0}(2, 2, 3, 3)$".

We set

$$u_0 = 2e_0 + e_1 + e_2 + e_3 + e_4, \quad v_0 = -u_0 - e_5, \quad e' = u_0 - e_6 \quad f = e_7 - e_5 - 2u_0.$$

$u_0^2 = v_0^2 = 0$, $u_0 \cdot v_0 = 1$, $P = P' \oplus (\mathbf{Z}u_0 + \mathbf{Z}v_0)$, and P' has a basis $e_0, e_1, e_2, e_3, f, e'$.

By Proposition 10.3 in Chapter 2 there exists a vector $u \in \Lambda_3$ satisfying the following conditions:

(1) u is isotropic.

(2) u is orthogonal to $Q(G)$.

(3) u is orthogonal to e_0, e_1, e_2, e_3, e_4, e_5 and e_6.

(4) $u \cdot e_7 > 0$.

Set $m = u \cdot e_7 = u \cdot f$. (Note that $m \neq 0$ since the I1-condition fails.)

The vector u is orthogonal to $\mathbf{Z}u_0 + \mathbf{Z}v_0$. Set $M = P' + \mathbf{Z}u$. M has signature $(6,1)$.

Lemma 25.4. *Assume that M is not primitive in Λ_3. Then, the primitive hull $\widetilde{M}$ of M in Λ_3 contains an element u' also satisfying the above conditions (1)–(4) and such that $u \cdot e_7 > u' \cdot e_7 > 0$.*

Proof. By $w_0, \ldots, w_3, z, w_4$ we denote the dual basis of $e_0, \ldots, e_3, f, e'$. Note $u \cdot e_i = 0$ $(0 \leq i \leq 3)$, $u \cdot f = m > 0$, and $u \cdot e' = 0$. Set $\xi = m(2e_0 + e_1 + e_2 + 2e_3 + 2f + e') - 5u$. We can check $\xi \in M$, $\xi \cdot e_i = 0$ $(0 \leq i \leq 3)$, $\xi \cdot f = 0$ and $\xi \cdot e' = 0$. Moreover, $\xi \cdot u = 2m^2$. $\xi^2 = -5\xi \cdot u = -10m^2$. The vector $y_0 = \xi/10m^2$ satisfies $y_0 \cdot \xi = -1$ and $y_0^2 = -1/10m^2$. We have $u = m(2e_0 + e_1 + e_2 + 2e_3 + 2f + e')/5 - \xi/5 = mz - 2m^2 y_0 = m(z - 2my_0)$.

Next, set $N = P' \oplus \mathbf{Z}\xi$. $N \subset M \subset \widetilde{M} \subset \widetilde{M}^* \subset M^* \subset N^*$. We consider the discriminant group $N^*/N = P'^*/P' \oplus (\mathbf{Z}y_0/\mathbf{Z}\xi)$. For $x \in N^*$ we denote $\overline{x} = x \bmod N \in N^*/N$. Set $I = M/N$. I is the cyclic group of order 5 generated by $\overline{u} = m\overline{z} - 2m^2\overline{y}_0$. $I^\perp = M^*/N$.

Here, we would like to show the following claim first:

Claim. If $M \neq \widetilde{M}$, then there is an element $\overline{x}_1 \in \widetilde{M}/N$ with $\overline{x}_1 \notin I$ satisfying either the following $\langle 1 \rangle$ or $\langle 2 \rangle$:

$\langle 1 \rangle$ $\overline{x}_1 = A(\overline{z} - 2m\overline{y}_0)$ for some integer A.

$\langle 2 \rangle$ $m \equiv 0 \pmod 5$ and $\overline{x}_1 = A(\overline{z} + 2m\overline{y}_0)$ for some integer A.

Some preparation is necessary. P'^*/P' is the orthogonal direct sum of the following 3 cyclic subgroups; the cyclic group of order 5 generated by $\overline{z}$, the cyclic group of order 2 generated by $\overline{g}_1 = 5\overline{w}_1$, the cyclic group of order 2 generated by $\overline{g}_2 = 5\overline{w}_2$. Let b_N be the discriminant bilinear form of N. $b_N(m\overline{z} - 2m^2\overline{y}_0, \overline{x}) \equiv 2ma/5 + c/5 \bmod \mathbf{Z}$ for $\overline{x} = a\overline{z} + b_1\overline{g}_1 + b_2\overline{g}_2 + c\overline{y}_0 \in N^*/N$. Thus, $\overline{x} \in I^\perp \Leftrightarrow 2ma + c \equiv 0 \bmod 5$.

Now, assume $M \neq \widetilde{M}$. We can pick an element $\overline{x}_0 \in \widetilde{M}/N$ with $\overline{x}_0 \notin I = M/N$. Since $\overline{x}_0 \in I^\perp$, we can write $\overline{x}_0 = a(\overline{z} - 2m\overline{y}_0) + b_1\overline{g}_1 + b_2\overline{g}_2 + 5c\overline{y}_0$.

Case 1. $5c \not\equiv 0 \pmod m$.

Set $\overline{x}_1 = 2m\overline{x}_0$. $\overline{x}_1 \in \widetilde{M}/N$. $\overline{x}_1 = 2ma(\overline{z} - 2m\overline{y}_0) + c(10m\overline{y}_0) = (2ma - 5c)(\overline{z} - 2m\overline{y}_0)$. Here, since $\overline{z}$ has order 5, $10m\overline{y}_0 = -5(\overline{z} - 2m\overline{y}_0)$. One knows $\overline{x}_1 \notin I$, since $\overline{z} - 2m\overline{y}_0$ has order $5m$, I is a cyclic group generated by $m(\overline{z} - 2m\overline{y}_0)$ and $2ma - 5c \not\equiv 0 \pmod m$ by assumption.

Case 2. $5c \equiv 0 \pmod m$.

Set $5c = dm$. $\overline{x}_0 = a\overline{z} + b_1\overline{g}_2 + b_2\overline{g}_2 + (-2a + d)m\overline{y}_0$. $0 \equiv q(\overline{x}_0) \equiv -(b_1^2 + b_2^2)/2 + (4ad - d^2)/10 \bmod 2\mathbf{Z} \iff 5(b_1^2 + b_2^2) - 4ad + d_2 \equiv 0 \pmod{20}$. In particular, $b_1^2 + b_2^2 + d^2 \equiv 0 \pmod 4$. Hence b_1, b_2 and d are all even. $b_1\overline{g}_1 + b_2\overline{g}_2 = 0$. We write $d = 2e$. $q(\overline{x}_0) \equiv 0 \iff e(e - 2a) \equiv 0 \pmod 5$.

When $e \equiv 0 \pmod 5$, setting $e = 5f$, we have $\overline{x}_0 = a(\overline{z} - 2m\overline{y}_0) + f(10m\overline{y}_0) = (a - 5f)(\overline{z} - 2m\overline{y}_0)$. Setting $\overline{x}_1 = \overline{x}_0$, $\overline{x}_1 \in \widetilde{M}/N$ and $\overline{x}_1 \notin I$. The condition $\langle 1 \rangle$ is satisfied.

If $e \not\equiv 0 \pmod 5$, we can write $e - 2a = 5f$. $\overline{x}_0 = a(\overline{z} - 2m\overline{y}_0) + 2(2a + 5f)m\overline{y}_0 = a(\overline{z} + 2m\overline{y}_0) + 10fm\overline{y}_0 = (a + 5f)(\overline{z} + 2m\overline{y}_0)$. Note here that $10m\overline{y}_0 = 5(\overline{z} + 2m\overline{y}_0)$. If $m \equiv 0 \pmod 5$, then $\overline{x}_1 = \overline{x}_0$ is the desired element satisfying $\langle 2 \rangle$. If $m \not\equiv 0 \pmod 5$, then $2m(a+5f) + 2m(a+5f) \equiv 4m(a+5f) \equiv 0 \pmod 5$ since $\overline{x}_0 \in I$. Thus, $a + 5f \equiv 0$

(mod 5). Since $\overline{z}$ has order 5, $\overline{x}_0 = (a+5f)(-\overline{z}+2m\overline{y}_0) = (-a-5f)(\overline{z}-2m\overline{y}_0)$. $\overline{x}_1 = \overline{x}_0$ is the desired element satisfying $\langle 1\rangle$. The above claim has been shown.

We continue the argument assuming the existence of $\overline{x}_1 \in \widetilde{M}/N$ with $\overline{x}_1 \notin I$ satisfying $\langle 1\rangle$. Set $A = Cm + B$ with $0 \leq B < m$. If $B = 0$, $\overline{x}_1 \in I$. Thus, $B \neq 0$. $\overline{x}_2 = \overline{x}_1 - Cm(\overline{z} - 2m\overline{y}_0) = B(\overline{z} - 2m\overline{y}_0)$ satisfies $\overline{x}_2 \in \widetilde{M}/N$ and $\overline{x}_2 \notin I$. Consider the vector $u' = Bu/m$ here. $u' \neq 0$. $u'^2 = 0$. $u' \cdot e_i = Bu \cdot e_i/m = 0$ $(0 \leq i \leq 3)$. $u' \cdot e' = 0$. $u' \cdot f = Bu \cdot f/m = B$. $u' \cdot \xi = 2mu' \cdot f = 2mB$. Thus, $u' \in N^*$. $m = u \cdot f > B = u' \cdot f$. Since $\overline{u}' = \overline{x}_2$, $u' \in \widetilde{M}$. This u' satisfies the desired condition.

The case where we have an element $\overline{x}_1 \in \widetilde{M}/N$ with $\overline{x}_1 \notin I$ satisfying $\langle 2\rangle$ is remaining. In this case, if A is a multiple of m, $\overline{x}_1 = A(\overline{z} + 2m\overline{y}_0) = A(-\overline{z} - 2m\overline{y}_0) = (-A)(\overline{z} - 2m\overline{y}_0) \in I$, a contradiction. We here used $5\overline{z} = -5\overline{x}$. Thus, we can write $A = Cm + B$ with $0 < B < m$. Set $\overline{x}_2 = B(\overline{z} + 2m\overline{y}_0) = \overline{x}_1 - Cm(\overline{z} + 2m\overline{y}_0) = \overline{x}_1 + Cm(\overline{z} - 2m\overline{y}_0)$. $\overline{x}_2 \in \widetilde{M}/N$ and $\overline{x}_2 \notin I$. Next, we consider the vector $u'' = B(z + 2my_0) = B(2z - u/m)$. $u''^2 = B^2(z^2 + 4m^2y_0^2) = 0$. $u'' \neq 0$. $u'' \cdot u = Bm(z^2 - 4m^2y_0^2) = Bm/5 \in \mathbf{Z}$. (Note $m \equiv 0 \pmod 5$.) $u'' \cdot e_i = B(z \cdot e_i + 2my_0 \cdot e_i) = 0$ $(0 \leq i \leq 6)$. $u'' \cdot e' = B(z \cdot e' + 2my_0 \cdot e') = 0$. $u'' \cdot u_0 = u'' \cdot v_0 = 0$. $u'' \cdot e_7 = u'' \cdot f = B(z \cdot f + 2my_0 \cdot f) = B$. $u'' \cdot \xi = 2mu'' \cdot f - 5u'' \cdot u = Bm$. Thus, $u'' \in N^*$. Since $\overline{u}'' = \overline{x}_2$, $u'' \in \widetilde{M}$. This u'' is the desired vector. Q.E.D.

By induction on $u \cdot e_7$ we can assume that u satisfies the following (5) in addition to (1)–(4):

(5) $M = P' + \mathbf{Z}u$ is primitive in Λ_3.

Then, of course, $M \oplus (\mathbf{Z}u_0 + \mathbf{Z}v_0) = P + \mathbf{Z}u$ is also primitive in Λ_3.

Proposition 25.5. *Assume that we have a vector $u \in \Lambda_3$ satisfying above (1)–(4) and (5).*

1. *If $u \cdot e_7 = 1$, then G is a subgraph of the Coxeter-Vinberg graph Γ of the lattice $Q(D_{12}) \oplus \mathbf{Z}\lambda$. ($\lambda^2 = -2$.)*
2. *If $u \cdot e_7 \geq 2$, then G can be obtained from a subgraph of Γ by one elementary transformation.*

Proof. **1.** Setting $v = f - 2u$, one has

$$M = (\mathbf{Z}e_0 + \mathbf{Z}e_1 + \mathbf{Z}e_2 + \mathbf{Z}(e_3 + u)) \oplus \mathbf{Z}(e' + u) \oplus (\mathbf{Z}u + \mathbf{Z}v),$$

and $u^2 = v^2 = 0$, $u \cdot v = 1$. Since $u \cdot e_i = u \cdot e' = 0$ $(0 \leq i \leq 3)$, $P + \mathbf{Z}u \cong Q(D_4 + A_1) \oplus H \oplus H$. For their discriminant quadratic forms we know $q_{P+\mathbf{Z}u} = q_{Q(D_4+A_1)}$.

Let L_1 be the orthogonal complement of $P + \mathbf{Z}u$ in Λ_3. L_1 has signature (12, 1). The discriminant quadratic form of L_1 is $-q_{P+\mathbf{Z}u} = -q_{Q(D_4+A_1)} = q_{Q(D_4)} \oplus (-q_{Q(A_1)})$. Since $Q(D_{12}) \oplus \mathbf{Z}\lambda$ $(\lambda^2 = -2)$ and L_1 have the same signature and the same discriminant quadratic form, they are isomorphic by Lemma 5.5 (4) and Lemma 12.4. In particular, the Coxeter-Vinberg graph of L_1 coincides with Γ. Since $Q(G)$ is full in L_1, G is a subgraph of Γ.

2. Assume that $m = u \cdot e_7 \geq 2$.

Let L be the orthogonal complement of $R = P + \mathbf{Z}u$ in Λ_3. We have a natural isomorphism $R^*/R \cong L^*/L$ preserving the discriminant quadratic forms up to sign.

Set $u_1 = u/m$ and $R_1 = R + \mathbf{Z}u_1$. R_1 is an even overlattice of R with index m, and is isomorphic to $P + \mathbf{Z}u$ in the case of $m = u \cdot e_6 = 1$. In particular, the discriminant quadratic form of R_1 is the same as that of $Q(D_4 + A_1)$.

By the above isomorphism one knows that L has an overlattice L_1 with index m whose discriminant quadratic form is $-q_{Q(D_4+A_1)}$. By reasoning in **1** $L_1 \cong Q(D_{12}) \oplus \mathbf{Z}\lambda$. ($\lambda^2 = -2$.)

Let $\widetilde{Q}_1$ (resp. $\widetilde{Q}$) be the primitive hull of $Q(G)$ in L_1 (resp. L). The Dynkin graph of $\widetilde{Q}_1$ is a subgraph of the Coxeter-Vinberg graph Γ of L_1. Since $\widetilde{Q}_1/\widetilde{Q} \subset L_1/L \cong \mathbf{Z}/m$ is cyclic, the Dynkin graph of $\widetilde{Q}$ is obtained from that of $\widetilde{Q}_1$ by one elementary transformation. Besides, by the fullness the Dynkin graph of $\widetilde{Q}$ is G. Q.E.D.

Remark. $Q(D_{10} + A_1) \oplus H \cong Q(D_{12}) \oplus \mathbf{Z}\lambda$ ($\lambda^2 = -2$).

Now, we would like to draw the Coxeter-Vinberg graph Γ of $Q(D_{12}) \oplus \mathbf{Z}\lambda$ ($\lambda^2 = -2$). To describe this lattice we use tools introduced in the case $Z_{1,0}$. (Section 9.) Now, by $K = \sum_{i=1}^{12} \mathbf{Z}v_i$, $v_i^2 = 1$ $(1 \le i \le 12)$, $v_i \cdot v_j = 0$ $(i \ne j)$, we define the lattice K. $Q(D_{12})$ is identified with a sublattice with index 2 as follows: $Q(D_{12}) = \{x = \sum x_i v_i \in K \mid \sum x_i \text{ is even.}\}$. In Vinberg's algorithm we use λ as the controlling vector. We can take

$$\begin{aligned} \gamma_i &= -v_i + v_{i+1} \quad (1 \le i \le 11) \\ \gamma_{12} &= -(v_{11} + v_{12}) \end{aligned}$$

as a root basis for the orthogonal complement of λ. In the next step of Vinberg's algorithm we get successively;

$$\begin{aligned} \gamma_{13} &= \lambda + 2v_1 \\ \gamma_{14} &= \lambda + v_1 + v_2 + v_3 + v_4 \\ \gamma_{15} &= 2\lambda + v_1 + v_2 + \cdots + v_{10} \\ \gamma_{16} &= 3\lambda + 3v_1 + v_2 + v_3 + \cdots \\ &\qquad + v_{11} - v_{12} \\ \gamma_{17} &= 3\lambda + 3v_1 + v_2 + \cdots + v_{12}. \end{aligned}$$

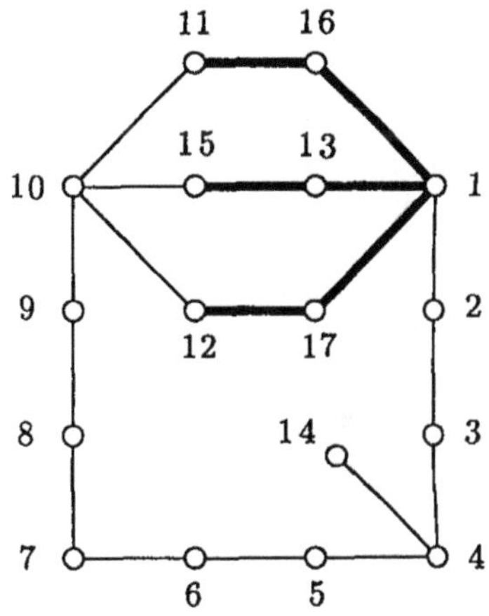

The Coxeter-Vinberg graph Γ for $Q(D_{12}) \oplus \mathbf{Z}\lambda$ ($\lambda^2 = -2$)

Drawing the graph for these 17 vectors, we get the above graph. This has no dotted edges, no Lannér subgraph and any extended Dynkin subgraph is a component of an extended Dynkin subgraph of rank 11. Thus, this is Γ.

Lemma 25.6. *There are 40 kinds of maximal Dynkin subgraphs with 12 vertices in Γ. Following is the list:*

(1) D_{12}	(2) $D_{11} + A_1$	(3) $A_{10} + 2A_1$
(4) $D_{10} + A_2$	(5) $D_{10} + 2A_1$	(6) $A_9 + 3A_1$
(7) $D_9 + 3A_1$	(8) $A_8 + D_4$	(9) $A_8 + 4A_1$
(10) $D_8 + A_3 + A_1$	(11) $D_8 + A_2 + 2A_1$	(12) $D_8 + 4A_1$
(13) $E_8 + D_4$	(14) $E_8 + A_3 + A_1$	(15) $E_8 + A_2 + 2A_1$

(16) E_8+4A_1	(17) $A_7+A_2+3A_1$	(18) D_7+A_5
(19) $D_7+A_4+A_1$	(20) $D_7+D_4+A_1$	(21) E_7+D_5
(22) $E_7+D_4+A_1$	(23) $E_7+A_3+2A_1$	(24) $E_7+A_2+3A_1$
(25) $A_6+A_4+2A_1$	(26) $A_6+D_4+A_2$	(27) $A_6+A_2+4A_1$
(28) $2D_6$	(29) $D_6+D_5+A_1$	(30) $D_6+D_4+A_2$
(31) $D_6+A_3+3A_1$	(32) $E_6+D_5+A_1$	(33) $E_6+A_4+2A_1$
(34) $E_6+A_3+3A_1$	(35) $A_5+D_5+2A_1$	(36) $A_5+A_4+3A_1$
(37) $A_5+D_4+A_3$	(38) $A_5+A_3+4A_1$	(39) $D_5+A_4+3A_1$
(40) $A_4+D_4+A_3+A_1$		

Lemma 25.7. *(1) Under the assumption $\rho=r+8$ only the case (A) in Proposition 25.3 takes place.*
(2) G has a component of type D, has 10, 11, or 12 vertices, and can be obtained from one of the 40 Dynkin graphs with 12 vertices in Lemma 25.6 by one elementary transformation.

Proof. (1) Assume $a=0$. Then, all elements in MW have finite order. In particular, the vector u in Proposition 10.3 is orthogonal also to the class of $C_7=s_1(C)$. Thus, u is orthogonal to the subgroup S generated by the classes in the union of the set of components of IF and the set of components of singular fibers disjoint from IF. Since u is isotropic, this contradicts the assumption that the I1-condition fails. Thus, $a\neq 0$. By Proposition 25.3 we get the conclusion.
(2) We can assume further $\rho=r+8$. Thus, by (1) and by Proposition 25.3 (A) only one component of F_3 of type I^* intersects IF, in particular. Note that the intersecting component has multiplicity 1. The dual graph of components of F_3 without intersection with IF is of type D and it is a component of G.

By Proposition 25.5 and by the theory of elementary transformations G can be made from a maximal Dynkin subgraph G' of Γ by an elementary transformation. By Lemma 8.7 G' has either 12 or 11 vertices. It has never 11 vertices, since the I1-condition fails.

Since the I1-condition fails, $r=10$, 11 or 12. Q.E.D.

For every singular fiber F_i with $3\leq i\leq t$, let G_i be the Dynkin graph defined as the dual graph of components of F_i disjoint from C_5. Note that by Lemma 25.7 (1) $G=\sum_{i=3}^{t}G_i$.

Lemma 25.8. *(1) $\nu(A)+2\nu(D)+2\nu(E)\leq 16-r$, where $\nu(T)$ denotes the number of components of G of type T.*
(2) $\nu(D_4)\leq 12-r$.

Proof. (1) In our situation the case (A) in Proposition 25.3 takes place, and we can substitute $\rho=8+r$, and $a=1$ into the equality (3) in the beginning of Section 10. Moreover, by the note just above one has $t-t_1=1+\nu(D)+\nu(E)+\nu(II)+\nu(III)+\nu(IV)$, $t_1=\nu(A)-\nu(III)-\nu(IV)+\nu(I_1)$. Note that these holds regardless of whether F_2 is of type I_2 or III. The inequality (1) follows from these.
(2) First, assume that the functional invariant J is constant. Then, $t_1=0$ in the equality (3) Section 10, since J has never poles. Since $\rho=20$, and $a=1$ under our

assumptions, we have $2t = 7$, which is a contradiction. Thus, J is not constant. We can apply the inequality (4) in the beginning of Section 10. Since we are in the case (A) in Proposition 25.3 $\nu(I_0^*) = 1 + \nu(D_4)$. We get (2). Q.E.D.

From here for a while we consider the case where G has 12 vertices.

Lemma 25.9. *(1) Let F be a Dynkin graph with 3 or more components of type A. Let F' be a Dynkin graph obtained from F by one elementary transformation. If F and F' have the same number of vertices and if F' has a component of type D, then for F' $\nu(A) + 2\nu(D) + 2\nu(E) \geq 5$.*
(2) Let F be a Dynkin graph with a D_4-component. Let F' be a Dynkin graph obtained from F by one elementary transformation. If F and F' have the same number of vertices and if F' has a component of type D_k with $k \geq 5$ and F' has no component of type D_4, then for F' $\nu(A) + 2\nu(D) + 2\nu(E) \geq 5$.

Among the 40 graphs in Lemma 25.6 F satisfies neither the condition of Lemma 25.9 (1) nor the condition of Lemma 25.9 (2) if, and only if, F is one of the following 12: (1), (2), (4), (5), (10), (14), (18), (19), (21), (28), (29), (32).

By Lemma 25.7 and Lemma 25.8 G satisfies the following conditions ⟨1⟩ –⟨5⟩ in our present situation.

Proposition 25.10. *A Dynkin graph F satisfies the following 5 conditions if, and only if, F belongs to one of the following 2 classes MPD12, NID12:*

The conditions:

⟨1⟩ *F has 12 vertices.*
⟨2⟩ *F can be made from one of the 40 graphs in Lemma 25.6 by one elementary transformation.*
⟨3⟩ *F has a component of type D.*
⟨4⟩ *F has no component of type D_4.*
⟨5⟩ *$\nu(A) + 2\nu(D) + 2\nu(E) \leq 4$ for F.*

MPD12: (6 items.)
D_{12}, $D_{10} + A_2$, $D_8 + A_3 + A_1$, $D_7 + A_5$,
$E_7 + D_5$, $2D_6$.

NID12: (9 items.)
$D_{11} + A_1$, $D_{10} + 2A_1$, $D_9 + A_3$, $A_7 + D_5$,
$D_7 + D_5$, $D_7 + A_4 + A_1$, $D_7 + A_3 + A_2$, $D_6 + 2A_3$,
$A_5 + D_5 + A_2$.

Proof. By Lemma 25.9 we can treat only graphs made from one of the above 12 by one elementary transformation. Thus, our proposition follows easily. Q.E.D.

Proposition 25.11. (MPD12) *Every graph belonging to MPD12 can be made from the basic graph $A_9 + BC_1$ by two tie transformations. More precisely, from $A_9 + BC_1$ we can make $D_{10} + BC_1$ by one tie transformation. From $D_{10} + BC_1$ we can make such a graph by a tie transformation.*

Graphs belonging to NID12 do not belong to $PC(S_{1,0})$. We will show this later.

We proceed to the case where the number of vertices in G is 11. By Lemma 25.7 (2) if we write $G = G' + G_D$ (G_D is the sum of components of type D.), G' has at most 7 vertices.

Lemma 25.12. *The I1-conition fails for $Q(G)$ if, and only if, $\epsilon_p(G') \neq (5, 2d(G'))_p$ for $p = 3$ or 5. Here, $\epsilon_p(G') = \epsilon_p(Q(G'))$ and $d(G') = d(Q(G'))$.*

Proof. Since $d(G) = 4^m d(G')$ and $\epsilon_p(G) = \epsilon_p(G')$, the I1-condition fails if, and only if, $\epsilon_p(G') \neq (5, 2d(G'))_p$ for some prime number p. Thanks to the product formula, we can assume $p \geq 3$. By Lemma 7.5 (3) and (4) we can assume moreover $p \leq 7$. Assume $\epsilon_7(G') \neq (5, 2d(G'))_7$. By Lemma 7.5 (4) $7 \mid d(G')$. Thus, $G' = A_6 + A_1$ or A_6. However, in both cases $\epsilon_7(G') = -1 = (5, 2d(G'))_7$. Thus, we can assume $p = 3$ or 5.
Q.E.D.

In our present situation G' satisfies the following 4 conditions:

⟨a⟩ Every component of G' is of type A or E.
⟨b⟩ G' has at most 7 vertices.
⟨c⟩ $\nu(A) + 2\nu(E) \leq 3$ for G'.
⟨d⟩ For $p = 5$ or 3 the following ⟨d_p⟩ holds:

⟨d_p⟩ $$\epsilon_p(G') \neq (5, 2d(G'))_p$$

Lemma 25.13. *A Dynkin graph F' with $5 \mid d(F')$ satisfies ⟨a⟩, ⟨b⟩, ⟨c⟩ and ⟨d_5⟩ if, and only if, F' is one of the following six:*

$$\begin{array}{lll} A_4 + A_3, & A_4 + A_2 + A_1, & A_4 + A_2, \\ A_4 + 2A_1, & A_4 + A_1, & A_4. \end{array}$$

Proof. We can write $F' = A_4 + G''$. By Lemma 7.5 (3) $5 \not| d'' = d(G'')$. We omit the lower index 5. $\epsilon = (-1, 5)(5, d'') = (5, d'')$. $(5, 2d) = (5, 2 \times 5 \times d'') = -(5, d'') \neq \epsilon$. We can choose an arbitrary G'' if F' satisfies ⟨a⟩, ⟨b⟩, ⟨c⟩.
Q.E.D.

Lemma 25.14. *A Dynkin graph F' with $5 \not| d(F')$ satisfies ⟨a⟩, ⟨b⟩, ⟨c⟩ and ⟨d_5⟩ if, and only if, F' is one of the following 13:*

$$\begin{array}{llll} A_6 + A_1, & E_6 + A_1, & A_5 + 2A_1, & A_3 + 2A_2, \\ 2A_3, & A_3 + A_2 + A_1, & A_5, & A_3 + 2A_1, \\ 2A_2 & A_3, & A_2 + A_1, & 2A_1, \\ \text{empty graph.} & & & \end{array}$$

Proof. Let G_0 be a connected Dynkin graph satisfying above ⟨a⟩ and ⟨b⟩ with $G_0 \neq A_4$. $\left(\dfrac{d(G_0)}{5}\right) = -1 \Leftrightarrow G_0 = A_7, E_7, A_6, E_6, A_2$, or A_1. $\left(\dfrac{d(G_0)}{5}\right) = +1 \Leftrightarrow G_0 = A_5, A_3$ or the empty graph. Under the assumption $\epsilon_5(F') = 1$ by Lemma 7.5 (4). $(5, 2d(F'))_5 = -\left(\dfrac{d(F')}{5}\right)$. By s we denote the total number of components of type A_7, E_7, A_6, E_6, A_2 or A_1. Under the assumption $5 \not| d(F')$ ⟨d_p⟩$\Leftrightarrow s$ is even. This implies the lemma.Q.E.D.

Lemma 25.15. *A Dynkin graph F' satisfies* ⟨a⟩, ⟨b⟩, ⟨c⟩ *and* ⟨d_3⟩ *if, and only if, F' is one of the following 9:*

$$\begin{array}{lllll} A_3+2A_2, & E_6, & A_5+A_1, & A_4+A_2, & A_3+A_2+A_1, \\ 3A_2, & 2A_2+A_1, & 2A_2, & A_2+A_1. & \end{array}$$

Proof. If $3 \nmid d(F')$, then $\epsilon_3(F') = (5,\, 2d(F'))_3 = 1$. Thus, $3 \mid d(F')$ is necessary. F' contains a component A_2, A_5 or E_6. If F' contains E_6 or A_5, it is easy to check that only E_6 and $A_5 + A_1$ satisfy the conditions. Assume that F' contains A_2 but does not contain E_6 and A_5. We can write $F' = kA_2 + G''$ with $1 \le k \le 3$ and $3 \nmid d'' = d(G'')$. We omit the lower index $p = 3$. We have $\epsilon = (-1)^{k(k-1)/2}(3,\, d'')^k$ and $(5,\, 2d) = (-1)^k$. If $k = 3$, then $F' = 3A_2$ and this satisfies the conditions. If $k = 2$, then for an arbitrary G'' ⟨d_3⟩ is satisfied. By the condition ⟨c⟩, $F' = A_3 + 2A_2$, $2A_2 + A_1$, or $2A_2$. In case $k = 1$ calculating $\epsilon = -(3,\, d'')$ we obtain the remaining 3 graphs. Q.E.D.

By Lemma 25.8 our graph G satisfies the following conditions ⟨A⟩ – ⟨D⟩.

Corollary 25.16. *A Dynkin graph F satisfies the following 4 conditions ⟨A⟩ – ⟨D⟩ if, and only if, F belongs one of the following 3 classes NPD11, MID11, NID11:*

⟨A⟩ *Every component of F is of type A, D or E. F has a component of type D.*
⟨B⟩ *F has 11 vertices.*
⟨C⟩ $\nu(A) + 2\nu(D) + 2\nu(E) \le 5$.
⟨D⟩ *The I1-condition fails.*

NPD11: (12 items.)

$$\begin{array}{lll} D_{11}, & D_8+A_3, & D_8+A_2+A_1, \\ D_7+A_4, & D_7+2A_2, & D_6+A_5, \\ D_6+D_5, & D_6+A_3+2A_1, & E_6+D_5, \\ A_5+D_5+A_1, & D_5+A_4+A_2, & D_5+A_3+A_2+A_1. \end{array}$$

MID11: (5 items.)

$$\begin{array}{lll} D_6+2A_2+A_1, & D_5+2A_3, & D_5+3A_2, \\ 2D_4+A_3, & D_4+A_3+2A_2. & \end{array}$$

NID11: (9 items.)

$$\begin{array}{lll} D_9+2A_1, & D_7+D_4, & A_6+D_4+A_1, \\ D_6+A_4+A_1, & E_6+D_4+A_1, & A_5+D_4+2A_1, \\ D_5+A_4+2A_1, & A_4+D_4+A_3, & A_4+D_4+A_2+A_1. \end{array}$$

Proposition 25.17. (NPD11) *Every graph F belonging to NPD11 is a subgraph of a graph belonging to MPD12. In particular, F can be made from the basic graph $A_9 + BC_1$ by tie or elementary transformation repeated twice.*

Proposition 25.18. (MID11) *Every graph in MID11 does not belong to $PC(S_{1,0})$.*

Proof. By F we denote a graph in MID11. Assuming that we have a full embedding $Q = Q(F) \hookrightarrow \Lambda_3/P$, we will deduce a contradiction. Let $S = P \oplus Q \hookrightarrow \Lambda_3$ be the induced embedding. $\widetilde{S}$ stands for the primitive hull of S in Λ_3. $I = \widetilde{S}/S$ is an isotropic subgroup of the discriminant group S^*/S of S. $T = C(S, \Lambda_3)$ has signature (1, 2). Thus, $d = d(T) > 0$.

• To $F = D_5 + 3A_2$ we can apply the method in Lemma 24.6. Assuming $I_3 = 0$, we get a contradiction $2 \equiv 1 \pmod 3$. If $I_3 \neq 0$, the primitive hull in Λ_3/P of the component $Q(3A_2)$ of Q contains a root system of type E_6, which contradicts fullness.

• For $F = D_5 + 2A_3$ the method in Lemma 24.7 is effective. We have a unique isotropic element $\overline{\alpha}_0$ of order 2 with the following property; if I_2 contains an isotropic element different from $\overline{\alpha}_0$, then the embedding $Q \hookrightarrow \Lambda_3/P$ is not full. Thus, I_2 is generated by $\overline{\alpha}_0$. One has $(\widetilde{S}^*/\widetilde{S})_2 \cong (I_2^{\perp}/I_2) \cong \mathbf{Z}/4 \oplus \mathbf{Z}/4 \oplus \mathbf{Z}/4$. Computing the discriminant form of $T \otimes \mathbf{Z}_2$ we get a contradiction $5 \equiv 1 \pmod 8$.

• To $D_6 + 2A_2 + A_1$ and $D_4 + A_3 + 2A_2$ we can apply the same method as in Lemma 24.8. Every isotropic element in $(S^*/S)_2$ corresponds to a root in the primitive hull of Q. On the other hand, $I_2 \neq 0$. Thus, we have a contradiction under the assumption of fullness.

• For the last case $F = 2D_4 + A_3$ the method is similar to that in Lemma 24.8. However, it is slightly different. In this case, $l(I_2) \geq 2$. On the other hand, every isotropic element in $(S^*/S)_2$ except a unique one $\overline{\alpha}_0$ corresponds to a root in the primitive hull of Q. Thus, by fullness I_2 is generated by $\overline{\alpha}_0$; thus, $l(I_2) \leq 1$, a contradiction. Q.E.D.

Later we will see that every graph in NID11 does not belong to $PC(S_{1,0})$, either.

The next case is the case where G has 10 vertices. We write $G = G' + G_D$ (G_D is the sum of components of type D.) as in the case of 11 vertices. G' has at most 6 vertices by Lemma 25.7 (2).

Lemma 25.19. *The I1-condition fails for G if, and only if, the following condition $\langle\langle d_p \rangle\rangle$ holds for $p = 2$, 3 or 5:*

$\langle\langle d_p \rangle\rangle$ $$5d(G') \in \mathbf{Q}_p^{*2} \text{ and } \epsilon_p(G') \neq (-2, 5)_p.$$

Proof. Since $d(G) = 4^m d(G')$ and $\epsilon_p(G) = \epsilon_p(G')$, the I1-condition fails if, and only if, $\langle\langle d_p \rangle\rangle$ holds for some prime number p. Assume that $\langle\langle d_p \rangle\rangle$ holds for some prime p with $p \geq 7$. Since $\epsilon_p(G') \neq (-2, 5)_p = 1$, $\epsilon_p(G') = -1$. By Lemma 7.5 (4) $p \mid d(G')$. Since $5d(G') \in \mathbf{Q}_p^{*2}$, $p^2 \mid d(G')$. By Lemma 7.5 (3) $12 \leq 2(p-1) \leq 6$, a contradiction. Q.E.D.

In our situation G satisfies the following conditions $\langle\langle A \rangle\rangle$–$\langle\langle D \rangle\rangle$.

Proposition 25.20. *A Dynkin graph F satisfies the following 4 conditions $\langle\langle A \rangle\rangle$–$\langle\langle D \rangle\rangle$ if, and only if, $F = D_6 + A_4$, $D_5 + 2A_2 + A_1$ or $A_4 + D_4 + 2A_1$:*

$\langle\langle A \rangle\rangle$ Every component of F is of type A, D or E.

$\langle\langle B \rangle\rangle$ F has 10 vertices.

$\langle\langle C \rangle\rangle$ $\nu(A) + 2\nu(D) + 2\nu(E) \leq 6$ for F.

$\langle\langle D \rangle\rangle$ The I1-condition fails for F.

Proof. Let F_D be the sum of components of type D. We set $F = F' + F_D$. First, we can classify F' with components A, E only and with at most 6 vertices and satisfying

$\langle\langle d_p \rangle\rangle$ for $p = 5$, 3 or 2. To treat $p = 2$ Lemma 7.7 is effective. Then, applying Lemma 25.19 one has 3 possibilities for F as above. Q.E.D.

Proposition 25.21. *(1)$D_6 + A_4$ is a subgraph of D_{12}. $D_5 + 2A_2 + A_1$ is a subgraph of $D_7 + A_5$. In particular, $D_6 + A_4$ and $D_5 + 2A_2 + A_1$ can be made from the basic graph $A_9 + BC_1$ by tie or elementary transformations repeated twice.*
(2) $A_4 + D_4 + 2A_1 \notin PC(S_{1,0})$.

Proof. (1) Obvious.
(2) To $F = A_4 + D_4 + 2A_1$ we can apply the method in Lemma 24.8. Every isotropic element in $(S^*/S)_2$ corresponds to a root in the primitive hull of $Q(F)$ in Λ_3/P. On the other hand, $I_2 \neq 0$. Thus, we get a contradiction. Q.E.D.

Proposition 25.22. (MID12, MID11) *(1) Every graph belonging to NID11 does not belong to $PC(S_{1,0})$.*
(2) Every graph belonging to NID12 does not belong to $PC(S_{1,0})$.

Proof. (1) Every graph F in NID11 contains $A_4 + D_4 + 2A_1$ as a subgraph. If $F \in PC(S_{1,0})$, then $A_4 + D_4 + 2A_1 \in PC(S_{1,0})$ by Proposition 13.5 (2). It contradicts Proposition 25.21 (2).
(2) Every graph in NID12 contains D_5+2A_3, D_5+3A_2, $D_4+A_3+2A_2$ or $A_4+D_4+2A_1$ as a subgraph. Since these 4 graphs do not belong to $PC(S_{1,0})$, we have the conclusion. Q.E.D.

Under the assumption ((1)) of this section and under the condition that the I1-condition fails, $G \in PC(S_{1,0})$ belongs to MPD12, or NPD11, or $G = D_6 + A_4$ or $D_5+2A_2+A_1$. Thus, by the above results, Proposition 25.1 has been established under ((1)).

26
A singular fiber of type II^*, III^*, IV^* in the case $S_{1,0}$

In this section we assume that the corresponding elliptic K3 surface $\Phi : Z \to C$ to $G \in PC(S_{1,0})$ has a singular fiber of type II^*, III^* or IV^*, i.e., we assume the condition ((2)) in the beginning of Section 25.

Proposition 26.1. *Under the assumption ((2)) one of the following (1), (2) holds for $G \in PC(S_{1,0})$:*
(1) G has a component of type E, and $G + A_1 \in PC(J_{3,0})$.
(2) We can write $G = D_5 + G_A$, where all components of G_A are of type A, and $E_7 + G_A \in PC(J_{3,0})$.

Proof. In the case (A) or (B) in Proposition 25.3 the above case (1) takes place. In the case (C) in Proposition 25.3 (2) takes place. Q.E.D.

By Section 11 if $G \in PC(J_{3,0})$ has a component of type E, then G is a subgraph of the Coxeter-Vinberg graph of the unimodular lattice with signature $(14, 1)$. (See the figure just before Proposition 11.2.)

Lemma 26.2. *If G has 12 vertices, then G belongs to one of the following 3 classes MPE12, MIE12, NIE12:*

MPE12:	(3 items.)	$E_7 + D_5$,	$E_7 + A_3 + A_2$,	$E_6 + A_6$.
MIE12:	(2 items.)	$E_8 + A_4$,	$E_7 + A_5$.	
NIE12:	(3 items.)	$E_8 + 2A_2$,	$2E_6$,	$E_6 + A_4 + A_2$.

Lemma 26.3. (MPE12)
(1) We can make $E_7 + D_5$ from the basic graph $A_9 + BC_1$ by two tie transformations.
(2) We can make $E_7 + A_3 + A_2$ from the sub-basic graph $B_6 + A_3 + A_2$ by one tie transformation.
(3) We can make $E_6 + A_6$ from the sub-basic graph $B_5 + A_6$ by one tie transformation.

Lemma 26.4. (MIE12) $E_8 + A_4 \notin PC(S_{1,0})$. $E_7 + A_5 \notin PC(S_{1,0})$.

Proof. To $G = E_8 + A_4$ we can apply the same method as in Lemma 24.5. Calculating the discriminant of the orthogonal complement of $S = P \oplus Q(G) \hookrightarrow \Lambda_3$ in 2 different ways, we have $2 \equiv x^2 \pmod 5$ for some integer x, a contradiction.

To $G = E_7 + A_5$ the method explained in Lemma 24.7 can be applied. Calculating the discriminant quadratic form of the primitive hull of $S = P \oplus Q(G) \hookrightarrow \Lambda_3$, and calculating the discriminant of the orthogonal complement of S in 2 different ways, we have $5 \equiv 1 \pmod 8$, a contradiction. Q.E.D.

Graphs in NIE12 do not belong to $PC(S_{1,0})$, either. We will see this later.

We proceed to the case where G has 11 vertices.

Lemma 26.5. *If G has 11 vertices, then G belongs to one of the following 3 classes MPE11, NPE11, MIE11:*

MPE11:	(2 items.)	$E_8 + A_3$,	$E_8 + A_2 + A_1$.	
NPE11:	(13 items.)			
	$E_7 + A_4$,	$E_7 + D_4$,	$E_7 + A_3 + A_1$,	$E_7 + 2A_2$,
	$E_7 + A_2 + 2A_1$,	$E_6 + A_5$,	$E_6 + D_5$,	$E_6 + A_4 + A_1$,
	$E_6 + A_3 + A_2$,	$D_5 + A_6$,	$D_5 + A_5 + A_1$,	$D_5 + A_4 + A_2$,
	$D_5 + A_3 + A_2 + A_1$.			
MIE11:	(1 item.)	$E_6 + 2A_2 + A_1$.		

Lemma 26.6. (MPE11) *We can make $E_8 + A_2$ from the basic graph $E_8 + BC_1$ by one tie transformation. From $E_8 + BC_1$ we can make the members $E_8 + A_3$, $E_8 + A_2 + A_1$ of MPE11 by one tie transformation.*

Lemma 26.7. (NPE11) *Every graph G in NPE11 is a subgraph of a graph in MPE12. In particular, either we can make G from a basic graph by tie or elementary transformations repeated twice, or we can make G from a sub-basic graph by one tie transformation or one elementary transformation.*

Lemma 26.8. (MIE11) $E_6 + 2A_2 + A_1 \notin PC(S_{1,0})$.

Proof. By the method explained in Lemma 24.6 we get this result.

Lemma 26.9. (MIE12) *Every graph in NIE 12 does not belong to $PC(S_{1,0})$.*

Proof. Any one G of 3 graphs in NIE12 contains a subgraph $E_6 + 2A_2 + A_1$. By Proposition 13.5 (2) and by Lemma 26.8 we get the lemma. Q.E.D.

The last remaining case is the case where G has 10 vertices.

Lemma 26.10. *(1) Let F be a Dynkin graph without vertices corresponding to a short root. We assume that F has 10 vertices and has a component of type E. F is a subgraph of the Coxeter-Vinberg graph of the unimodular lattice with signature $(14,1)$ if, and only if, $F \neq E_6 + 4A_1$.*
(2) If G has 10 vertices, then G belongs to the following class NPE10:

NPE10: (16 items.)

$E_8 + A_2$,	$E_8 + 2A_1$,	$E_7 + A_3$,	$E_7 + A_2 + A_1$,
$E_7 + 3A_1$,	$E_6 + A_4$,	$E_6 + D_4$,	$E_6 + A_3 + A_1$,
$E_6 + 2A_2$,	$E_6 + A_2 + 2A_1$,	$D_5 + A_5$,	$D_5 + A_4 + A_1$,
$D_5 + A_3 + A_2$,	$D_5 + A_3 + 2A_1$,	$D_5 + 2A_2 + A_1$,	$D_5 + A_2 + 3A_1$.

Lemma 26.11. (NPE10) *Every graph G in NPE10 is a subgraph of a graph in MPE11 or NPE11. In particular, either we can make G from a basic graph by 2 of tie or elementary transformations, or we can make G from a sub-basic graph by one tie or elementary transformation.*

Under the assumption that the I1-condition fails, our G has 12, 11 or 10 vertices. Thus, we can complete the proof of Proposition 25.1 under the assumption ((2)).

27
Combinations of graphs of type A in the case $S_{1,0}$

In this section we treat the case ((3)). Let $G \in PC(S_{1,0})$ be a Dynkin graph with r vertices. We assume that every singular fiber except F_1 of the elliptic K3 surface $\Phi : Z \to C$ associated with G is of type I, II, III or IV. F_1 is of type I_0^*. As before, ρ denotes the Picard number of Z, and a denotes the rank of the Mordell-Weil group MW of Φ.

Lemma 27.1. *(1) If $\rho = r + 8$ and $a = 0$, then we can write $G = 2A_4 + G'$.*
(2) $r \leq 12$. If $r = 12$, then $a = 1$.
(3) The number of components of G is at most $16 - r$.

Proof. (1) Under our assumptions only the case (B) in Proposition 25.3 can take place.

Let S_i be the subgroup of Pic(Z) generated by components of F_i disjoint from C_5. Let F be a smooth fiber of Φ. Set $H_0 = \mathbf{Z}[F] + \mathbf{Z}[F + C_5]$ and $\overline{S} = \oplus_{i=1}^{t} S_i$. $H_0 \oplus \overline{S}$ is a subgroup of Pic(Z). Also, C_7 is a section of Φ. The assumption $a = 0$ implies $[C_7] \in H_0 \oplus \widetilde{\overline{S}} \subset H_0 \oplus \overline{S}^*$. Thus, we can write $[C_7] = b[F] + c[F + C_5] + \sum_{i=1}^{t} \chi_i$ with $b, c \in \mathbf{Z}$, $\chi_i \in S_i^*$. $1 = [C_7] \cdot [F] = b$, $0 = [C_7] \cdot [C_5] = b - c$. Hence $b = c = 1$.

Next, we consider a number $i \geq 4$. Let D be an arbitrary component of F_i disjoint from C_5. By the description in Proposition 25.3 (B) D is also disjoint from C_7. $0 = [C_7] \cdot [D] = \chi_i \cdot [D]$. It implies $\chi_i = 0$.

Moreover, by the description in (B)

$$S_1 \cong Q(-D_4), \quad S_2 \cong Q(-A_1), \quad S_3 \cong Q(-A_{2k-1}).$$

Here, $Q(-X)$ denotes the negative definite root lattice of type X. Let $\omega_0, \ldots, \omega_3$ be the dual basis of the basis $[C_0], \ldots, [C_3]$ of S_1. One has $\chi_1 = \omega_3$. In particular, $\chi_1^2 = -1$. Let D_1 be the component of F_2 intersecting C_7. The class $[D_1]$ is the generator of S_2 and $1 = [C_7] \cdot [D_1] = \chi_2[D_1]$. Thus, $\chi_2^2 = -1/2$. Finally, we consider F_3. χ_3 is the fundamental weight corresponding to the central vertex of the Dynkin graph A_{2k-1}. Thus, $\chi_3^2 = -k/2$. We have $-2 = [C_7]^2 = \{[F]+[F+C_5]\}^2 + \sum_{i=1}^{3} \chi_i^2 = 2-1-1/2-k/2$. One knows $k = 5$ and $S_3 \cong Q(-A_9)$. G contains A_9 minus the central vertex, i.e., $2A_4$.
(2) The number of positive eigen values of the bilinear form on Λ_3 equals to 19. That of the form on P equals to 7. Thus, $r \leq 19 - 7 = 12$. Assume $r = 12$ and $a = 0$. We will deduce a contradiction. Since $20 \geq \rho \geq 8 + r = 20$, the equality $\rho = r + 8$ holds. Thus, the case (B) in Proposition 25.3 takes place under our assumption. Let $J : C \to \mathbf{P}^1$ be the functional invariant of the elliptic surface $\Phi : Z \to C$. By the above proof of (1) F_3 is of type I_{10} and J has a pole with order 10 at c_3. Therefore, J is not constant. By the inequality (4) in Section 10 we have $0 = 20 - \rho + a \geq \nu(I_0^*) \geq 1$, a contradiction.
(3) We can assume $\rho = 8 + r$ without losing generality. On the other hand, by the equality (3) in Section 10, $\rho = 26 + a - 2(t - t_1) - t_1$. By our assumption on singular fibers, the case (C) in Proposition 25.3 never takes place. We consider case (A) and (B) separately.

In case (A) $a = 1$, $t_1 = 1 + \nu(A) - \nu(III) - \nu(IV) + \nu(I_1)$, $t - t_1 = 1 + \nu(II) + \nu(III) + \nu(IV)$, regardless of whether F_2 is of type I_2 or III. Here, $\nu(A)$ is the number of components of G of type A. Under the assumption $\nu(A)$ equals to the number of components of G. Thus, $16 - r = \nu(A) + 2\nu(II) + \nu(III) + \nu(IV) + \nu(I_1) \geq \nu(A)$.

In case (B) by (1) F_3 is of type I_{10}. $a = 0$, $t_1 = 1+\nu(A)-1-\nu(III)-\nu(IV)+\nu(I_1)$, $t-t_1 = 1+\nu(II)+\nu(III)+\nu(IV)$, regardless of whether F_2 is of type I_2 or III. Thus, we get the same result. Q.E.D.

Lemma 27.2. $G + A_1 \in PC(J_{3,0})$.

Proof. We consider $\Phi : Z \to C$ associated with $G \in PC(S_{1,0})$. By definition the dual graph of the union $\mathcal{E}$ of components of singular fibers disjoint from $IF = F_1 \cup C_5 \cup C_7 \cup C_6$ coincides with G. We can assume the equality $\rho = r + 8$ without losing generality here. Under our assumption, case (C) in Proposition 25.3 never takes place and F_2 is of type I_2 or III. Let D_1 be the component of F_2 intersecting C_7. D_1 is disjoint from C_5. Moreover, since the case (A) or (B) takes place, $\mathcal{E} \subset \cup_{i=3}^{t} F_i$. Thus, the dual graph of

$\overline{\mathcal{E}} = \{D_1\} \cup \mathcal{E}$ is $G + A_1$, and $\overline{\mathcal{E}}$ is disjoint from $\overline{IF} = F_1 \cup C_5$. Since $\overline{IF}$ is isomorphic to the curve at infinity defined in the case $J_{3,0}$, the subgroup in $H^2(Z, \mathbf{Z})$ generated by the components of $\overline{\mathcal{E}} \cup \overline{IF}$ defines an embedding $P(2,2,2,3) \oplus Q(G + A_1) \hookrightarrow \Lambda_3$ satisfying Looijenga's condition (L1) and (L2). Thus, $G + A_1 \in PC(J_{3,0})$. Q.E.D.

Recall that P has a basis $e_0, \ldots, e_7$. By $\overline{P}$ we denote the sublattice of P generated by $e_0, e_1, \ldots, e_5$ and $e' = u_0 - e_6$. $\overline{P}$ is isomorphic to the lattice $P = P(2,2,2,4)$ defined for $Z_{1,0}$.

Lemma 27.3. *Every element of order 2 in $P^*/P \cong \mathbf{Z}/5 \oplus \mathbf{Z}/2 \oplus \mathbf{Z}/2$ belongs to the subgroup $(P^* \cap \overline{P}^*) + P/P \cong (P^* \cap \overline{P}^*)/\overline{P}$. Here, $P^* \cap \overline{P}^* = \{x \in \overline{P} \otimes \mathbf{Q} \mid$ For every $y \in P$ $x \cdot y \in \mathbf{Z}\}$.*

Proof. We use the notations in Section 4 "The case of $S_{1,0}$". The elements of order 2 in P^*/P are $5\overline{w}_1$, $5\overline{w}_2$ and $5(\overline{w}_1 + \overline{w}_2)$. Set $x_1 = (e_1 + e_3 + e')/2$. We can check $x_1 \in P^* \cap \overline{P}^*$ and $5w_1 - x_1 \in P$. Thus, $5\overline{w}_1 \in (P^* \cap \overline{P}^*) + P/P$. Similarly $5\overline{w}_2 \in (P^* \cap \overline{P}^*) + P/P$. Since the right-hand side is a group, $5(\overline{w}_1 + \overline{w}_2) \in (P^* \cap \overline{P}^*) + P/P$. Q.E.D.

Let G_i with $i \geq 3$ be the dual graph of the components of the singular fiber F_i disjoint from C_5. Under our assumption ((3)) G_i is of type A. Assume that G_i is of type A_{n_i}.

Under the assumption $\rho = r + 8$ and $a = 1$ the case (A) in Proposition 25.3 takes place, and we have $G = \sum_{i=3}^{t} G_i$.

Consider the induced embedding $S = P \oplus Q(G) \hookrightarrow \Lambda_3$ here. Let $\widetilde{S}$ be the primitive hull of S in Λ_3. Set $I = \widetilde{S}/S$ $(\subset S^*/S)$.

Proposition 27.4. *Assume $\rho = r + 8$ and $a = 1$.*
(1) $Q(G)$ is primitive in Λ_3.
(2) Let $\overline{\alpha} \in I$ be an element of order 2. Since $S^*/S = P^*/P \oplus \bigoplus_{i=3}^{t} Q(A_{n_i})^*/Q(A_{n_i})$, we can write $\overline{\alpha} = \overline{\omega} + \sum_{i=1}^{t} \overline{\chi}_i$ with $\overline{\omega} \in P^*/P$, $\overline{\chi}_i \in Q(A_{n_i})^*/Q(A_{n_i})$. One of the following ⟨a⟩, ⟨b⟩, and only one of them holds. We set $M = \{i \mid 3 \leq i \leq t,\ \overline{\chi}_i \neq 0\}$ and $N(2) = \{i \mid 3 \leq i \leq t,\ n_i$ is odd.$\}$:
  ⟨a⟩ $\overline{\omega} = 5\overline{w}_1$ or $\overline{\omega} = 5\overline{w}_2$. $M \subset N(2)$ and $\sum_{i \in M}(n_i + 1) = 10$.
  ⟨b⟩ $\overline{\omega} = 5(\overline{w}_1 + \overline{w}_2)$. $M \subset N(2)$ and $\sum_{i \in M}(n_i + 1) = 12$.
(3) Assume $l(I_2) \geq 2$. Then, I_2 is an abelian group of type $(2,2)$. Besides, there exist sets M_1 and M_2 satisfying the following conditions ⟨1⟩–⟨4⟩:
  ⟨1⟩ $M_1 \subset N(2)$. $M_2 \subset N(2)$.
  ⟨2⟩ For $k = 1$ and 2, $\sum_{i \in M_k}(n_i + 1) = 10$.
  ⟨3⟩ $\sum_{i \in M_1 \cap M_2}(n_i + 1) = 4$.
  ⟨4⟩ For $i \in M_k$ by $\overline{\chi}_i$ we denote the unique element of order 2 in $Q(A_{n_i})^*/Q(A_{n_i})$. Then,

$$\overline{\alpha}_1 = 5\overline{w}_1 + \sum_{i \in M_1} \overline{\chi}_i, \quad \text{and} \quad \overline{\alpha}_2 = 5\overline{w}_2 + \sum_{i \in M_2} \overline{\chi}_i$$

  are generators of I_2.
(4) $I_3 = 0$.

Proof. (1) Let $\overline{\beta} \in \widetilde{Q}(G)/Q(G)$ be an arbitrary element. Assuming $\overline{\beta} \neq 0$, we will deduce a contradiction.

Under the assumption the group of sections MW of Φ is identified with $\widetilde{S}/(\overline{P} \oplus Q(G))$. ($C_5$ is the unit element of MW.) Under the natural inclusion $\widetilde{Q}(G)/Q(G) \subset \widetilde{S}/(\overline{P} \oplus Q(G))$, $\overline{\beta}$ defines a section $C' \in \operatorname{Tor} MW$ with a finite order. Since $\overline{\beta} \neq 0$, $C' \neq C_5$.

We here consider the group $F_a^{\#}$ consisting of non-singular points of the fiber $F_a = \Phi^{-1}(a)$ over a point $a \in C$. The intersection point of F_a and C_5 is the unit element. Recall that the natural homomorphism $\operatorname{Tor} MW \to \operatorname{Tor} F_a^{\#}$ is injective. Thus, in particular, the intersection point of C' and F_1 is a point of finite order different from the unit element.

Since F_1 is of type I_0^*, $F_1^{\#}$ has only 4 points with finite order, and these 4 points lie on different components of F_1. Thus, C' and C_5 hit different components of F_1.

On the other hand, if we regard $\overline{\beta} \in \overline{P}^*/\overline{P} \oplus (Q(G)^*/Q(G))$, the $\overline{P}^*/\overline{P}$-component of $\overline{\beta}$ is zero. This implies that C' and C_5 hit the same component of F_1, a contradiction.
(2) By (1) $\overline{\omega} \neq 0$. Thus, $\overline{\omega} = 5\overline{w}_1$, $5\overline{w}_2$, or $5(\overline{w}_1 + \overline{w}_2)$. By Lemma 27.3 $\overline{\omega} \in (P^* \cap \overline{P}^*)/\overline{P}$. We can regard

$$\overline{\alpha} \in P(\overline{P} \oplus Q(G), \Lambda_3)/(\overline{P} \oplus Q(G)) \subset \widetilde{S}/(\overline{P} \oplus Q(G)).$$

Corresponding to $\overline{\alpha}$ a section $C' \in MW$ with order 2 is defined. By S_i we denote the subgroup of $\operatorname{Pic}(Z)$ generated by the components of F_i disjoint from C_5. We can write

$$[C'] = [F] + [F + C_5] + \chi_1 + \chi_2 + \sum_{i=3}^{t} \chi_i$$

with $\chi_i \in S_i^*$. Here, F denotes a smooth fiber of Φ. Under the isomorphism $H^2(Z, \mathbf{Z}) \to \Lambda_3$, $V = \mathbf{Z}[F] + \mathbf{Z}[F + C_5] \oplus \bigoplus_{i=1}^{t} S_i$ corresponds to $\overline{P} \oplus Q(G)$, $[C']$ mod V corresponds to $\overline{\alpha}$, $\chi_1 + \chi_2$ mod V corresponds to $\overline{\omega}$, and χ_i mod V corresponds to $\overline{\chi}_i$ for $i \geq 3$. Every χ_i is the fundamental weight corresponding to the vertex of G_i corresponding to the component of F_i hit by C', if $\chi_i \neq 0$. $\chi_i = 0$ if, and only if, C' and C_5 hit the same component of F_i. Since G_i is of type A, for $i \geq 3$, $\chi_i \neq 0 \Leftrightarrow \chi_i \bmod V \neq 0 \Leftrightarrow \overline{\chi}_i \neq 0$. Thus, $M = \{i \mid 3 \leq i \leq t,\ \chi_i \neq 0\}$.

We fix $i \in M$. Since C' has order 2, n_i is odd and χ_i is the weight corresponding to the central vertex of A_{n_i}. Thus, $M \subset N(2)$, and $\chi_i^2 = -(n_i + 1)/4$.

Next, we consider χ_1. Since F_1 is of type I_0^*, $\chi_1 = 0$ implies $C' = C_5$. Thus, $\chi_1 \neq 0$. Since C' hits a component of F_1 with multiplicity 1, χ_1 corresponds to one of the three vertices in D_4 with only 1 edge issuing from it. In particular, $\chi_1^2 = -1$.

Finally, we consider χ_2. $\chi_1 + \chi_2$ mod V corresponds to $\overline{\omega}$. First, assume $\overline{\omega} = 5\overline{w}_1$. Since $5\overline{w}_1 = (e_1 + e_2)/2 + e'/2 \bmod P$, and since χ_1 mod V corresponds to $(e_1 + e_2)/2 \bmod P$, χ_2 mod V corresponds to $e'/2 \bmod P \neq 0$. In particular, $\chi_2 \neq 0$. $\chi_2^2 = -1/2$. In the case $\overline{\omega} = 5\overline{w}_2$, $5\overline{w}_2 = (e_2 + e_3)/2 + e'/2 \bmod P$. By the same reason we have $\chi_2^2 = -1/2$. If $\overline{\omega} = 5(\overline{w}_1 + \overline{w}_2)$, $\chi_2 = 0$.

By the argument so far, if $\overline{\omega} = 5\overline{w}_1$, or $5\overline{w}_2$, then we have $-2 = [C']^2 = 2 - 1 - 1/2 - \sum_{i \in M}(n_i + 1)/4 \Leftrightarrow \sum_{i \in M}(n_i + 1) = 10$. Similarly if $\overline{\omega} = 5(\overline{w}_1 + \overline{w}_2)$, one knows $\sum_{i \in M}(n_i + 1) = 12$.

(3) By (1) the restriction to I of the projection $P^*/P \oplus Q(G)^*/Q(G) \to P^*/P$ is injective. Thus, I_2 is isomorphic to a subgroup of $(P^*/P)_2 \cong \mathbf{Z}/2+\mathbf{Z}/2$. In particular, if $l(I_2) \geq 2$, $I_2 \cong \mathbf{Z}/2+\mathbf{Z}/2$. The latter half follows from (2).
(4) By the injectivity mentioned in (3) I_3 is isomorphic to a subgroup of $(P^*/P)_3 = 0$.
Q.E.D.

From here, for a while, we consider case $r = 12$. Thanks to Lemma 27.2, we can consider only graphs F with 12 vertices and with only components of type A such that $F + A_1$ corresponds to one item in the class [1] or [4] in Section 13 in the case $J_{3,0}$ of 13 vertices. Once we have chosen such graphs F in Section 13 for $Z_{1,0}$. We know that there are 15 such graphs.

Lemma 27.5. *A Dynkin graph F with 12 vertices and with only components of type A satisfies $F + A_1 \in PC(J_{3,0})$ if, and only if, F belongs to one of the following 4 classes MPA12, MOA12, MIA12, NIA12:*

MPA12:	(3 items.)	A_{12},	$A_{11} + A_1$,	$A_{10} + A_2$.
MOA12:	(2 items.)	$A_9 + A_3$,	$A_9 + A_2 + A_1$.	
MIA12:	(5 items.)	$A_8 + A_4$,	$2A_6$,	$A_7 + A_4 + A_1$,
	$A_6 + A_4 + A_2$,	$A_5 + A_4 + A_3$.		
NIA12:	(5 items.)	$A_7 + A_3 + A_2$,	$A_7 + 2A_2 + A_1$,	$2A_5 + 2A_1$,
	$A_5 + 2A_3 + A_1$,	$2A_4 + 2A_2$.		

Lemma 27.6. (MPA12) *Three members A_{12}, $A_{11}+A_1$, $A_{10}+A_2$ of MPA12 can be made from the basic graph $A_9 + BC_1$ by two tie transformations. Thus, they belong to $PC(S_{1,0})$.*

Lemma 27.7. (MOA12) *(1) Two members $A_9 + A_3$, $A_9 + A_2 + A_1$ of MOA12 can be made from the basic graph $A_9 + BC_1$ by two tie transformations. However, in these cases the component A_9 remains as an obstruction component.*
(2) $A_9 + A_3 \notin PC(S_{1,0})$. $A_9 + A_2 + A_1 \notin PC(S_{1,0})$.

Proof. (1) is obvious. We will show (2). Let F be the Dynkin graph under consideration. Assuming that we have a full embedding $Q = Q(F) \hookrightarrow \Lambda_3/P$ without an obstruction component A_9, we will deduce a contradiction. Let $S = P \oplus Q \hookrightarrow \Lambda_3$ be the induced embedding, T be the orthogonal complement of S in Λ_3. T is a lattice with signature $(0,2)$. By $d = d(T)$ we denote the discriminant of T. $d > 0$. Let $\widetilde{S}$ be the primitive hull of S in Λ_3. Set $I = \widetilde{S}/S$. We regard $I \subset S^*/S$. The restriction of the projection $S^*/S = P^*/P \oplus Q^*/Q \to Q^*/Q$ to I is denoted by π. It induces a homomorphism between p-Sylow subgroups $\pi_p : I_p \to (Q^*/Q)_p$. Since in our case we can write $F = A_9 + F_1$, $(Q^*/Q)_p = (Q(A_9)^*/Q(A_9))_p \oplus (Q(F_1)^*/Q(F_1))_p$. Note that the component A_9 is an obstruction component if, and only if, $\pi_p(I_p) \supset (Q(A_9)^*/Q(A_9))_p$ for $p = 5$ and $p = 2$.

We consider $F = A_9 + A_3$ first. $S^*/S = (S^*/S)_5 \oplus (S^*/S)_2$. We consider $p = 5$. $(S^*/S)_5 \cong \mathbf{Z}/5 \oplus \mathbf{Z}/5$. The first component of the right-hand side is induced by P and the second is induced by the component A_9. The discriminant quadratic form on $(S^*/S)_5$ is given by $q_5(x,y) \equiv 2(x^2 - y^2)/5 \bmod 2\mathbf{Z}$. Assume $I_5 = 0$. Then, $d =$

$|d(P)||d(G)|/[\widetilde{S}:S]^2 = 2^5 \times 5^2/m^2$ with $5 \not| m$. On the other hand, the discriminant form of $T \otimes \mathbf{Z}_5$ coincides with $-q_5$ and $T \otimes \mathbf{Z}_5$ is isomorphic to the 5-adic lattice of rank 2 defined by the 2×2 diagonal matrix whose diagonal entries are 2×5 and -2×5. Thus, $d \equiv -2^2 \times 5^2 \bmod \mathbf{Z}_5^{*2}$. Comparing 2 expressions of d, we get $-2 \in \mathbf{Z}_5^{*2}$. It implies $-2 \equiv x^2 \pmod 5$ for some integer x, a contradiction. Thus, $I_5 \neq 0$. It is easy to check that for any non-zero element $(x,y) \in I_5$, $y \neq 0$. Thus, $\pi_5 : I_5 \to (Q^*/Q)_5 = (Q(A_9)^*/Q(A_9))_5 \cong \mathbf{Z}/5$ is surjective.

Next, we consider $p = 2$. $(S^*/S)_2 \cong \mathbf{Z}/2 \oplus \mathbf{Z}/2 \oplus \mathbf{Z}/4 \oplus \mathbf{Z}/2$. The first component of the right-hand side and the second are induced by P, the third $\mathbf{Z}/4$ corresponds to A_3, and the last fourth $\mathbf{Z}/2$ to A_9. Since $l((S^*/S)_2) = 4 > \operatorname{rank} T = 2$, $I_2 \neq 0$. By Lemma 27.4 (2) any non-zero element $\overline{\alpha}$ in I_2 coincides with $(0,1,0,1)$ or $(1,0,0,1)$. Anyway, the third coordinate of $\overline{\alpha}$ is 0, and the fourth coordinate is not 0. This implies $\pi_2(I_2)$ contains $(Q(A_9)^*/Q(A_9))_2$. We conclude that for every full embedding $Q(A_9 + A_3) \hookrightarrow \Lambda_3/P$ A_9 is an obstruction component. Thus, $A_9 + A_3 \notin PC(S_{1,0})$.

We proceed to the case $A_9 + A_2 + A_1$. $d(S) = d(P)d(G) = -5^2 \times 3 \times 2^4$. $S^*/S = (S^*/S)_5 \oplus (S^*/S)_3 \oplus (S^*/S)_2$.

We consider $p = 5$ first. Assume $I_5 = 0$. $(S^*/S)_5$ and the discriminant quadratic form on it is same as in the above case $A_9 + A_3$. Thus, by the same reasoning we have $-3 \in \mathbf{Z}_5^{*2}$. It implies that $-3 \equiv x^2 \pmod 5$ for some integer x, a contradiction. By the same reason as in the above $\pi_5 : I_5 \to (Q^*/Q)_5 = (Q(A_9)^*/Q(A_9))_5 \cong \mathbf{Z}/5$ is surjective.

Next, we consider $p = 2$. $(S^*/S)_2 \cong \mathbf{Z}/2 \oplus \mathbf{Z}/2 \oplus \mathbf{Z}/2 \oplus \mathbf{Z}/2$. The first component of the right-hand side and the second are induced by P, the third is induced by A_9, and the fourth by A_1. The discriminant quadratic form on $(S^*/S)_2$ is given by $q_2(a_1, a_2, b, c) \equiv -(a_1^2 + a_2^2)/2 + (b^2 + c^2)/2 \bmod 2\mathbf{Z}$. Since $l((S^*/S)_2) = 4 > \operatorname{rank} T = 2$, $I_2 \neq 0$. By Lemma 27.4 (2) any non-zero element $\overline{\alpha}$ coincides with either $\overline{\alpha}_1 = (1,0,1,0)$, $\overline{\alpha}_2 = (0,1,1,0)$ or $\overline{\alpha}_3 = (1,1,1,1)$. Note that for $\overline{\alpha}_1$ and $\overline{\alpha}_2$ the last coordinate is 0 and the third coordinate is non-zero. Thus, unless I_2 is the group of order 2 generated by $\overline{\alpha}_3$, $\pi_2(I_2)$ contains $(Q(A_9)^*/Q(A_9))_2$.

Finally, assuming that I_2 is generated by $\overline{\alpha}_3$ we deduce a contradiction. In this case, $m = [\widetilde{S}:S] = 2n$ with an odd integer n. Hence $d = 5^2 \times 3 \times 2^2/n^2$. Since $(1,1,1,1)$, $(1,0,1,0)$ and $(0,1,1,0)$ generate $I_2^{\perp}$, and since $q_2(1,0,1,0) \equiv q_2(0,1,1,0) \equiv 0$, and $q_2((1,0,1,0) + (0,1,1,0)) \equiv 1 \bmod 2\mathbf{Z}$, The discriminant quadratic form on $(\widetilde{S}^*/\widetilde{S})_2 \cong I_2^{\perp}/I_2 \cong \mathbf{Z}/2 + \mathbf{Z}/2$ can be written $q_2'(e_1, e_2) \equiv e_1 e_2 \bmod 2\mathbf{Z}$. Thus, $T \otimes \mathbf{Z}_2$ is isomorphic to the 2-adic lattice defined by the matrix $A = \begin{pmatrix} 0 & 2 \\ 2 & 0 \end{pmatrix}$. It implies $d \equiv -2^2 \bmod \mathbf{Z}_2^{*2}$. By the 2 expressions of d we have $-3 \in \mathbf{Z}_2^{*2}$, which implies $-3 \equiv x^2 \pmod 8$ for some integer x, a contradiction.

In conclusion, we know that for every full embedding $Q(A_9 + A_2 + A_1) \hookrightarrow \Lambda_3/P$ A_9 is an obstruction component. Thus, $A_9 + A_2 + A_1 \notin PC(S_{1,0})$. Q.E.D.

Lemma 27.8. (MIA12) *Every graph in MIA12 does not belong to $PC(S_{1,0})$.*

Proof. We use the same notations S, $\widetilde{S}$, I, T, d, etc. as in the verification of Lemma 27.7 (2).

First, we deal with $2A_6$. We have $d = d(T) = |d(P)||d(2A_6)|/m^2 = 7^2 \times 5 \times 2^2/m^2$ with $m = \#I = [\widetilde{S} : S]$. We apply the p-adic method for $p = 7$. Now, $(S^*/S)_7 \cong \mathbf{Z}/7 \oplus \mathbf{Z}/7$. The discriminant form on it can be written $q_7(x,y) \equiv 6(x^2+y^2)/7 \bmod 2\mathbf{Z}$ for $(x,y) \in \mathbf{Z}/7 \oplus \mathbf{Z}/7$. We can check that $q_7(x,y) \equiv 0$ implies $(x,y) = (0,0)$. Thus, $I_7 = 0$. $T \otimes \mathbf{Z}_7$ is isomorphic to the 7-adic lattice defined by the diagonal 2×2 matrix whose diagonal entries are -6×7 and -6×7. Thus, $d \equiv 6^2 \times 7^2 \bmod \mathbf{Z}_7^{*2}$. By the 2 expressions of d we have $5 \in \mathbf{Z}_7^{*2}$, which implies $5 \equiv x^2 \pmod 7$ for some integer x, a contradiction.

Next, we treat 4 graphs $F = A_8 + A_4$, $A_7 + A_4 + A_1$, $A_6 + A_4 + A_2$, $A_5 + A_4 + A_3$ at the same time. The method applied to them is same as in Lemma 24.5. For these cases we can write $d(S) = d(P)d(F) = -5^2 d_1$. Here, $d_1 = 3^2 \times 2^2$, 2^6, $7 \times 3 \times 2^2$, 3×2^5 in the respective case. Here, always $d_1 \equiv x^2 \pmod 5$ has an integral solution; thus $d_1 \in \mathbf{Z}_5^{*2}$. Under this notation $d = 5^2 d_1/m^2$ with $m = \#(I) = [\widetilde{S} : S]$.

On the other hand, always in 4 cases $(S^*/S)_5 \cong \mathbf{Z}/5 \oplus \mathbf{Z}/5$ and the discriminant form on it is given by $q_5(x,y) \equiv (2x^2 + 4y^2)/5 \bmod 2\mathbf{Z}$ for $(x,y) \in \mathbf{Z}/5 \oplus \mathbf{Z}/5$. Since $q_5(x,y) \equiv 0$ implies $(x,y) = (0,0)$, $I_5 = 0$. $T \otimes \mathbf{Z}_5$ is isomorphic to the 5-adic lattice defined by the 2×2 diagonal matrix whose diagonal entries are -2×5 and $-2^2 \times 5$. Thus, $d \equiv 2^3 \times 5^2 \bmod \mathbf{Z}_5^{*2}$. By the 2 expressions of d we have $2 \in \mathbf{Z}_5^{*2}$, which implies the existence of an integral solution $2 \equiv x^2 \pmod 5$, a contradiction. Q.E.D.

5 graphs in NIA12 do not belong to $PC(S_{1,0})$, either. We will see this later.

Remark. Note that in Lemma 27.2 the condition ((3)) was assumed. Thus, it does not imply the claim $F + A_1 \notin PC(J_{3,0}) \Rightarrow F \notin PC(S_{1,0})$ for any Dynkin graph F with only components of type A. However, this claim seems to hold. If the number of vertices of F is at least 11, we can show this claim by case-by-case calculation.

We proceed to the case $r = 11$. In this case, by Lemma 27.1 (3) we can write $G = \sum_{i=1}^{5} A_{k_i}$ with $k_1 \geq k_2 \geq \cdots \geq k_5 \geq 0$, $\sum_{i=1}^{5} k_i = 11$. (A_0 stands for the empty graph.) Thus, G corresponds to a division of 11 into a sum of 5 non-negative integers. There are 37 such divisions as follows. We omit 0.

⟨1⟩ 11	⟨2⟩ 10+1	⟨3⟩ 9+2	⟨4⟩ 8+3
⟨5⟩ 7+4	⟨6⟩ 6+5	⟨7⟩ 9+1+1	⟨8⟩ 8+2+1
⟨9⟩ 7+3+1	⟨10⟩ 7+2+2	⟨11⟩ 6+4+1	⟨12⟩ 6+3+2
⟨13⟩ 5+5+1	⟨14⟩ 5+4+2	⟨15⟩ 5+3+3	⟨16⟩ 4+4+3
⟨17⟩ 8+1+1+1	⟨18⟩ 7+2+1+1	⟨19⟩ 6+3+1+1	⟨20⟩ 6+2+2+1
⟨21⟩ 5+4+1+1	⟨22⟩ 5+3+2+1	⟨23⟩ 5+2+2+2	⟨24⟩ 4+4+2+1
⟨25⟩ 4+3+3+1	⟨26⟩ 4+3+2+2	⟨27⟩ 3+3+3+2	⟨28⟩ 7+1+1+1+1
⟨29⟩ 6+2+1+1+1	⟨30⟩ 5+3+1+1+1	⟨31⟩ 5+2+2+1+1	⟨32⟩ 4+4+1+1+1
⟨33⟩ 4+3+2+1+1	⟨34⟩ 4+2+2+2+1	⟨35⟩ 3+3+3+1+1	⟨36⟩ 3+3+2+2+1
⟨37⟩ 3+2+2+2+2			

We divide these 37 items into 4 classes:

NPA11: ⟨1⟩–⟨16⟩, ⟨18⟩, ⟨20⟩, ⟨22⟩, ⟨24⟩.
MPA11: ⟨21⟩, ⟨25⟩, ⟨26⟩, ⟨30⟩, ⟨36⟩.
MIA11: ⟨27⟩, ⟨32⟩, ⟨34⟩, ⟨35⟩.
E11 : ⟨17⟩, ⟨19⟩, ⟨23⟩, ⟨28⟩, ⟨29⟩, ⟨32⟩, ⟨33⟩, ⟨37⟩.

Lemma 27.9. (NPA11) *Every graph F in NPA11 is a subgraph of a graph F' belonging to MPA12 or MOA12. Moreover, if F has a component of type A_9, we can take F' without a component of type A_9.*

Thus, F can be made from the basic graph $A_9 + BC_1$ by 2 of elementary or tie transformations. We can take the procedure of transformations such that the component A_9 of F is not an obstruction component. In particular, $F \in PC(S_{1,0})$.

Lemma 27.10. (MPA11)
(1) We can make $D_{10} + BC_1$ from the basic graph $A_9 + BC_1$ by one tie transformation. From $D_{10} + BC_1$ we can make graphs $\langle 21\rangle$ $A_5 + A_4 + 2A_1$, $\langle 25\rangle$ $A_4 + 2A_3 + A_1$, $\langle 30\rangle$ $A_5 + A_3 + 3A_1$ by one tie transformation.
(2) $\langle 26\rangle$ $A_4 + A_3 + 2A_2$ can be made from the basic graph $E_8 + BC_1$ by two tie transformations. (For example, by the first transformation we can make $E_6 + A_4$ from $E_8 + BC_1$.)
(3) $\langle 36\rangle$ $2A_3 + 2A_2 + A_1$ can be made from the sub-basic graph $B_6 + A_3 + A_2$ by one tie transformation.

Lemma 27.11. (MIA11) *Graphs $\langle 27\rangle$ $3A_3 + A_2$, $\langle 31\rangle$ $A_5 + 2A_2 + 2A_1$, $\langle 34\rangle$ $A_4 + 3A_2 + A_1$, $\langle 35\rangle$ $3A_3 + 2A_1$ do not belong to $PC(S_{1,0})$.*

Proof. Let F denote one of the above 4 graphs. Assuming the existence of a full embedding $Q = Q(F) \hookrightarrow \Lambda_3/P$, we deduce a contradiction. By $S = P \oplus Q \hookrightarrow \Lambda_3$ be the induced embedding. $\widetilde{S}$ denotes the primitive hull if S in Λ_3. $I = \widetilde{S}/S$ can be regarded as an isotropic subgroup of S^*/S. We write $m = \#(I) = [\widetilde{S} : S]$. $T = C(S, \Lambda_3)$ is a lattice with signature $(1, 2)$. $d = d(T)$ denotes the discriminant of T. $d > 0$.

For $\langle 31\rangle$ $A_5 + 2A_2 + 2A_1$ and $\langle 34\rangle$ $A_4 + 3A_2 + A_1$ we apply the method explained in Lemma 24.6. We consider $p = 3$. By Lemma 27.4 (4) $I_3 = 0$. Thus, $3 \not| m$. For $A_5 + 2A_1$ $d = 5 \times 2^5 \times 3^3/m^2$. On the other hand, calculating the discriminant form on $(\widetilde{S}^*/\widetilde{S})_3 = (S^*/S)_3 \cong \mathbf{Z}/3 \oplus \mathbf{Z}/3 \oplus \mathbf{Z}/3$, one knows that $d \equiv 2^3 \times 3^3 \bmod \mathbf{Z}_3^{*2}$. Comparing both expressions of d one has $5 \in \mathbf{Z}_3^{*2}$, a contradiction.

For $A_4 + 3A_2 + A_1$ $d = 5^2 \times 3^3 \times 2^3/m^2$. On the other hand, one has $d \equiv -2^3 \times 3^3 \bmod \mathbf{Z}_3^{*2}$. Thus, one has $-1 \in \mathbf{Z}_3^{*2}$, a contradiction.

For $\langle 27\rangle$ $3A_3 + A_2$ and $\langle 35\rangle$ $3A_3 + 2A_1$ we apply the method in Lemma 24.7. In these cases we can apply Lemma 27.4 to look for overlattices and the calculation is slightly easier than in Lemma 24.7.

Consider $3A_3 + A_2$. In this case $d = 5 \times 3 \times 2^8/m^2$. $(S^*/S)_2 \cong \mathbf{Z}/4 \oplus \mathbf{Z}/4 \oplus \mathbf{Z}/4 \oplus \mathbf{Z}/2 \oplus \mathbf{Z}/2$. The first component of the right-hand side, the second and the third correspond to A_3-components respectively. The fourth and the fifth $\mathbf{Z}/2$ are induced by P. The discriminant form on it can be written $q_2(\overline{\alpha}) \equiv 3(a_1^2 + a_2^2 + a_3^2)/4 - (b_1^2 + b_2^2)/2 \bmod 2\mathbf{Z}$. Since $5 = l((S^*/S)_2) > 3 = \operatorname{rank} T$, $I_2 \neq 0$. By Lemma 27.4 (2) I_2 is the cyclic group of order 2 generated by $(2, 2, 2, 1, 1)$. Thus, $m = 2n$ with an odd integer n and $d = 5 \times 3 \times 2^6/n^2$. Consider the subgroup J generated by $\overline{\beta}_1 = (1, 0, 0, 1, 0)$, $\overline{\beta}_2 = (0, 1, 0, 0, 1)$, $\overline{\beta}_3 = (2, 0, 1, 1, 0) \in (S^*/S)_2$ here. J is the direct sum of 3 cyclic groups of order 4 generated by $\overline{\beta}_i$'s. $I_2^\perp = J + I_2$ and $J \cap I_2 = 0$. Thus, we have isomorphisms $(\widetilde{S}^*/\widetilde{S})_2 \cong I_2^\perp/I_2 \cong J$ preserving the finite quadratic forms. Since $\overline{\beta}_i$'s are mutually orthogonal and since $q_2(\overline{\beta}_1) \equiv q_2(\overline{\beta}_2) \equiv 1/4 \bmod 2\mathbf{Z}$,

$q_2(\overline{\beta}_3) \equiv -3/4 \bmod 2\mathbf{Z}$, one can conclude that $T \otimes \mathbf{Z}_2$ is isomorphic to the 2-adic lattice defined by the diagonal 3×3 matrix whose diagonal entries are -2^2, -2^2 and 3×2^2. It implies $d \equiv 3 \times 2^6 \bmod \mathbf{Z}_2^{*2}$. By the 2 expressions of d we have $5 \in \mathbf{Z}_2^{*2}$, a contradiction.

Next, we consider $3A_3 + 2A_1$. $S^*/S = (S^*/S)_2 \oplus \mathbf{Z}/5$. $(S^*/S)_2 \cong (\mathbf{Z}/4)^{\oplus 3} \oplus (\mathbf{Z}/2)^{\oplus 4}$. The first component, the second and the third correspond to A_3-components respectively. The 4-th and the 5-th are induced by P, and the 6-th and the 7-th are induced by A_1-components respectively. The discriminant form can be given for $\overline{\alpha} = (a_1, a_2, a_3, b_1, b_2, c_1, c_2) \in (\mathbf{Z}/4)^{\oplus 3} \oplus (\mathbf{Z}/2)^{\oplus 4}$ by $q_2(\overline{\alpha}) \equiv 3(a_1^2 + a_2^2 + a_3^2)/4 - (b_1^2 + b_2^2)/2 + (c_1^2 + c_2^2)/2 \bmod 2\mathbf{Z}$. On the other hand, $3 = \operatorname{rank} T \geq l((S^*/S)_2) - 2l(I_2) = 7 - 2l(I_2)$. Thus, $l(I_2) \geq 2$. By Lemma 27.4 (3) I_2 is an abelian group of type $(2, 2)$, and after exchanging the order of 3 of A_3 and 2 of A_1, we can assume that I_2 is generated by $(2, 2, 0, 1, 0, 1, 0)$ and $(2, 0, 2, 0, 1, 0, 1)$. Consider here the subgroup J in $(S^*/S)_2$ generated by the following 3 elements; $\overline{\beta}_1 = (1, 0, 0, 0, 0, 1, 1)$, $\overline{\beta}_2 = (0, 1, 0, 1, 0, 0, 0)$, $\overline{\beta}_3 = (0, 0, 1, 0, 1, 0, 0)$. $J + I_2 = (S^*/S)_2$ and $J \cap I_2 = 0$. Thus, $(\widetilde{S}^*/\widetilde{S})_2 \cong J$. $\overline{\beta}_i$'s are mutually orthogonal, and $q_2(\overline{\beta}_1) \equiv -1/4$, $q_2(\overline{\beta}_2) \equiv q_2(\overline{\beta}_3) \equiv 1/4 \bmod 2\mathbf{Z}$. Thus, $T \otimes \mathbf{Z}_2$ is isomorphic to the 2-adic lattice defined by the diagonal 3×3 matrix whose diagonal entries are 2^2, -2^2 and -2^2. It implies $d \equiv 2^6 \bmod \mathbf{Z}_2^{*2}$. On the other hand, $\widetilde{S}^*/\widetilde{S} = (\widetilde{S}^*/\widetilde{S})_2 \oplus \mathbf{Z}/5$ and $d = 5 \times 2^6$. By the 2 expressions we have $5 \in \mathbf{Z}_2^{*2}$, a contradiction. Q.E.D.

Lemma 27.12. (E11) *For every graph F in E11 $F + A_1 \notin PC(J_{3,0})$.*

Indeed, $F + A_1$ corresponds to one of the 13 graphs in paragraph **A** in Section 13 just after Proposition 13.5. By Lemma 27.2 we need not consider graphs in E11.

Remark. We can show that every graph in E11 does not belong to $PC(S_{1,0})$ by case-by-case calculation.

Lemma 27.13. (NIA12) *Every graph in NIA 12 contains a subgraph belonging to MIA11. Thus, every graph in NIA12 does not belong to $PC(S_{1,0})$.*

The last remaining case is the case $r = 10$.

Lemma 27.14. *If a Dynkin graph F with components of type A only has 10 vertices, then F does not satisfy the I1-condition if, and only if, F is one of the following graphs:*

$$A_4 + 2A_3, \quad A_4 + A_3 + A_2 + A_1, \quad A_4 + 2A_2 + 2A_1, \quad A_3 + 2A_2 + 3A_1,$$
$$A_4 + 6A_1.$$

Proof. Recall that the I1-condtion fails if, and only if, $5d(F) \in \mathbf{Q}_p^{*2}$ and $\epsilon_p(F) \neq (-2, 5)_p$ for some prime number p. With the aid of lemmas in Section 7 we can show the above lemma. Q.E.D.

Here, $A_4 + 6A_1$ does not satisfy the condition in Lemma 27.1 (3). Thus, we need not consider $A_4 + 6A_1$.

Lemma 27.15. *A_4+2A_3 is a subgraph of a graph in MOA12. So do $A_4+A_3+A_2+A_1$ and $A_4+2A_2+2A_1$. $A_3+2A_2+3A_1$ is a subgraph of ⟨30⟩ $A_5+A_3+3A_1$ in MPA11. Thus, they 4 can be made from the basic graph A_9+BC_1 by 2 of tie or elementary transformations. In particular, they belong to $PC(S_{1,0})$.*

We have studied all cases and we can complete the verification of Proposition 25.1 under assumption ((3)).

By the results in Section 25, 26 and 27, Proposition 25.1 and Theorem 0.6 have been established.

Chapter 4

28
Concept of co-root modules

In this last chapter we study the last remaining case $U_{1,0}$. This case has 2 characteristics. The first is that the curve IF at infinity contains 3 of sections. Therefore, we cannot apply the theory of transcendental cycles in Section 10 and 11. The second is that the associated lattice P has an even proper overlattice $\hat{P}$. For this reason we have to treat a non-regular root module $(\Lambda_N/P, F)$. Moreover, since there exists an isotropic vector $u \in \Lambda_N/P$ with $u \notin F$, the theory of elementary transformations and tie transformations in Section 6 is not sufficient to treat $U_{1,0}$.

Therefore, in this section we would first like to give the theoretical ground of dual elementary transformations. To formulate them we introduce the concept of co-root modules.

Definition 28.1. The pair $(L, U) = (L, (\ ,\ ), U)$ of a lattice $L = (L, (\ ,\ ))$ and a subset $U \subset L$ satisfying the following 3 conditions $\langle 1\rangle$, $\langle 2\rangle$, $\langle 3\rangle$ is called a *co-root module*:

$\langle 1\rangle$ For every $\alpha \in U$ $\alpha^2 = 2$, 4 or 6.

$\langle 2\rangle$ For every $\alpha \in U$ and for every $x \in L$ $(\alpha^\vee, x) \in \mathbf{Z}$, where $\alpha^\vee = 2\alpha/\alpha^2$.

$\langle 3\rangle$ Let $s_\alpha : L \to L$ denotes the reflection associated with $\alpha \in U$, which is defined by $s_\alpha(x) = x - (x, \alpha^\vee)\alpha$ for $x \in L$. For every $\alpha \in U$ $s_\alpha(U) = U$.

We call U the *root system* of (L, U), $\alpha \in U$ a *root*, and $\alpha^\vee$ the *co-root* of α.

Let M be a submodule of L. By $(\ ,\ )_M$ we denote the restriction of the bilinear form to M. Then, $(M, (\ ,\ )_M, M \cap U)$ is a co-root module. It is called a *co-root submodule* of (L, U).

Definition 28.2. We define that a homomorphism $f : (M, (\ ,\ ), U_M) \to (L, (\ ,\ ), U_L)$ between 2 co-root modules is a homomorphism $f : M \to L$ of abelian groups satisfying the following $\langle\!\langle 1\rangle\!\rangle$ and $\langle\!\langle 2\rangle\!\rangle$:

$\langle\!\langle 1\rangle\!\rangle$ f preserves the bilinear forms.

$\langle\!\langle 2\rangle\!\rangle$ $f^{-1}(U_L) = U_M$.

We say that f is an *embedding*, if f is injective. When f is an embedding, f is *full*, if $f(M) \cap U_L = \widetilde{f(M)} \cap U_L$, where $\widetilde{f(M)}$ denotes the primitive hull of $f(M)$ in L.

Example 28.3. Let $L = (L, (\ ,\ ))$ be a lattice. Set

$$U = \{\alpha \in L \mid \alpha^2 = 2,\ 4 \text{ or } 6. \text{ For every } x \in L\ 2(x, \alpha) \in \alpha^2\mathbf{Z}.\}.$$

Then, $(L, (\ ,\ ), U)$ is a co-root module.

Example 28.4. Let L be a non-degenerate lattice. Let L^* be the dual quasi-lattice of L. We assume that for some submodule FL with $L \subset FL \subset L^*$ the pair (L^*, FL) is a root module. By $\hat{R}$ we denote the *reduced* root system of (L^*, FL). That is,

$$\hat{R} = \{\alpha \in FL \mid \alpha^2 = 2\} \cup \{\beta \in L^* \mid \beta^2 = 2/3 \text{ or } 1\}.$$

We write $\alpha^\vee = 2\alpha/\alpha^2$ for $\alpha \in L \otimes \mathbf{Q}$ and $X^\vee = \{\alpha^\vee \mid \alpha \in X\}$ for a subset $X \subset L \otimes \mathbf{Q}$. Note that $(\alpha^\vee)^\vee = \alpha$.

On the other hand, as in Example 28.3, we can define a co-root module (L, U). It is easy to see $\hat{R}^\vee = U \subset L$ here.

Let $T \subset L^*$ be a submodule. For any $\alpha \in \hat{R} \cap T$ we can write $\alpha^\vee = k\alpha$ with $k = 1$, 2 or 3. Thus, $\alpha^\vee \in \hat{R}^\vee \cap T$. We have $\hat{R}^\vee \cap T \supset (\hat{R} \cap T)^\vee$.

Lemma 28.5. *(1) For any submodule $T \subset L^*$ $(T \cap L, (\ ,\ )_{T\cap L}, (\hat{R} \cap T)^\vee)$ is a co-root module.*
(2) If $T \subset L^*$ is full as a root submodule, then

$$\hat{R}^\vee \cap T = (\hat{R} \cap T)^\vee.$$

Proof. (1) We check that $U_T = (\hat{R} \cap T)^\vee$ satisfies the conditions of the root system of a co-root module. Obviously, $U_T \subset T$ and $\langle 1 \rangle$ is satisfied. For any $\alpha \in \hat{R} \cap T$ and for any $x \in T \cap L$ $(\alpha^{\vee\vee}, x) = (\alpha, x) \in \mathbf{Z}$, since $\alpha \in L^*$. Thus, $\langle 2 \rangle$ holds. We check $\langle 3 \rangle$. For any $\alpha \in \hat{R} \cap T$ $s_{\alpha^\vee}(\hat{R} \cap T) = s_\alpha(\hat{R} \cap T) = \hat{R} \cap T$, since $s_{\alpha^\vee} = s_\alpha$. Thus, $s_{\alpha^\vee}(U_T) = U_T$.
(2) By the above $\hat{R}^\vee \cap T \supset U_T$.

Next, assume that $\alpha \in \hat{R}$ and $\alpha^\vee \in T$. Since $\alpha^\vee = k\alpha$ for $k = 1$, 2 or 3, $\alpha \in \widetilde{T}$. Since T is full, we have $\alpha \in T$. Thus, $\hat{R}^\vee \cap T \subset (\hat{R} \cap T)^\vee = U_T$. Q.E.D.

Let (L, U) be a co-root module, and (M, U_M) be a co-root submodule. We assume that M is positive definite. The root system U_M is a finite set satisfying the axioms of a root system in Bourbaki [3] except that U_M need not span M. Thus, we can choose a root basis $\Delta \subset U_M$. Let $U_M = \bigoplus_{i=1}^m U_i$ be the irreducible decomposition. Let $\zeta_i \in U_i$ be the maximal root of the irreducible component U_i associated with the irreducible root basis $\Delta_i = \Delta \cap U_i$. The extended root basis $\Delta^+ = \bigcup_{i=1}^m \Delta_i^+$ where $\Delta_i^+ = \Delta_i \cup \{-\zeta_i\}$ is also defined. We would like to draw a graph associated with Δ or Δ^+.

Now, associated with a finite subset S of L consisting of vectors with length $\sqrt{2}$, 2 or $\sqrt{6}$, we can draw a graph Γ by the following rules:

(1) Vertices in Γ have one-to-one correspondence with vectors in S.
(2) Any vertex in Γ has one of 4 different expressions and an expression has the following meaning depending on the corresponding vector $\alpha \in S$:

expression	$\circ$	$\bullet$	$\circledcirc$	$\otimes$
meaning	$\alpha^2 = 2$ and $\alpha \cdot x$ is an odd integer for some $x \in L$.	$\alpha^2 = 4$	$\alpha^2 = 6$	$\alpha^2 = 2$ and $\alpha \cdot x$ is an even integer for any $x \in L$.

(3) If two vectors α, $\beta \in S$ are orthogonal, then we do not connect the corresponding two vertices in Γ. $\alpha \ast \qquad \ast\ \beta$. ($\ast$ denotes one of $\circ$, $\bullet$, $\circledcirc$, and $\otimes$.)
(4) If α and $\beta \in S$ are not orthogonal, then the corresponding two vertices in Γ are connected by a single edge. $\alpha \ast\!\!-\!\!-\!\!-\!\!\ast\ \beta$.
(5) If α and $\beta \in S$ are proportional, i.e., $\beta = t\alpha$ for some real number t, then the edge connecting the corresponding two vertices is replaced by a bold edge. $\alpha \ast\!\!\blacksquare\!\!\blacksquare\!\!\ast\ \beta$.
Note that the graph Γ depends not only on S but also on the embedding $S \subset L$.

If S is a root basis Δ, then the graph Γ is called the Dynkin graph of the pair (M, L). It does not depend on the choice of Δ. When we need not mention L, we say that Γ is the Dynkin graph of the co-root module (M, U_M).

Proposition 28.6. *We consider the situation in Example 28.4. Assume that a submodule $T \subset L^*$ is positive definite and full as a root submodule.*

(1) The Dynkin graph of a root module $(T, T \cap FL)$ and the Dynkin graph of a pair $(T \cap L, L)$ are equal.

(2) The graph associated with the extended root basis Δ^+ of the root system $T \cap U$ is the dual extended Dynkin graph associated with the Dynkin graph of $(T \cap L, T \cap U)$. (Recall the definition of the dual extended Dynkin graph in the Introduction just before Definition 0.7.)

Proof. (1) From Lemma 28.5 (2) and from the definitions of Dynkin graphs the claim follows. When a graph of type BC is concerned, note that for a vector $\alpha \in L$, $\alpha^2 = 2$ and $\alpha \cdot x \in 2\mathbf{Z}$ for any $x \in L$ if, and only if, $(\alpha/2)^2 = 1/2$ and $\alpha/2 \in L^*$.
(2) Obvious by the definitions. Q.E.D.

Lemma 28.7. *Let (L, U) be a positive definite co-root module. Let $M \subset L$ be a submodule.*

(1) If M is primitive in L, then any root basis of $U \cap M$ can be extended to a root basis of U.

(2) If the subgroup of the quotient L/M consisting of elements of finite order is cyclic, then there are a root basis Δ_M of $U \cap M$ and a root basis Δ_L of U such that $\Delta_M \subset \Delta_L^+$. That is, the root system $U \cap M$ is obtained from the root system U by one elementary transformation.

Proof. (1) The verification is same as that of Proposition 6.1 (1) for root modules.
(2) It is easy to check the corresponding claim to Lemma 5.3, i.e., the equality $\mathbf{R}\alpha^\vee \cap Q(U^\vee) = \mathbf{Z}\alpha^\vee$ for every $\alpha \in U$. Thus, it follows from results in Urabe [16]. (Prop. 2.5, Cor. 2.6, Prop. 2.9 (4), Lemma 2.10 in [16].) Q.E.D.

Theorem 28.8. *Let $n = 2$ or 3, and l be an integer. Let $K = \mathbf{Z}s + \mathbf{Z}t$ be a lattice with a basis s, t satisfying $s^2 = 0$, $s \cdot t = n$ and $t^2 = l$. Let L be a lattice. We define root systems U and V by the following:*

$$U = \{\alpha \in L \mid \alpha^2 = 2 \text{ or } 2n. \text{ For every } x \in L\ 2(x, \alpha) \in \alpha^2\mathbf{Z}\}$$
$$V = \{\beta \in L \oplus K \mid \beta^2 = 2 \text{ or } 2n. \text{ For every } x \in L \oplus K\ 2(x, \beta) \in \beta^2\mathbf{Z}\}$$

We consider the co-root modules (L, U) and $(L \oplus K, V)$.

(1) Let M be a positive definite full co-root submodule of $L \oplus K$ contained in the orthogonal complement $L \oplus \mathbf{Z}s$ of the vector s. Let $Q = Q(V \cap M)$ be the sublattice of M generated by roots $V \cap M$ in M. The image J_0 of Q under the composition of homomorphisms $\pi : Q \subset L \oplus \mathbf{Z}s \to (L \oplus \mathbf{Z}s)/\mathbf{Z}s \cong L$ satisfies the following condition $\langle * \rangle$:

$\langle * \rangle$ *For any positive definite full co-root module J with $J_0 \subset J \subset L$ the root system of $V \cap M$ can be obtained from the root system $U \cap J$ by one elementary transformation.*

(2) Let $J \subset L$ be a positive definite full co-root submodule. Let U' be a root system obtained from the root system $U \cap J$ by one elementary transformation. Then, there exists a full embedding of co-root modules $\phi : (Q(U'), U') \to (L \oplus K, V)$ whose image is contained in the orthogonal complement $L \oplus \mathbf{Z}s$ of s.
(3) We consider the case $n = 2$ in particular. Set

$$U_\times = \{\alpha \in U \mid \alpha^2 = 2. \text{ For every } x \in L\ (x, \alpha) \in 2\mathbf{Z}\}$$
$$V_\times = \{\beta \in V \mid \beta^2 = 2. \text{ For every } x \in L \oplus K\ (x, \beta) \in 2\mathbf{Z}\}.$$

In the situation of (1) the induced mapping $\overline{\pi} : V \cap M \to U \cap J$ satisfies $\overline{\pi}^{-1}(U_\times \cap J) = V_\times \cap M$.

Remark. The above (3) is necessary to distinguish the vertex ⊗ from the vertex o.

Proof. (1) By $\widetilde{Q}$ we denote the primitive hull of the root lattice $Q = Q(V \cap M)$ of M in $L \oplus K$. We have $\widetilde{Q} \subset \widetilde{M}$, and $V \cap \widetilde{Q} = V \cap M$. Note here that $Q \subset \widetilde{Q} \subset L \oplus \mathbf{Z}s$. Let $\pi : L \oplus \mathbf{Z}s \to L$ be the projection. Since s is isotropic, π preserves the bilinear forms. Moreover, $\pi \mid \widetilde{Q}$ is injective, since $\widetilde{Q}$ is positive definite. Set $J_0 = \pi(Q)$. Let $\omega_1, \ldots, \omega_m$ be a basis of L^*. Then, we have

$$\widetilde{Q} = \{x \in Q \otimes \mathbf{Q} \mid x \cdot \omega_i \in \mathbf{Z}\ (1 \le i \le m),\ x \cdot t \in n\mathbf{Z}\}$$
$$\widetilde{J_0} \cong \{x \in Q \otimes \mathbf{Q} \mid x \cdot \omega_i \in \mathbf{Z}\ (1 \le i \le m)\}.$$

Therefore, $\widetilde{J_0}/\pi(\widetilde{Q})$ is a cyclic group.

Now, let J be a positive definite full co-root submodule with $J_0 \subset J \subset L$. $\widetilde{J_0} \subset \widetilde{J}$. By fullness $U \cap \widetilde{J} = U \cap J$. Since $\widetilde{J}/\widetilde{J_0}$ is a free module, the subgroup of $\widetilde{J}/\pi(\widetilde{Q})$ consisting of elements of finite order is cyclic. Thus, by Lemma 28.7 (2) the root system $\pi(V \cap \widetilde{Q}) = \pi(V \cap M) \cong V \cap M$ is obtained from $U \cap \widetilde{J} = U \cap J$ by one elementary transformation.
(2) Let $U \cap J = U_1 \cup U_2 \cup \cdots \cup U_m$ be the irreducible decomposition of the root system. Let $\Delta_i \subset U_i$ be a root basis, $\zeta_i \in U_i$ be the associated maximal root. Set $\Delta_i^+ = \Delta_i \cup \{-\zeta_i\}$. We choose a proper subset $\Delta_i' \subset \Delta_i^+$ for each i. We can assume that U' is the root system whose root basis is $\bigcup_{i=1}^m \Delta_i'$. Here, $\bigcup_{i=1}^m \Delta_i' \subset L$. We define a homomorphism $\phi : Q(U') \to L \oplus K$ by setting for a vector $\alpha \in \Delta_i'$ in $\bigcup_{i=1}^m \Delta_i'$

$$\phi(\alpha) = \begin{cases} \alpha \oplus 0 & (\text{if } \alpha \neq -\zeta_i) \\ \alpha \oplus s & (\text{if } \alpha = -\zeta_i). \end{cases}$$

Obviously, the image is contained in the orthogonal complement of s, and ϕ preserves the bilinear forms. Since $Q(U')$ is non-degenerate, ϕ is an embedding.

Let $\alpha \in U'$ be an arbitrary root. If $\alpha^2 = 2$, then for every $x \in L \oplus K$ $(x, \phi(\alpha)) \in \mathbf{Z} \Leftrightarrow 2(x, \phi(\alpha)) \in 2\mathbf{Z}$. Next, assume $\alpha^2 = 2n$. We can write $\phi(\alpha) = \alpha \oplus ms$ with $m \in \mathbf{Z}$. An arbitrary vector $x \in L \oplus K$ can be written $x = y + as + bt$ with $y \in L$, a,

$b \in \mathbf{Z}$. $(x, \phi(\alpha)) = (y, \alpha) + mbn$. Here, $y \cdot \alpha \in n\mathbf{Z}$, since $\alpha \in U$. Thus, $x \cdot \phi(\alpha) \in n\mathbf{Z}$. Consequently, we have $\phi(U') \subset V$.

The equality $P(\phi(Q(U')), L \oplus K) \cap V = \phi(U')$ follows from the same argument as in the proof to show fullness in Proposition 4.2 of Urabe [18]. Thus, in particular, we have $\phi^{-1}(V) = U'$.

(3) It is easy to see that if $\alpha \in V_\times \cap M$, then $\overline{\pi}(\alpha) \in U_\times \cap J$. Thus, $\overline{\pi}^{-1}(U_\times \cap J) \supset V_\times \cap M$. Conversely, any element in $\overline{\pi}^{-1}(U_\times \cap J) \subset V \cap M$ can be written in the form $\alpha + ms$ with $\alpha \in U_\times \cap J$, $m \in \mathbf{Z}$. For any $y \in L$, a, $b \in \mathbf{Z}$ we have $(\alpha + ms, y + as + bt) = (\alpha, y) + mbn \in 2\mathbf{Z}$, since $n = 2$ and $\alpha \in U_\times$. Thus, $\alpha + ms \in V_\times \cap M$. Q.E.D.

Proposition 28.9. *In addition to the situation in Theorem 28.8 we assume either the following* $\langle a \rangle$ *and* $\langle b \rangle$*, or the following* $\langle a \rangle$ *and* $\langle c \rangle$*:*

$\langle a \rangle$ *The pair* $(L^* \oplus K^*, L \oplus K)$ *is a root module.*

$\langle b \rangle$ $n = 2$ *and* $\alpha^2 \neq 2/3$ *for every* $\alpha \in L^* \oplus K^*$.

$\langle c \rangle$ $n = 3$, $\alpha^2 \neq 1$ *for every* $\alpha \in L^* \oplus K^*$, *and* $2\beta \notin L \oplus K$ *for every* $\beta \in L^* \oplus K^*$ *with* $\beta^2 = 1/2$.

(1) For every positive definite full root submodule $S \subset L^* \oplus K^*$ *orthogonal to* $s \in K$, *there exists a positive definite root submodule* T_0 *of* L^* *with* $\operatorname{rank} T_0 =$ *(rank of the root system of* S*) satisfying the following* $\langle *' \rangle$*:*

$\langle *' \rangle$ *For any positive definite full root module* T *with* $T_0 \subset T \subset L^*$ *the Dynkin graph of* S *can be obtained from the Dynkin graph of* T *by one dual elementary transformation.*

(2) Let $T \subset L^*$ *be a positive definite full root submodule with the Dynkin graph* G. *Let* G' *be a Dynkin graph obtained from* G *by one dual elementary transformation. Then, there is a full embedding of root modules* $Q(G') \hookrightarrow L^* \oplus K^*$ *whose image is orthogonal to* $s \in K$.

Proof. It follows from Lemma 28.5, Proposition 28.6 and Theorem 28.8. Q.E.D.

Next, we consider the behavior of an obstruction component under a dual elementary transformation. We assume the situation in Theorem 28.8 and Proposition 28.9. We can go along the line in the former half of Section 15. Let $k \geq 4$ and $K = \mathbf{Z}s + \mathbf{Z}t$ $(s^2 = 0,\ s \cdot t = 2 \text{ or } 3,\ t^2 = l)$ be the lattice introduced in Theorem 28.8. Let G' be a Dynkin graph. We consider the case where a full embedding of root modules

$$Q(G') \hookrightarrow L^* \oplus K^*$$

into the orthogonal complement of s is given. We assume moreover that G' has an obstruction component G'_0 of type A_k. By definition

$$[P(Q(G'_0),\ L^* \oplus K^*) : Q(G'_0)] = k + 1.$$

By $\pi : L^* \oplus K^* \to L^*$ we denote the projection. Let $T \subset L^*$ be a positive definite full root submodule containing the image $\pi(Q(G'))$, and $Q(T)$ be the root quasi-lattice of T. Let $\Delta' \subset Q(G')$ and $\Delta \subset Q(T)$ be the respective root bases. Let $\Delta = \bigcup_{i=1}^{m} \Delta_i$ be the irreducible decomposition. When Δ_i is not of type BC_r for any $r \geq 2$, we pick the maximal short root $\zeta_i \in Q(T)$ associated with Δ_i. In the case Δ_i is of type BC_r for some $r \geq 2$ we pick the maximal element ζ_i in the set of roots with length

1 associated with Δ_i. (The maximality is defined with respect to the lexicographical order associated with a total order on Δ_i. ζ_i does not depend on the total order on Δ_i and it depends only on Δ_i.) Set $\Delta_i^\bullet = \Delta_i \cup \{-\zeta_i\}$ and $\Delta^\bullet = \bigcup_{i=1}^m \Delta_i^\bullet$. Under this notation we can assume $\pi(\Delta') \subset \Delta^\bullet$ by Lemma 28.5, Proposition 28.6 and Theorem 28.8. Now, replacing Δ^+ by $\Delta^\bullet$, we can follow the lines in the verification of Lemma 15.1. We get the following:

Lemma 28.10. *The component Q_1 of $Q(T)$ containing the image $\pi(Q(G_0'))$ is also of type A_k and $[P(Q_1, L^*) : Q_1] = k+1$.*

Next, we consider the situation when we go up from L^* to $L^* \oplus K^*$.

Let R be a finite root system and $R = \bigoplus_{i=1}^m R_i$ be the irreducible decomposition. Assume that R_1 is of type A_k with $k \geq 4$. Assume further that a full embedding $Q(R) \hookrightarrow L^*$ of root modules such that $Q_1 = Q(R_1)$ and $\widetilde{Q}_1 = P(Q_1, L^*)$ satisfy $[\widetilde{Q}_1 : Q_1] = k+1$ is given. Let U' be a root system obtained from $U = \hat{R}^\vee$ by one elementary transformation. We assume moreover $U'^\vee \cap R_1 = R_1 \Leftrightarrow U' \cap R_1 = R_1$.

Lemma 28.11. *There exists a full embedding $\phi : Q(U'^\vee) \hookrightarrow L^* \oplus K^*$ of root modules satisfying either the following conditions (1), (2) and (3), or (1), (2) and (4). We denote here $Q_1' = \phi(Q_1)$ and $\widetilde{Q}_1' = P(Q_1', L^* \oplus K^*)$:*
(1) The image $\phi(Q(U'^\vee))$ of ϕ is orthogonal to $s \in K$.
(2) The composition of ϕ and the projection $L^* \oplus K^* \to L^*$ coincides with the given embedding $Q(U'^\vee) \hookrightarrow Q(R) \hookrightarrow L^*$.
(3) $[\widetilde{Q}_1' : Q_1'] < k+1$
(4) $[\widetilde{Q}_1' : Q_1] = k+1$.

Proof. Let $\Delta \subset R$ be a root basis, and $\Delta = \bigcup_{i=1}^m \Delta_i$ with $\Delta_i = \Delta \cap R_i$ be the irreducible decomposition. We consider the set $\Delta^\bullet = \bigcup_{i=1}^m \Delta_i^\bullet$ defined above. By definition $\Delta_i^\bullet = \Delta_i \cup \{-\zeta_i\}$.

For every i with $1 \leq i \leq m$ we have a proper subset $\Delta_i' \subset \Delta_i^\bullet$ and $\bigcup_{i=1}^m \Delta_i'$ is a root basis of $U'^\vee$. Here, $\Delta_1 = \{\alpha_1, \ldots, \alpha_k\}$ is of type A_k (We assign numbers of α_i's from the end of the Dynkin graph A_k in order.) and $\Delta_1^\bullet$ consists of $k+1$ elements. We have $k+1$ ways of choosing Δ_1', and under any choice the root basis Δ_1' generates the same root system $U'^\vee \cap R_1$.

In order to define the embedding ϕ satisfying (3), we choose $\Delta_1' = \{\alpha_2, \alpha_3, \ldots, \alpha_k, -\zeta_i\}$. To define ϕ satisfying (4) we choose $\Delta_1' = \Delta_1$. Note that $\Delta' = \bigcup_{i=1}^m \Delta_i'$ is a basis of $Q(U'^\vee)$. We define the embedding ϕ by setting for $\alpha \in \Delta_i'$

$$\phi(\alpha) = \begin{cases} \alpha \oplus 0 & (\text{if } \alpha \neq -\zeta_i) \\ \alpha \oplus s & (\text{if } \alpha = -\zeta_i \text{ and } \zeta_i^2 = 2) \\ \alpha \oplus s/n & (\text{if } \alpha = -\zeta_i \text{ and } \zeta_i^2 < 2). \end{cases}$$

(Note that $s/n \in K^*$.) Obviously, it defines an embedding of quasi-lattices satisfying (1) and (2). The fullness follows from Lemma 28.5, Proposition 28.6 and Theorem 28.8.

We show the condition (3). Assume $\Delta_1' = \{\alpha_2, \alpha_3, \ldots, \alpha_k, -\zeta_1\}$. Since α_2 corresponds to an end of the Dynkin graph A_k made from Δ_1', the fundamental weight ω corresponding α_2 can be written $\omega = \{k\alpha_2 + (k-1)\alpha_3 + \cdots + 2\alpha_k - \zeta_1\}/(k+1)$. Now, if

$[\widetilde{Q}'_1 : Q_1] \geq k+1$, then we have $\phi(\omega) \in L^* \oplus K^*$. However, $\phi(\omega) = \omega \oplus (n/k+1)(s/n) \notin L^* \oplus K^*$, a contradiction. (Notice that $s/n \in K^*$ is primitive in K^*.) We have (3).

If $\Delta'_1 = \Delta_1$, then $\widetilde{Q}'_1 = \widetilde{Q}_1$, and we have (4), since $[\widetilde{Q}_1 : Q_1] = k+1$. Q.E.D.

By Lemma 28.10 and Lemma 28.11 one knows that obstruction components behave like in Definition 0.4 (Addition) under a dual elementary transformation.

Now, we would like to apply the above general theory to our singularity $U_{1,0}$. The following is the key for the application:

Proposition 28.12. *Let $P = P(2,3,3,3)$ be the lattice in the case $U_{1,0}$. We fix a primitive embedding $P \hookrightarrow \Lambda_N$, $N \geq 2$. The orthogonal complement F of P is identified with its image under the canonical surjective homomorphism $\Lambda_N \to \Lambda_N/P$. Assume that a primitive isotropic vector $u \in \Lambda_N/P$ does not belong to F. Then, $s = 3u \in F$ and there exists a vector $t \in F$ with $s \cdot t = 3$, $t^2 = -2$ such that $K = \mathbf{Z}s + \mathbf{Z}t$ is a direct summand of F. Thus, in particular, $F = L \oplus K$ and $\Lambda_N/P \cong L^* \oplus K^*$ where $L = C(K, F)$. Moreover, for some primitive embedding $\hat{P} \hookrightarrow \Lambda_{N-1}$ $L^* \cong \Lambda_{N-1}/\hat{P}$ and L is isomorphic to the orthogonal complement of $\hat{P}$ in Λ_{N-1}.*

Proof. $F^*/F \cong \mathbf{Z}/9 \oplus \mathbf{Z}/3$. By q we denote the discriminant quadraric form of F and by b we denote the discriminant bilinear form of F. For $\overline{\alpha} = (x, y) \in \mathbf{Z}/9 \oplus \mathbf{Z}/3$ we can write

$$q(\overline{\alpha}) \equiv \frac{2}{9}x^2 + \frac{2}{3}y^2 \bmod 2\mathbf{Z}.$$

In the following we denote $\overline{x} = x \bmod F \in F^*/F$ for $x \in \Lambda_N/P \cong F^*$.

By assumption $\overline{u} \neq 0$. We have $\overline{u} = (\pm 3, 0)$ since $q(\overline{u}) \equiv u^2 = 0$. Thus, $3\overline{u} = 0$. Namely, $s = 3u \in F$.

Since u is primitive in F^*, we have a vector $\alpha_1 \in F$ with $u \cdot \alpha_1 = 1$. On the other hand, for a vector $\alpha_2 \in F^*$ with $\overline{\alpha}_2 = (1, 0) \in \mathbf{Z}/9 \oplus \mathbf{Z}/3$ we have $u \cdot \alpha_2 \equiv b(\overline{u}, \overline{\alpha}_2) \equiv b((\pm 3, 0), (1, 0)) \equiv \pm 2/3 \bmod \mathbf{Z}$. Thus, we can conclude $\frac{1}{3}\mathbf{Z} \subset \{u \cdot \alpha \mid \alpha \in F^*\}$.

In particular, we have a vector $\tau \in F^*$ with $u \cdot \tau = 1/3$. Setting $\overline{\tau} = (x_0, y_0)$, $x_0 \not\equiv 0 \pmod 3$, since $1/3 \equiv \pm 2x_0/3 \bmod \mathbf{Z}$. Thus, $\overline{\tau}$ has order 9 in F^*/F, and $\tau^2 \equiv q(\overline{\tau}) \equiv 2/9, 8/9$ or $-4/9 \bmod 2\mathbf{Z}$. Namely, $9\tau^2 = m \in \mathbf{Z}$ and $m \equiv 2 \pmod 6$.

Now, let $\alpha \in F^*$ be an arbitrary vector. We can write $\alpha \cdot u = a/3$ with $a \in \mathbf{Z}$. Set $\beta = \alpha - a\tau$. $\beta \cdot u = 0$. Thus, we can write $\overline{\beta} = (3z_1, y_1) \in \mathbf{Z}/9 \oplus \mathbf{Z}/3$. $\tau \cdot \beta \equiv b(\overline{\tau}, \overline{\beta}) \equiv 2x_0z_1/3 + 2y_0y_1/3 \bmod \mathbf{Z}$. Thus, we can write $\beta \cdot \tau = b/3$ with $b \in \mathbf{Z}$. Set $\gamma = \beta - bu = \alpha - a\tau - bu$. We can check $\gamma \cdot \tau = \gamma \cdot u = 0$.

The argument so far implies $F^* = \Gamma \oplus \Xi$ where $\Xi = \mathbf{Z}u + \mathbf{Z}\tau$ and $\Gamma = C(\Xi, F^*)$. Here, we write $X^* = \{x \in X \otimes \mathbf{Q} \mid \text{For every } y \in X\ x \cdot y \in \mathbf{Z}\}$ for a non-degenerate quasi-lattice X. We have $(F^*)^* = F \subset F^*$. We get the orthogonal decomposition $F = \Gamma^* \oplus \Xi^*$.

Setting $9\tau^2 = m = 6k+2$, $s = 3u$ and $t = -(3k+2)u + 3\tau$, we can check that these s and t satisfy the desired conditions.

Here, $L = \Gamma^*$ is a lattice with $L^*/L \cong \mathbf{Z}/3$ and the discriminant quadratic form on it is given by $x \in \mathbf{Z}/3 \mapsto 2x^2/3 \bmod 2\mathbf{Z}$. Thus, by Proposition 4.5 and Lemma 5.5 we get the remaining parts of the claim. Q.E.D.

Lemma 28.13. *Let $K = \mathbf{Z}s + \mathbf{Z}t$ be a lattice satisfying $s^2 = 0$, $s \cdot t = 3$ and $t^2 = -2$. For any primitive embedding $\hat{P} \hookrightarrow \Lambda_{N-1}$ with $N \geq 2$ and any primitive embedding $P \hookrightarrow \Lambda_N$, $(\Lambda_{N-1}/\hat{P}) \oplus K^* \cong \Lambda_N/P$.*

Proof. Let $\hat{F}$ be the orthogonal complement of $\hat{P}$ in Λ_{N-1}. We can check that the discriminant quadratic form of $\hat{F} \oplus K$ is equal to (-1) times that of P. By Lemma 5.5 one knows that the above claim holds for some primitive embedding $P \hookrightarrow \Lambda_N$. By Proposition 5.2 we can replace the word "some" by "any". Q.E.D.

Lattice L and K in Proposition 28.12 correspond to the case $n = 3$, $l = -2$ in Theorem 28.8. Moreover, L and K satisfy $\langle a \rangle$ and $\langle c \rangle$ in Proposition 28.9. Therefore, the claim (2) and (3) in Proposition 28.9 hold for $L^* \cong \Lambda_{N-1}/\hat{P}$ and $L^* \oplus K^* \cong \Lambda_N/P$, $N \geq 2$. (We assume here that primitive embeddings $\hat{P} \hookrightarrow \Lambda_{N-1}$ and $P \hookrightarrow \Lambda_N$ are given.)

Theorem 28.14. *Let G' be a Dynkin graph, N be an integer with $N \geq 2$. The following are equivalent:*
(1) There exist a primitive embedding $P \hookrightarrow \Lambda_N$ and a full embedding $Q(G') \hookrightarrow (\Lambda_N/P, F)$ such that the orthogonal complement of $Q(G')$ in Λ_N/P contains a primitive isotropic vector u with $u \notin F$.
(2) There exist a primitive embedding $\hat{P} \hookrightarrow \Lambda_{N-1}$ and a positive definite submodule $T_0 \subset \Lambda_{N-1}/\hat{P}$ such that for any positive definite full root submodule T with $T_0 \subset T \subset \Lambda_{N-1}/\hat{P}$ G' is obtained from the Dynkin graph of T by one dual elementary transformation.

Remark. The lattice K in Lemma 29.13 contains neither a vector $\alpha \in K$ with $\alpha^2 = 2$ nor a vector $\beta \in K$ with $\beta^2 = 6$, $\beta \cdot s = 3$. Therefore, we cannot define the concept of "dual tie transformation" in our present case.

29 Theorems with the Ik-conditions for $U_{1,0}$

In this section we would like to draw the Coxeter-Vinberg graph of Λ_2/P for $U_{1,0}$.

In what follows we fix a primitive embedding $P \hookrightarrow \Lambda_2$. By F we denote the orthogonal complement of P. F is identified with its image under $\Lambda_2 \to \Lambda_2/P$.

Recall that in this case, the discriminant group $P^*/P \cong P'^*/P'$ of $P = P' \oplus H_0$ is isomorphic to $\mathbf{Z}/9 \oplus \mathbf{Z}/3$. The discriminant quadratic form was given in Proposition 4.3. Special elements in it are of type G. Thus, any root in Λ_2/P has length $\sqrt{2}$ or $\sqrt{2/3}$.

However, Λ_2/P does not satisfy the equivalent conditions in Proposition 8.4. Thus, it is impossible to draw the Coxeter-Vinberg graph down. We can regard that this is because we do not have enough roots. We try to increase roots.

Let $\alpha \in \Lambda_2/P$ be a vector with $\alpha^2 = 2/9$. It is easy to see $9\alpha \in F$. Thus, the reflection $s_\alpha(x) = x - 2(x, \alpha)\alpha/\alpha^2 = x - (x, 9\alpha)\alpha$ associated with α is well-defined. In the following we regard vectors with length $\sqrt{2}/3$ also as roots in addition to those with length $\sqrt{2}$ or $\sqrt{2/3}$. A root with length $\sqrt{2}/3$ can be called a *dummy* root. We will see that under this increment of roots Λ_2/P satisfies the conditions in Proposition 8.4.

Let us draw the Coxeter-Vinberg graph of Λ_2/P. We will indicate a vertex corresponding to a dummy root with length $\sqrt{2}/3$ by the symbol: ⊙.

F and $Q(A_8+A_2)\oplus H$ have the same discriminant quadratic form and the same signature. Thus, by Proposition 5.2 (2) and by Lemma 5.5 (4) $F \cong Q(A_8+A_2)\oplus H$.

Here we introduce lattices $K_1 = \sum_{i=0}^{4} \mathbf{Z}v_i$ satisfying $v_0^2 = -1$, $v_1^2 = v_2^2 = v_3^2 = v_4^2 = +1$, $v_i \cdot v_j = 0$ $(i \neq j)$ and $K_2 = \sum_{i=0}^{8} \mathbf{Z}w_i$ satisfying $w_i^2 = +1$ $(0 \le i \le 8)$, $w_i \cdot w_j = 0$ $(i \neq j)$. Set $\omega = v_0 + v_1 + v_2 + v_3 + v_4 \in K_1$, $\chi = \sum_{i=0}^{8} w_i \in K_2$, and set $M_1 \cong C(\mathbf{Z}\omega, K_1)$, $M_2 = C(\mathbf{Z}\chi, K_2)$. Then, we have $M_1 \cong Q(A_2)\oplus H$, and $M_2 \cong Q(A_8)$.

Let $p : K_1 \otimes \mathbf{Q} \to M_1 \otimes \mathbf{Q}$ and $r : K_2 \otimes \mathbf{Q} \to M_2 \otimes \mathbf{Q}$ be the projections. $p(x) = x - (x\cdot\omega)\omega/3$. $r(y) = y - (y\cdot\chi)\chi/9$. We here consider the vectors $p(v_0) = 4v_0/3 + (v_1+v_2+v_3+v_4)/3$ and $r(w_0) = 8w_0/9 - (w_1+w_2+\cdots+w_9)/9$ in particular.

$p(v_0)^2 = -4/3$. $r(w_0)^2 = 8/9$. Moreover, $M_1^* = M_1 + \mathbf{Z}p(v_0)$, and $M_2^* = M_2 + \mathbf{Z}r(w_0)$. Thus, we have

$$\Lambda_2/P \cong [M_1 + \mathbf{Z}p(v_0)] \oplus [M_2 + \mathbf{Z}r(w_0)] .$$

We apply Vinberg's algorithm to this expression. We choose $p(v_0)$ as the controlling vector.

Noting that M_2^* contains no short roots, we can choose the following root basis of type $A_8 + A_3$ as the root basis orthogonal to $p(v_0)$:

$$e_1 = -v_1+v_2,\quad e_2 = -v_2+v_3,\quad e_3 = -v_3+v_4,$$
$$e_j = -w_{j-4}+w_{j-3} \quad (4 \le j \le 11)$$

By the algorithm we get successively

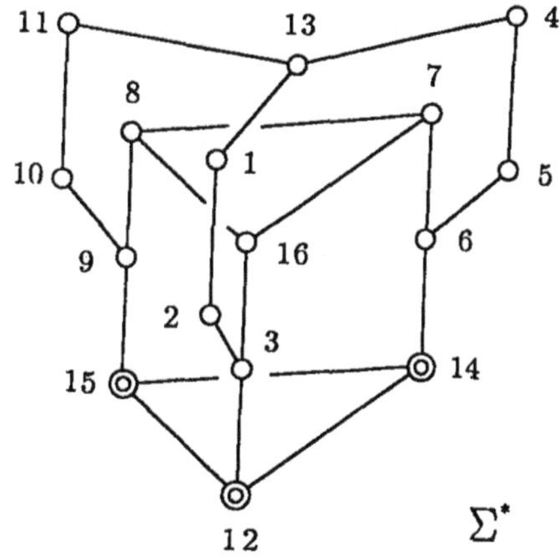

$$e_{12} = v_0/3 + (v_1+v_2+v_3)/3 - 2v_4/3$$
$$e_{13} = v_0 + v_1 + w_0 - w_8$$
$$\begin{aligned} e_{14} &= 4v_0/3 + (v_1+v_2+v_3+v_4)/3 \\ &\quad + 2(w_0+w_1+w_2)/3 \\ &\quad - (w_3+w_4+\cdots+w_8)/3 \end{aligned}$$
$$\begin{aligned} e_{15} &= 4v_0/3 + (v_1+v_2+v_3+v_4)/3 \\ &\quad + (w_0+w_1+\cdots+w_5)/3 \\ &\quad - 2(w_6+w_7+w_8)/3 \end{aligned}$$
$$\begin{aligned} e_{16} &= 3v_0 + (v_1+v_2+v_3) \\ &\quad + (w_0+w_1+w_2+w_3) \\ &\quad - (w_5+w_6+w_7+w_8) \end{aligned}$$
$$e_{17} = 4v_0 + (v_1+v_2+v_3+v_4)/3 + 2(w_0+w_1+\cdots+w_6)/9 - 7(w_7+w_8)/9$$
$$e_{18} = 4v_0/3 + (v_1+v_2+v_3+v_4)/3 + 7(w_0+w_1)/9 - 2(w_2+w_3+\cdots+w_8)/9.$$

Here, $e_{13}^2 = e_{16}^2 = 2$, $e_{12}^2 = e_{14}^2 = e_{15}^2 = 2/3$, $e_{17}^2 = e_{18}^2 = 2/9$. Drawing the graph for the 16 vectors e_i $(1 \le i \le 16)$ excluding the last two, we get the graph Σ^* on the above figure.

This graph Σ^* has the automorphism group isomorphic to the symmetric group S_3 of degree 3. Note here that the following 2 facts: first, $e_1, e_2, \ldots, e_{11}$ and e_{13} generate $F = M_1 \oplus M_2$; second, the sum of long roots among e_i $(1 \le i \le 16)$

$$g = \sum_{j=1}^{11} e_j + e_{13} + e_{16} = 4v_0 + (v_1+v_2+v_3+v_4) + (w_0+w_1+w_2+w_3) - (w_5+w_6+w_7+w_8)$$

satisfies $g^2 = -4 < 0$.

Thus, we get the following lemma by the argument developed in the latter half of Section 21:

Lemma 29.1. *(1) Every automorphism of Σ^* is induced by an integral orthogonal transformation on Λ_2/P.*
(2) The Coxeter-Vinberg graph Σ of Λ_2/P contains Σ^* as a subgraph and Σ has the action of S_3 extending the action on Σ^*.

In particular, roots conjugate to e_{17} or e_{18} are fundamental roots. We get the following 4 fundamental roots in addition:

$$\begin{aligned} e_{19} &= 2v_0 + v_1 + v_2 + 5(w_0 + w_1 + w_2 + w_3)/9 - 4(w_4 + \cdots + w_8)/9 \\ e_{20} &= 2v_0 + v_1 + v_2 + 4(w_0 + w_1 + \cdots + w_4)/9 - 5(w_5 + w_6 + w_7 + w_8)/9 \\ e_{21} &= 8v_0/9 + 2(v_1 + v_2 + v_3 + v_4)/3 + 7(w_0 + w_1 + w_2 + w_3)/9 \\ &\qquad -2(w_4 + w_5 + w_6)/9 - 11(w_7 + w_8)/9 \\ e_{22} &= 8v_0/9 + 2(v_1 + v_2 + v_3 + v_4)/3 + 11(w_0 + w_1)/9 \\ &\qquad +2(w_2+w_3+w_4)/9 - 7(w_5+w_6+w_7+w_8)/9. \end{aligned}$$

Drawing the graph for these 22 vectors we get the graph Σ described on the following 2 pages.

Here we give several remarks on Dynkin graphs.

In the original situation for $U_{1,0}$ we had to treat two kinds of vertices o and ◎. However, we added the third kind of vertices ⊙. For this reason a connected Dynkin graph all of whose vertices are ⊙ and a Dynkin graph of type $G_2(3)$ ⊙——◎ can appear in our theory. (The following is the extended Dynkin graph of type $G_2(3)$: ⊙——◎——◎)

Moreover, we have to give a new definition of Lannér graphs. (See Proposition 8.5.) The table below on the next page but one shows the new Lannér graphs.

Finally, remember that in our case $U_{1,0}$ dual extended graphs may appear in our theory. We have to count the following dual graphs corresponding to G_2 and $G_2(3)$:

◎——◎——o, ⊙——⊙——◎

We check the condition $\langle a \rangle$ and $\langle b \rangle$ in Proposition 8.5 for the graph Σ.

The condition $\langle a \rangle$ is easy. We have the maximal extended Dynkin graphs $E_8 + G_2(3)$, $E_7 + G_2 + A_1(9)$, $E_6 + A_2 + A_2(3)$, $A_8 + G_2$ and the maximal dual extended Dynkin graph $E_8 + G_2$.

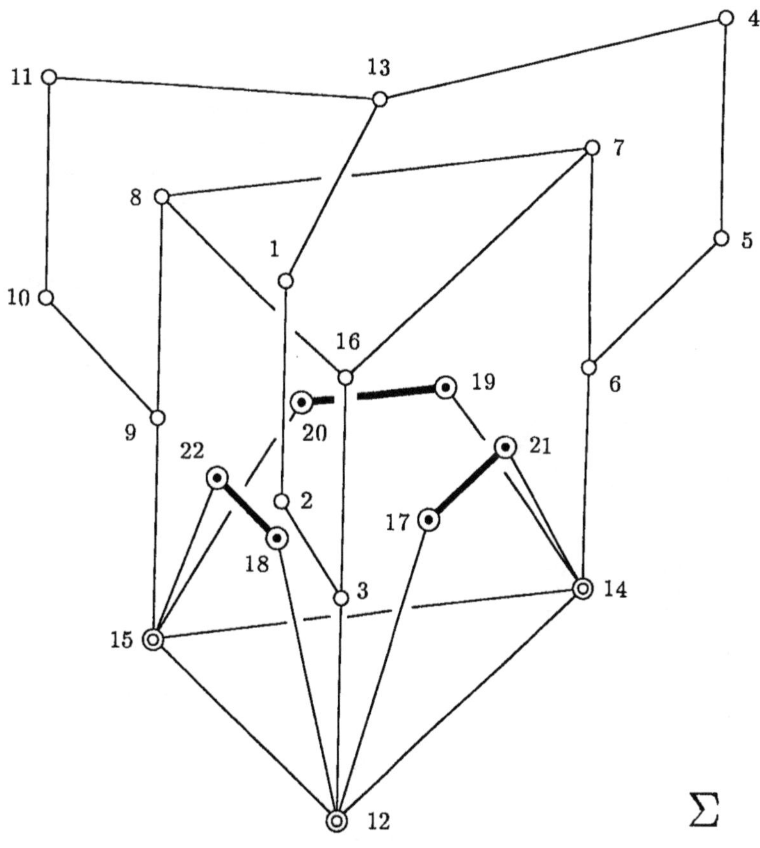

In the above figure we have omitted all dotted edges to make it easy to see. As for the dotted edges, see the next page.

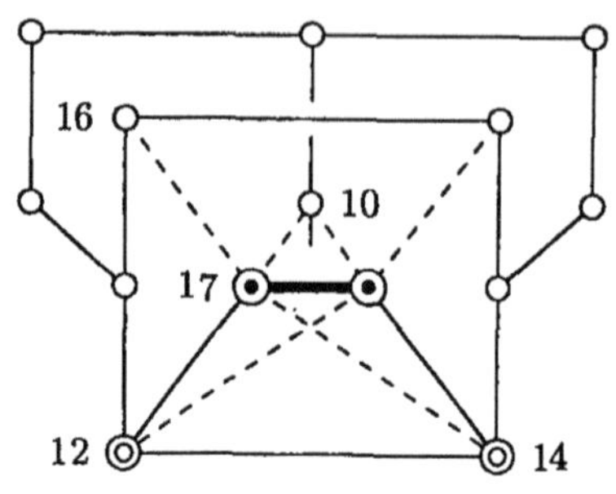

Every dotted edge has a vertex ⊙ corresponding to a root with length $\sqrt{2}/3$ at least at one of two ends. The left figure shows the connection of the vertex 17 ⊙ with other vertices with length $\sqrt{2}$ or $\sqrt{2/3}$. Thanks to the S_3-action, we can indicate only this case. Vertices corresponding to a long root connected with 17 are 10 and 16. Vertices corresponding to length $\sqrt{2/3}$ connected with 17 are 12 and 14. Vertices 17 and 12 are connected by a solid edge, while 17 and 10, 17 and 16, 17 and 14 are by a dotted edge.

The left figure below shows the connection among vertices ⊙ corresponding to length $\sqrt{2}/3$. It contains a bold edge and a dotted edge.

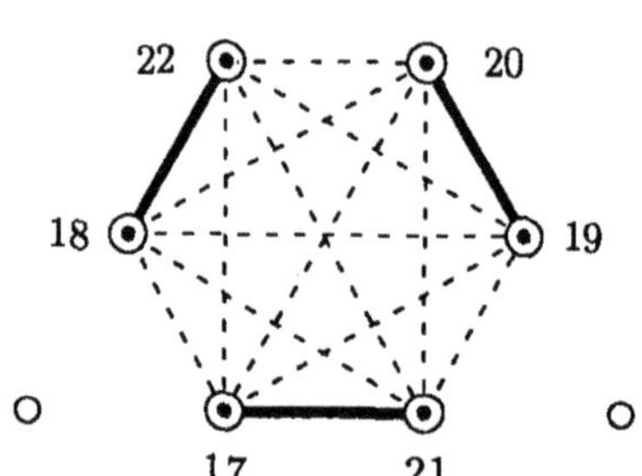

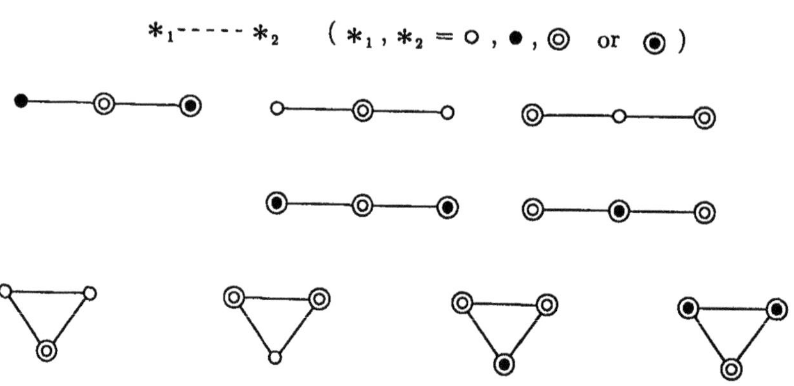

Lannér graphs in the case $U_{1,0}$

We proceed to the condition $\langle b \rangle$. We have to check it for the Lannér subgraphs ⊙——◎——o and dotted edges. However, thanks to the S_3-action, we can check it only for the representatives of the conjugate subgraphs. Thus, we can check it only for the following 7 cases. By Γ we denote the Lannér subgraph under consideration:

(1) The Lannér subgraph 3, 12, 17.

In this case, the graph Ξ defined in Lemma 8.6 is a Dynkin graph of type D_9. Since the number of elements in $S(\Gamma) \cup S(\Xi)$ is equal to $12 = \operatorname{rank} \Lambda_2/P$, $\langle b \rangle$ holds by Lemma 8.6.

(2) The dotted edge 14, 17.

Ξ is a Dynkin graph of type $E_7 + A_3$. $\#(S(\Gamma) \cup S(\Xi)) = 12$.

(3) The dotted edge 10, 17.

Ξ is a Dynkin graph of type $A_9 + G_1$. $\#(S(\Gamma) \cup S(\Xi)) = 12$.

(4) The dotted edge 16, 17.

In this case, the graph Ξ consists of vertices 1, 2, 4, 5, 6, 9, 11, 13, 15. Ξ is a Dynkin graph of type $E_7 + G_2$. Thus, $\#(S(\Gamma) \cup S(\Xi)) = 11$. This number is less than the expected number 12 by 1. The same argument as above cannot be applied. We apply the method explained in the last part of Section 21 just before Corollary 21.9. Consider the subset $S_1 = \{e_{16}\} \cup S(\Xi) \subset S(\Gamma) \cup S(\Xi)$. The graph of S_1 is a Dynkin graph of type $E_7 + G_2 + A_1$. S_1 corresponds to a 2-dimensional facet F_1 of the fundamental polyhedron. Next, consider $\{e_{21}\} \cup S_1$. This defines a Dynkin graph of type $E_7 + G_2 + A_1 + A_1(9)$, and corresponds to a 1-dimensional facet adjacent to F_1. Finally, we consider $\{e_{12}, e_{17}\} \cup S_1$. The graph of this set is a dual extended Dynkin graph of type $E_8 + G_2$. Thus, it corresponds to a 1-dimensional facet at infinity adjacent to F_1. Since F_1 has only 2 adjacent 1-dimensional facets, $S(\Gamma) \cup S(\Xi)$ does not define a facet. This implies the condition in Lemma 8.6 holds. Thus, we get the condition $\langle b \rangle$.

(5) The dotted edge 17, 18.

Ξ is a Dynkin graph of type $D_6 + A_4$.

(6) The dotted edge 17, 20.

Ξ consists of vertices 1, 3, 4, 5, 6, 7, 9, 11, 13, and has fewer vertices than the expected number 10 by 1. We consider $S_1 = \{e_{17}\} \cup S(\Xi) \subset S(\Gamma) \cup S(\Xi)$. The graphs of S_1, $\{e_{15}\} \cup S_1$ and $\{e_8\} \cup S_1$ are Dynkin graphs of type $D_7 + 2A_1 + A_1(9)$, $D_7 + A_1 + G_2 + A_1(9)$ and $D_9 + A_1 + A_1(9)$ respectively. By the same argument as in the case (4) we get $\langle b \rangle$.

(7) The dotted edge 17, 19.

Ξ consists of only 10 vertices 1, 3, 4, 5, 6, 8, 9, 11, 13. We consider $S_1 = \{e_{17}\} \cup S(\Xi) \subset S(\Gamma) \cup S(\Xi)$. The graphs of S_1, $\{e_7\} \cup S_1$ and $\{e_2, e_{15}, e_{21}\} \cup S_1$ are a Dynkin graph of type $D_6 + A_2 + A_1 + A_1(9)$, a Dynkin graph of type $D_9 + A_1 + A_1(9)$ and an extended Dynkin graph of type $E_7 + G_2 + A_1(9)$ respectively. Thus, we get $\langle b \rangle$.

By Lemma 8.6 and Proposition 8.5 the graph Σ described above is the Coxeter-Vinberg graph of Λ_2/P.

Lemma 29.2. *Consider the following 5 vectors in $M_1^* \oplus M_2^* \cong \Lambda_2/P$:*

$$u_1 = -(4e_1 + 2e_2 + 5e_4 + 4e_5 + 3e_6 + 2e_7 + e_8 + 3e_{11} + 6e_{13})$$
$$u_2 = -(3e_1 + 2e_2 + e_3 + 3e_4 + 2e_5 + e_6 + 2e_{11} + 4e_{13})$$

$$u_3 = -(e_7 + e_8 + e_{16})$$

$$u_4 = -(\sum_{i=4}^{11} e_i + e_{13})$$

$$u_5 = -(e_6 + e_{12} + 2e_{14})$$

Set $Z(j) = \{i \mid 1 \leq i \leq 22,\ u_j \cdot e_i = 0\}$. Recall $M_1 \oplus M_2 \cong F$.

(1) $Z(1) = \{1,2,4,5,6,7,8,11,13\} \cup \{12,15,17\}$
$Z(2) = \{1,2,3,4,5,6,11,13\} \cup \{8,9,15\} \cup \{17,21\}$
$Z(3) = \{1,2,4,5,10,11,13\} \cup \{7,8,16\} \cup \{12,14,15\}$
$Z(4) = \{4,5,6,7,8,9,10,11,13\} \cup \{2,3,12\}$
$Z(5) = \{1,2,4,8,9,10,11,13,16\} \cup \{6,12,14\}$.

(2) $u_1 \cdot e_9 = 1$. $u_2 \cdot e_7 = 1$. $u_3 \cdot e_3 = 1$, $u_4 \cdot e_1 = 1$.

(3) For $1 \leq j \leq 4$ u_j is a primitive isotropic vector with $u_j \in F$. The generalized root system $\widetilde{R} = \{\alpha \in (\mathbf{Z}u_j)^\perp/\mathbf{Z}u_j \mid \alpha^2 = 2,\ 2/3$ or $2/9$, If $\alpha^2 = 2$, then $\alpha \in F \cap (\mathbf{Z}u_j)^\perp + \mathbf{Z}u_j/\mathbf{Z}u_j\}$ of $(\mathbf{Z}u_j)^\perp/\mathbf{Z}u_j$ is of type $E_8 + G_2(3)$ (when $j = 1$), $E_7 + G_2 + A_1(9)$ (when $j = 2$), $E_6 + A_2 + A_2(3)$ (when $j = 3$), or $A_8 + G_2$ (when $j = 4$).

(4) $u_5 \cdot e_5 = 1$. $u_5 \cdot e_{18} = 1/3$.

(5) u_5 is a primitive isotropic vector with $u_5 \notin F$. The generalized root system $\widetilde{R}$ of $(\mathbf{Z}u_5)^\perp/\mathbf{Z}u_5$ is of type $E_8 + G_2$.

Proof. Claim (1), (2) and (4) follow immediately.
(3) Assume $1 \leq j \leq 4$. Obviously, $u_j \neq 0$. It is easy to check $u_j^2 = 0$. Since u_j is a sum of long roots, we have $u_j \in F$. Assume $u_j = aw$ for some $a \in \mathbf{Z}$ and $w \in M_1^* \oplus M_2^*$. By (2) we have a vector $e \in M_1 \oplus M_2 = F$ with $1 = u_j \cdot e = aw \cdot e$. Since $w \cdot e \in \mathbf{Z}$, we have $a = \pm 1$. Thus, u_j is primitive. The type of $\widetilde{R}$ coincides with the type of the extended Dynkin graph formed by the vertices whose number belongs to $Z(j)$. (See Lemma 8.2.)
(5) Since $e_5 \in F = M_1 \oplus M_2$, by the same argument as above that u_5 is primitive. Since $u_5 \cdot e_{18} \notin \mathbf{Z}$, $u_5 \notin F$.

Next, we consider $L = (\mathbf{Z}u_5)^\perp/\mathbf{Z}u_5$. It has the induced structure of a root module $(L,\ FL)$. By Proposition 28.12 one knows $L = FL^*$ and $L/FL \cong \mathbf{Z}/3$. Thus, the root system of $(L,\ FL)$ is equal to that of the regular root module L.

We here note the following fact: Let Q_1 be a quasi-lattice of rank 2 defined by the Dynkin graph of type $A_2(3)$, that is, $Q_1 = \mathbf{Z}x + \mathbf{Z}y$ with $x^2 = y^2 = 2/3$, $x \cdot y = -1/3$. The generalized root system of a regular root module Q_1 is not of type $A_2(3)$ but of type G_2.

The last claim follows from these facts. Q.E.D.

Proposition 29.3. *Fixing a primitive embedding $P \hookrightarrow \Lambda_2$, we consider a root module $(\Lambda_2/P,\ F)$. Let $L \subset \Lambda_2/P$ be a positive definite full root submodule. Assume that the root system of $(L,\ F \cap L)$ has no component of type G_1. Then, there exist a primitive isotropic vector $u \in \Lambda_2/P$ belonging to F satisfying the following ⟨1⟩, ⟨2⟩:*

⟨1⟩ u is in a nice position with respect to L.

⟨2⟩ The root system of the positive definite root module $(I^\perp/I,\ (I^\perp \cap F) + I/I)$ where $I = \mathbf{Z}u$ is of type either $E_8 + A_2(3)$, $E_7 + G_2$, $E_6 + A_2 + A_2(3)$ or $A_8 + G_2$.

Proof. Replacing L by a conjugate one, we can assume that a root basis $\widetilde{\Delta}_L$ of the generalized root system $\widetilde{R}(L) = \{\alpha \in F \cap L \mid \alpha^2 = 2\} \cup \{\beta \in L \mid \beta^2 = 2/3 \text{ or } 2/9\}$ is a subset of the above set $\{e_i \mid 1 \leq i \leq 22\}$ of fundamental roots.

Let $R(L) = \{\alpha \in F \cap L \mid \alpha^2 = 2\} \cup \{\beta \in L \mid \beta^2 = 2/3\}$ be the root system of L. Obviously, $R(L) \subset \widetilde{R}(L)$. Note that a root basis Δ_L of $R(L)$ can be obtained by replacing a component of $\widetilde{\Delta}_L$ of type $G_2(3)$ by a component of type $A_2(3)$ and by discarding all components of $\widetilde{\Delta}_L$ consisting of only short roots with length $\sqrt{2}/3$. In particular, any component of $\widetilde{\Delta}_L$ containing a long root can be regarded as a component of Δ_L.

Case 1. $\widetilde{\Delta}_L$ contains a short root with length $\sqrt{2}/3$.

By Lemma 29.1 it suffices to consider only the case where $e_{17} \in \widetilde{\Delta}_L$.

Since the graph of $\widetilde{\Delta}_L$ is a Dynkin graph and contains no dotted edge and no bold edge. Thus, we can conclude

$$e_{10}, e_{14}, e_{16}, e_{18}, e_{19}, e_{20}, e_{21}, e_{22} \notin \widetilde{\Delta}_L.$$

Moreover, since vertex 3, 12 and 17 form a Lannér subgraph, either $e_3 \notin \widetilde{\Delta}_L$ or $e_{12} \notin \widetilde{\Delta}_L$.

First, we consider the case $e_3 \notin \widetilde{\Delta}_L$. We consider the primitive isotropic vector u_1 in Lemma 29.2. It satisfies $u_1 \in F$. Since the generalized root system of $(\mathbf{Z}u_1)^{\perp}/\mathbf{Z}u_1$ is of type $E_8 + G_2(3)$, the root system of it is of type $E_8 + A_2(3)$. The condition $\langle 2 \rangle$ is satisfied.

On the other hand, under the assumption u_1 is orthogonal to $\widetilde{\Delta}_L - \{e_9\}$ by Lemma 29.2 (1). Thus, if $e_9 \notin \widetilde{\Delta}_L$, u_1 is orthogonal to $\widetilde{R}(L)$ and $R(L)$. The condition $\langle 1 \rangle$ holds in this case.

If $e_9 \in \widetilde{\Delta}_L$, then the component of $\widetilde{\Delta}_L$ containing e_9 is a component of Δ_L, and u_1 is orthogonal to $\Delta_L - \{e_9\}$. Since $u_1 \cdot e_9 = 1$, the condition $\langle 1 \rangle$ holds also in this case.

Second, we consider the case $e_{12} \notin \widetilde{\Delta}_L$. In this case, the vector u_2 in Lemma 29.2 is orthogonal to $\widetilde{\Delta}_L - \{e_{12}\}$ and $u_2 \cdot e_{12} = 1$. Thus, the condition $\langle 1 \rangle$ holds in this case. The generalized root system of $(\mathbf{Z}u_2)^{\perp}/\mathbf{Z}u_2$ is of type $E_7 + G_2 + A_1(9)$ and the root system of it is of type $E_7 + G_2$.

Case 2. $\widetilde{\Delta}_L$ contains no short root with length $\sqrt{2}/3$.

In this case, $\widetilde{R}(L) = R(L)$ and $\widetilde{\Delta}_L = \Delta_L$. Moreover, by assumption

$$e_{17}, e_{18}, e_{19}, e_{20}, e_{21}, e_{22} \notin \Delta_L.$$

In what follows we do not repeat these facts.

Vertex 12, 14, 15 form an extended Dynkin graph of type $A_2(3)$ and at least one of e_{12}, e_{14}, e_{15} does not belong to Δ_L.

First, we consider the case where only one of e_{12}, e_{14}, e_{15} does not belong to Δ_L. By Lemma 29.1 we can consider only the case where $e_{12} \notin \Delta_L$ and $e_{14}, e_{15} \in \Delta_L$. Since vertex 9, 14, 15 form a dual extended Dynkin graph of type G_2, we know $e_9 \notin \Delta_L$. By the same reason $e_6 \notin \Delta_L$. Thus, the vector u_3 in Lemma 29.2 is orthogonal to $\Delta_L - \{e_3\}$ and $u_3 \cdot e_3 = 1$. Thus, the condition $\langle 1 \rangle$ is satisfied. Both of the generalized root system of $(\mathbf{Z}u_3)^{\perp}/\mathbf{Z}u_3$ and the root system of it are of type $E_6 + A_2 + A_2(3)$ and we have the condition $\langle 2 \rangle$.

Second, we assume that at least 2 of e_{12}, e_{14}, e_{15} do not belong to Δ_L. By Lemma 28.1 we can consider only the case where $e_{14}, e_{15} \notin \Delta_L$.

If $e_{16} \notin \Delta_L$, then the vector u_4 in Lemma 29.2 is orthogonal to $\Delta_L - \{e_1\}$ and $u_4 \cdot e_1 = 1$. The generalized root system of $(\mathbf{Z}u_4)^\perp/\mathbf{Z}u_4$ and the root system of it are of type $A_8 + G_2$, and we have the condition $\langle 1 \rangle$ and $\langle 2 \rangle$.

In what follows we assume $e_{16} \in \Delta_L$. Since vertex 7, 8, 16 form an extended Dynkin graph of type A_2, one of e_7 and e_8 does not belong to Δ_L. By Lemma 29.1 we can consider only the case $e_7 \notin \Delta_L$. In what follows we assume $e_7, e_{14}, e_{15} \notin \Delta_L$ and $e_{16} \in \Delta_L$.

Vertex 3, 12, 16 form an extended Dynkin graph of type G_2, and one of e_3 and e_{12} does not belong to Δ_L. First, we consider the case $e_{12} \notin \Delta_L$. Set $u_4' = -(e_1+e_2+e_3+e_8+e_9+e_{10}+e_{11}+e_{13}+e_{16})$. This is conjugate to u_4 with respect to the S_3-action, and is orthogonal to $\Delta_L - \{e_4\}$. Since $u_4' \cdot e_4 = 1$, we have the condition $\langle 1 \rangle$. The generalized root system of $(\mathbf{Z}u_4')^\perp/\mathbf{Z}u_4'$ and the root system of it is of type $A_8 + G_2$, and we have the condition $\langle 2 \rangle$.

The last remaining case is the case where $e_3, e_7, e_{14}, e_{15} \notin \Delta_L$. By assumption Δ_L has no component of type G_1 and we have $e_{12} \notin \Delta_L$. Thus, this case is reduced to the case just above. Q.E.D.

Proposition 29.4. *We consider the root module $(\Lambda_2/P, F)$ for a fixed primitive embedding $P \hookrightarrow \Lambda_2$. By G we denote a Dynkin graph. The following 3 conditions are equivalent:*
(1) There exists a primitive isotropic vector $u \in \Lambda_2/P$ belonging to F such that the root system of the root module $((\mathbf{Z}u)^\perp/\mathbf{Z}u, (F \cap (\mathbf{Z}u)^\perp) + \mathbf{Z}u/\mathbf{Z}u)$ is of type G.
(2) There exists a primitive embedding $P \hookrightarrow \Lambda_1$ such that the root system of the root module $(\Lambda_1/P, F)$ is of type G.
(3) $G = E_8 + A_2(3)$, $E_7 + G_2$, $E_6 + A_2 + A_2(3)$ or $A_8 + G_2$.

Moreover, if the root system of $(\Lambda_1/P, F)$ is of type $A_8 + G_2$ for some primitive embedding $P \hookrightarrow \Lambda_1$, then $\Lambda_1/P \cong Q(A_8 + A_2)^*$. In particular, in this case, A_8 is an obstruction component.

Proof. (1)$\Rightarrow$(2) It follows from Lemma 6.2.
(2)$\Rightarrow$(1) By Proposition 5.2 for some integral orthogonal transformation $\phi : \Lambda_1 \oplus H \xrightarrow{\sim} \Lambda_2$ the composition $P \hookrightarrow \Lambda_1 \hookrightarrow \Lambda_1 \oplus H \xrightarrow{\phi} \Lambda_2$ coincides with the given embedding $P \hookrightarrow \Lambda_2$, where $\Lambda_1 \hookrightarrow \Lambda_1 \oplus H$ is the embedding into the direct summand. The image of $u \in H$ by ϕ satisfies the condition of (1).
(1)$\Leftrightarrow$(3) Note that any extended Dynkin subgraph G with the maximal rank in the Coxeter-Vinberg graph Σ of Λ_2/P satisfies (3), and if 2 such subgraphs in Σ have the same type, then they are conjugate under the S_3-action. Thus, the claim follows from Lemma 29.1 and Lemma 29.2.

Let us consider the last claim on A_8+G_2. Since the discriminant quadratic form of P coincides with (-1) times that of $Q(A_8+A_2)$, by Lemma 5.5 $\Lambda_1/P \cong Q(A_8+A_2)^*$ for some primitive embedding $P \hookrightarrow \Lambda_1$. Since the Dynkin graph of $(Q(A_8+A_2)^*, Q(A_8+A_2))$ is $A_8 + G_2$, by (2)$\Rightarrow$(1) $Q(A_8+A_2)^* \cong (\mathbf{Z}u)^\perp/\mathbf{Z}u$ for some primitive isotropic vector u. Since extended Dynkin subgraphs of type $A_8 + G_2$ in the Coxeter-Vinberg graph are all conjugate with respect to S_3, such u's are all conjugate with respect to the orthogonal transformation group of Λ_2/P by Lemma 29.1. Q.E.D.

Proposition 29.5. *We consider the same situation as in Proposition 29.4 above. The following 3 conditions are equivalent:*
(1) There exists a primitive isotropic vector $u \in \Lambda_2/P$ with $u \notin F$ such that the root system of the root module $((\mathbf{Z}u)^\perp/\mathbf{Z}u, (F \cap (\mathbf{Z}u)^\perp) + \mathbf{Z}u/\mathbf{Z}u)$ is of type G.
(2) There exists a primitive embedding $\hat{P} \hookrightarrow \Lambda_1$ such that the root system of the regular root module $\Lambda_1/\hat{P}$ is of type G.
(3) $G = E_8 + G_2$.

Proof. (1)$\Rightarrow$(2) It follows from Proposition 28.12.
(2)$\Rightarrow$(1) By Lemma 28.13 we have an isomorphism $\Lambda_1/\hat{P} \oplus K^* \cong \Lambda_2/P$. The image of $u = s/3 \in K^*$ under this isomorphism satisfies the condition in (1).
(1)$\Leftrightarrow$(3). A maximal generalized extended Dynkin subgraph in Σ which did not appear in Proposition 29.4 is a dual extended Dynkin graph of type $E_8 + G_2$. It is easy to see that any 2 such dual extended graphs are conjugate with respect to S_3. Thus, by Lemma 29.1 we can consider only the graph consisting vertices whose number is in $Z(5)$ in Lemma 29.2. By Lemma 29.2 we know that this subgraph actually corresponds to an isotropic vector satisfying (1). Q.E.D.

We complete the study of Λ_2/P in the case $U_{1,0}$.

Now, note that we have also the case where in the reduction from Λ_3 to Λ_2 the dual elementary transformation is applied. To treat this case we have to study $\Lambda_2/\hat{P}$, too.

We fix a primitive embedding $\hat{P} \hookrightarrow \Lambda_2$. By $\hat{F}$ we denote the orthogonal complement of $\hat{P}$ in Λ_2. Recall that $\hat{F}^*/\hat{F} \cong \mathbf{Z}/3$ and the discriminant quadratic form on it is given by $x \in \mathbf{Z}/3 \mapsto 2x^2/3 \bmod 2\mathbf{Z}$. (See Proposition 4.5.) Since $\hat{F}^*/\hat{F}$ contains no isotropic element, we can treat the regular root module $\Lambda_2/\hat{P}$ instead of $(\Lambda_2/\hat{P}, \hat{F})$.

Since $\hat{F}$ and $Q(E_8 + A_2) \oplus H$ has the same discriminant quadratic form and the same signature, they are isomorphic by Lemma 5.5 and Proposition 5.2. Here we use the tools when we drew the Coxeter-Vinberg graph of Λ_2/P, i.e., $K_1 = \sum_{i=0}^{4} \mathbf{Z}v_i$, $K_2 = \sum_{i=0}^{8} \mathbf{Z}w_i$, $\omega = v_0 + \cdots + v_4 \in K_1$, $\chi = \sum_{i=0}^{8} w_i \in K_2$, $M_1 = C(\mathbf{Z}\omega, K_1)$, $M_2 = C(\mathbf{Z}\chi, K_2)$, projections p, r, $p(v_0) = (4v_0 + v_1 + v_2 + v_3 + v_4)/3$, $r(w_0) = (8w_0 - \sum_{i=1}^{9} w_i)/9$, etc. Recall that $M_1 \cong Q(A_2) \oplus H$, $M_2 \cong Q(A_8)$. Thus, we have

$$\hat{F} = M_1 \oplus [M_2 + \mathbf{Z}(3r(w_0))].$$
$$\Lambda_2/\hat{P} \cong \hat{F}^* = [M_1 + \mathbf{Z}p(v_0)] \oplus [M_2 + \mathbf{Z}(3r(w_0))].$$

We use this expression to draw the Coxeter-Vinberg graph. We choose $p(v_0)$ as the controlling vector. We take the following vectors as a root basis orthogonal to $p(v_0)$:

$$\begin{aligned} f_1 &= -v_1 + v_2, \quad f_2 = -v_2 + v_3, \quad f_3 = -v_3 + v_4, \\ f_j &= -w_{j-4} + w_{j-3} \quad (4 \le j \le 10), \\ f_{11} &= (w_0 + w_1 + \cdots + w_4)/3 - 2(w_5 + w_6 + w_7)/3 + w_8/3. \end{aligned}$$

By Vinberg's algorithm we get further:

$$f_{12} = v_0/3 + (v_1 + v_2 + v_3)/3 - 2v_4/3$$
$$f_{13} = v_0 + v_1 + w_0 - w_8.$$

Here, $f_j^2 = 2$ for $1 \le j \le 11$, $f_{12}^2 = 2/3$, $f_{13}^2 = 2$.

Drawing the graph for these 13 vectors we get the following:

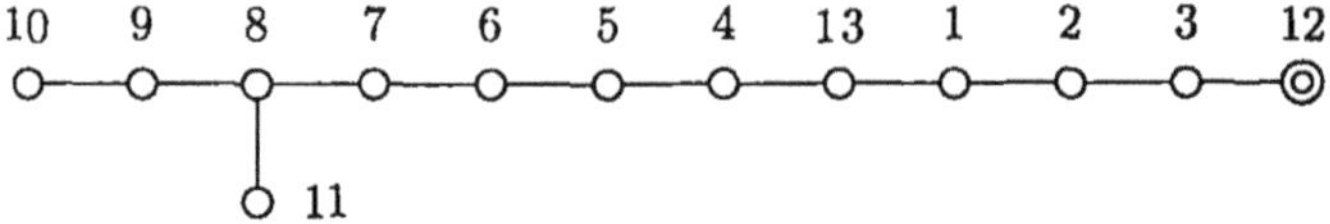

By Proposition 8.5 this is the Coxeter-Vinberg graph of $\Lambda_2/\hat{P}$.

Corollary 29.6. *For every positive definite full root submodule $L \subset \Lambda_2/\hat{P}$ there exists a primitive isotropic vector $u \in \Lambda_2/\hat{P}$ in a nice position with respect to L such that the root system of $(\mathbf{Z}u)^\perp/\mathbf{Z}u$ is of type $E_8 + G_2$.*

Proof. Exchanging L for a conjugate one we can assume that a root basis Δ_L is a subset of $\{f_j \mid 1 \le j \le 13\}$. Set $u = -(f_2 + 2f_3 + 3f_{12})$. We can check $u^2 = 0$, $u \cdot f_j = 0$ $(2 \le j \le 13)$ and $u \cdot f_1 = 1$. This u is the desired vector. Q.E.D.

By the results in this section we can show the following theorem by standard arguments:

Theorem 29.7. *We consider the Ik-conditions in the case $U_{1,0}$. (See Section 7.) By r we denote the number of vertices in the Dynkin graph G. $Q = Q(G)$ stands for the root (quasi-)lattice of type G. $r = \text{rank } Q$.*

[I] The following two conditions (a) and (b) are equivalent:

(a) $G \in PC(U_{1,0})$ and the I2-condition holds for the root lattice $Q = Q(G)$ of type G.

(b) G contains no vertex corresponding to a short root and either the following (b-1) or (b-2) holds:

(b-1) G can be obtained from one of the 4 basic Dynkin graphs in Theorem 0.8 by elementary transformations repeated twice.

(b-2) G can be made from the dual basic Dynkin graph $E_8 + G_2$ by a combination of one elementary transformation and one dual elementary transformation. (Both of two kinds of combinations — i.e., "dual elementary" after "elementary", and "elementary" after " dual elementary" are permitted.)

[II] The following two conditions (A) and (B) are also equivalent:

(A) $G \in PC(U_{1,0})$ and the I1-condition holds for the root lattice $Q = Q(G)$ of type G.

(B) G contains no vertex corresponding to a short root and either the following (B-1) or (B-2) holds:

(B-1) G can be obtained from one of the 4 basic Dynkin graphs in Theorem 0.8 by one of the following 3 kinds of procedures:
The procedures:
⟨1⟩ *elementary transformations repeated twice*
⟨2⟩ *an elementary transformation following after a tie transformation*
⟨3⟩ *a tie transformation following after an elementary transformation.*

(B-2) G can be obtained from the dual basic Dynkin graph $E_8 + G_2$ by two transformations one of which is dual elementary transformation and the other is an elementary or a tie transformation. (Thus, only four kinds of combinations of transformations — i.e., "dual elementary" after "elementary," "elementary" after "dual elementary," "dual elementary" after "tie," and "tie" after "dual elementary" — are permitted.)

The reader may notice that Proposition 29.3 contains a condition "The root system of L contains no component of type G_1". However, the influence of this condition does not appear in the above theorem because of the following lemma similar to Lemma 16.3 and Lemma 21.10:

Lemma 29.8. *Let $G'+G''$ be a Dynkin graph containing a component G'' of type G_1. Let G be a Dynkin graph obtained from $G'+G''$ by one tie or elementary transformation. If G contains no vertex corresponding to a short root, then G can be obtained even from G' by the same transformation.*

30
Elliptic K3 surfaces with 3 sections

The most important remaining part of the verification of Theorem 0.8 is the following proposition:

Proposition 30.1. *Assume that $G \in PC(U_{1,0})$, that G does not belongs to the exception list in Theorem 0.8 (1) and that $Q = Q(G)$ does not satisfy the I1-condition. Then, with respect to some full embedding $Q(G) \hookrightarrow \Lambda_3/P$ without an obstruction component A_8, there exists a primitive isotropic vector u in Λ_3/P in a nice position, i.e., such that either u is orthogonal to $Q(G)$, or u belongs to F and there is a root basis $\Delta \subset Q(G)$ and a long root $\theta \in \Delta$ such that $\alpha \cdot u = 0$ for every $\alpha \in \Delta$ with $\alpha \neq \theta$ and $\theta \cdot u = 1$.*

Recall that corresponding to $G \in PC(U_{1,0})$ we have a lattice embedding $P \oplus Q(G) \hookrightarrow \Lambda_3$ satisfying Looijenga's condition (L1) and (L2), a full embedding $Q(G) \hookrightarrow \Lambda_3/P$ without an obstruction component A_8, and the associated elliptic K3 surface $\Phi : Z \to C\ (\cong \mathbf{P}^1)$. The critical values of Φ are denoted by $\{c_1, c_2, \dots, c_t\} \subset C$. The fiber over a point $a \in C$ is denoted by $F_a = \Phi^{-1}(a)$. We write $F_i = F_{c_i}$ for simplicity. Φ satisfies the following:

1. F_1 is a singular fiber of type I_0^* and it has the irreducible decomposition $F_1 = 2C_0 + C_1 + C_2 + C_3 + C_4$ with multiplicities.
2. Φ has 3 disjoint sections C_5, C_6, C_7.
3. C_i $(5 \leq i \leq 7)$ intersects only one component of F_1. C_5 intersects C_4, C_6 intersects C_2 and C_7 intersects C_3. The union $IF = \bigcup_{i=0}^{7} C_i$ is called *the curve at infinity.*

4. The dual graph of the set of components of singular fibers of Φ disjoint from IF coincides with G.
5. The Picard number ρ of Z satisfies $\rho = r + 8$, where r is the number of vertices of G.

The group structure with the unit element C_5 is defined on the set MW of sections. For every $a \in C$ the set $F_a^{\#}$ of non-singular points on F_a has the group structure whose unit element is the intersection point with C_5.

Lemma 30.2. *The rank a of the group MW satisfies $a = 1$ or 2.*

Proof. Let $m(F_i)$ denote the number of components of F_i. We have $\rho = 2 + a + \sum_{i=1}^{t}(m(F_i) - 1)$. (See Section 10. The equality (2).) By $n(F_i)$ we denote the number of components of F_i disjoint from IF. By above **4** $r = \sum_{i=1}^{t} n(F_i)$ and $m(F_1) = 5$, $n(F_1) = 0$. By **5** we have

$$2 - a = \sum_{i=2}^{t}(m(F_i) - 1 - n(F_i)). \qquad (**)$$

Since $m(F_i) \geq n(F_i) + 1$, we have $0 \leq a \leq 2$.

Assume $a = 0$. We will deduce a contradiction. Under this assumption MW is a finite group. $MW = \operatorname{Tor} MW$, where by $\operatorname{Tor} X$ we denote the subgroup of X consisting of elements with finite order. $\operatorname{Tor} F_1^{\#} \cong \mathbf{Z}/2 + \mathbf{Z}/2$ and the images of C_6, $C_7 \in MW$ on $F_1^{\#}$ generate $\operatorname{Tor} F_1^{\#}$. Since the natural homomorphism $\operatorname{Tor} MW \to \operatorname{Tor} F_1^{\#}$ is injective, we know $\operatorname{Tor} MW \cong \mathbf{Z}/2 + \mathbf{Z}/2$.

We consider a singular fiber F_i with $i \geq 2$. Since $\operatorname{Tor} MW \to \operatorname{Tor} F_i^{\#}$ is injective, we know $\mathbf{Z}/2 + \mathbf{Z}/2 \subset \operatorname{Tor} F_i^{\#}$ and F_i is either of type I_{2b} for some b with $b \geq 1$ or of type I_{2b}^* for some b with $b \geq 0$. We know also the following:

$$(*) \begin{cases} \text{If } F_i \text{ is of type } I_{2b}, \text{ then } m(F_i) = n(F_i) + 2. \\ \text{If } F_i \text{ is of type } I_{2b}^*, \text{ then } m(F_i) = n(F_i) + 3. \end{cases}$$

Case 1. For some j with $2 \leq j \leq t$ F_j is of type I_{2b}^*.

Exchanging the numbers $i \geq 2$, we can assume that $j = 2$ and $m(F_2) = n(F_2) + 3$. By (**) $m(F_i) = n(F_i) + 1$ for $i \geq 3$. By (*) we can conclude $t = 2$. Substituting this in the equality $\rho = 26 + a - 2t + t_1$ (t_1 is the number of singular fibers of type I.) one has $\rho \geq 22$. On the other hand, $20 = \dim H^2(Z, \Omega_Z^1) \geq \rho$, a contradiction.

Case 2. For every i with $2 \leq i \leq t$ F_i is of type I_{2b}.

By (*) $m(F_i) = n(F_i) + 2$ for $2 \leq i \leq t$. By (**) we have $t = 3$ and $t_1 = 2$. Thus, $\rho = 22$ by the above equality, a contradiction. Q.E.D.

Lemma 30.3. *$D_{11} + A_1 \notin PC(U_{1,0})$.*

Proof. Assume $D_{11} + A_1 \in PC(U_{1,0})$. Let $S = P \oplus Q(A_{11} + A_1) \hookrightarrow \Lambda_3$ be the corresponding embedding. Under this embedding P is primitive in Λ_3. Let T be the orthogonal complement of S. T is negative definite with rank $T = 2$.

Set $I = \widetilde{S}/S$, where $\widetilde{S}$ is the primitive hull of S. The 3-Sylow subgroup I_3 of I is an isotropic subgroup of $(S^*/S)_3 \cong \mathbf{Z}/9 \oplus \mathbf{Z}/3$. The discriminant quadratic form on it can be written $q(\overline{\alpha}) \equiv 4x^2/9 - 2y^2/3 \bmod 2\mathbf{Z}$ for $\overline{\alpha} = (x, y) \in \mathbf{Z}/9 \oplus \mathbf{Z}/3 \cong (S^*/S)_3$. Thus, $q(\overline{\alpha}) \equiv 0 \Leftrightarrow \overline{\alpha} = (0, 0)$ or $(\pm 3, 0)$. Thus, if $I_3 \neq 0$, then $(\pm 3, 0) \in I_3$. This implies that P is not primitive in Λ_3, a contradiction. Therefore, $I_3 = 0$. $3 \not| m = [\widetilde{S} : S]$.

Now, the discriminant quadratic form of the 3-adic lattice $T \otimes \mathbf{Z}_3$ is equal to (-1) times that on $(\widetilde{S}^*/\widetilde{S})_3 = (S^*/S)_3$, and, thus, coincides with $-4x^2/9 + 2y^2/3$. By Lemma 12.3 $T \otimes \mathbf{Z}_3$ is isomorphic over $\mathbf{Z}_3$ to the 3-adic lattice of rank 2 defined by the 2×2 diagonal matrix whose diagonal entries are $-2^2 \times 3^2$ and 2×3. Thus, $d(T) \equiv -2^3 \times 3^3 \bmod \mathbf{Z}_3^{*2}$. On the other hand, the discriminant of T $d(T) = |d(T)| = |d(\widetilde{S})| = |d(S)|/m^2 = 2^3 \times 3^3/m^2$. Comparing the 2 expressions of $d(T)$, we have $-1 \in \mathbf{Z}_3^{*2}$. This implies that $z^2 \equiv -1 \pmod 3$ has an integral solution z, a contradiction. Q.E.D.

Proposition 30.4. *If $a = 1$, then either $G = D_{11}$ or G is a subgraph of $A_{11} + A_1$.*

Proof. Recall that $C_5, C_6, C_7 \in MW$. C_5 is the unit element of MW. By assumption $a = 1$, we have integers m, l with $(m, l) \neq (0,0)$ such that $C' = lC_6 + mC_7 \in MW$ has finite order, where the sum is taken in MW. By dividing l and m by their common divisor we can assume moreover that l and m are relatively prime. $C' \in \operatorname{Tor} MW$.

We denote the intersection point of C' and F_1 by c', the intersection point of C_6 and F_1 by c_6, the intersection point of C_7 and F_1 by c_7, and the intersection point of C_5 and F_1 by e. Under the group structure of $F_1^\#$ we have $c' = lc_6 + mc_7$. Let F_1^0 be the connected component of $F_1^\#$ containing e. $F_1^\#/F_1^0 \cong \mathbf{Z}/2 + \mathbf{Z}/2$. c_6 and c_7 are generators of $F_1^\#/F_1^0$. Since at least one of l and m is odd, $c' \notin F_1^0$. Thus, $C' \neq C_5$.

Since $\operatorname{Tor} MW \to \operatorname{Tor} F_a^\#$ is injective for all $a \in C$, C' and C_5 are disjoint. $C' \cdot C_5 = 0$.

The image of C' on $\operatorname{Tor} F_a^\#$, i.e., the intersection point of C' and $F_a^\#$ is a point on $F_a^\#$ with order 2.

Now, by (**) we can assume $m(F_2) = n(F_2)+2$ and $m(F_i) = n(F_i)+1$ for $3 \leq i \leq t$, after exchanging the numbers $i \geq 2$ if necessary.

Let S_i be the subgroup of $\operatorname{Pic}(Z)$ generated by components of F_i disjoint from C_5. Set $S = \bigoplus_{i=1}^t S_i$ and $H_0 = \mathbf{Z}[F]+\mathbf{Z}[F+C_5]$, where F is a smooth fiber of Φ. By Lemma 10.5 $\operatorname{Tor} MW \cong P(H_0 \oplus S, \operatorname{Pic}(Z))/(H_0 \oplus S)$. Thus, $[C'] \in P(H_0 \oplus S, \operatorname{Pic}(Z)) \subset H_0 \oplus S^*$. We can write $[C'] = b[F] + c[F + C_5] + \sum_{i=1}^t \chi_i$ with $b, c \in \mathbf{Z}$, $\chi_i \in S_i^*$. It follows from $C_5 \cdot C' = 0$ and $C' \cdot F = 1$ that $b = c = 1$.

Consider a number i with $3 \leq i \leq t$. C_5, C_6 and C_7 hit the same component of F_i. Thus, also $C' = lC_6 + mC_7$ and C_5 hit the same component of F_i. This implies $\chi_i = 0$.

Consider χ_1. χ_1 is the fundamental weight of $S_1 \cong Q(-D_4)$ corresponding to the component of $F_1^\#$ hit by C'. Thus, $\chi_1^2 = -1$. We have $-2 = C'^2 = 2 - 1 + \chi_2^2$. Thus, $\chi_2^2 = -3$.

Let G^* be the dual graph of components of F_2 disjoint from C_5. χ_2 is a fundamental weight of $S_2 \cong Q(-G^*)$ corresponding to the component of F_2 hit by C'. Note that this component has multiplicity (=the coefficient of the maximal root) 1.

Since F_2 contains a point with order 2 which is the intersection point of F_2 and C', it is of type either III, III^* I_b or I_b^*.

If F_2 is of type III, then $G^* = A_1$ and $\chi_2^2 = -1/2 \neq -3$, a contradiction.

If F_2 is of type III^*, then $G^* = E_7$ and $\chi_2^2 = -3/2 \neq -3$, a contradiction.

We consider the case where F_2 is of type I_b. Since C' has order 2, the component hit by C' corresponds to the central vertex of $G_2 = A_{b-1}$. Thus, b is even and $-3 = \chi_2^2 = -b/4$. Thus, $b = 12$. F_2 is of type I_{12}. Now, our Dynkin graph can be written $G = G' + G''$ where G' corresponds to components in F_2 and G'' corresponds to components in $F_3, \ldots, F_t$. This G' has 10 ($= n(F_2) = m(I_{12}) - 2$) vertices and is a subgraph of A_{11}. Since $r \leq 12$, G'' has at most 2 components.

Note here that $\bigcup_{i=0}^{6} C_i$ and $\bigcup_{i=0}^{5} C_i \cup C_7$ are isomorphic to the curve at infinity in the case $W_{1,0}$. Thus, $A_{11} + G'' \in PC(W_{1,0})$. On the other hand, since $A_{11} + A_2 \notin PC(W_{1,0})$ and $A_{11} + 2A_1 \notin PC(W_{1,0})$, G'' has only at most one vertex. Namely, G is a subgraph of $A_{11} + A_1$.

Finally, we consider the case where F_2 is of type I_b^*. Since the fundamental weight χ_2 corresponds to a vertex with multiplicity 1 and since $S_2 \cong Q(-D_{b+4})$, $\chi_2^2 = -(b+4)/4 = -3$. Thus, $b = 8$, F_2 is of type I_8^* and $G^* = D_{12}$. We can write $G = G' + G''$ as before. Note that $n(F_2) = m(I_8^*) - 2 = 11$, and that the unique vertex of G^* not lying on G' corresponds to a component of F_2 with multiplicity 1. Thus, $G' = D_{11}$ or A_{11}. Since $r \leq 12$, G'' has only at most one component. Since $D_{11} + A_1 \notin PC(U_{1,0})$ by the above lemma, we get the conclusion. Q.E.D.

Lemma 30.5. *Graphs $A_{11} + A_1$ and D_{11} can be made from the basic graph $A_8 + G_2$ by two elementary transformations. Also, any subgraph of them can be made from $A_8 + G_2$ by a combination of 2 of tie or elementary transformations. In particular, any subgraph of $A_{11} + A_1$ or D_{11} belongs to $PC(U_{1,0})$.*

Proof. From $A_8 + G_2$ we can make A_{11}. From A_{11} we can make $A_{11} + A_1$ and D_{11}. The latter half follows from Proposition 13.5. Q.E.D.

Proposition 30.6. *Assume $a = 2$.*
(1) Let G_i with $2 \leq i \leq t$ be the dual graph of components of F_i disjoint from C_5. Then, $G = \sum_{i=2}^{t} G_i$.
(2) For the corresponding lattice embedding $S = P \oplus Q(G) \hookrightarrow \Lambda_3$, $Q(G)$ is primitive in Λ_3.
(3) Set $I = \widetilde{S}/S$, where $\widetilde{S}$ is the primitive hull of S in Λ_3. Then, $I = I_3$. Moreover, I is isomorphic to a subgroup of $\mathbf{Z}/9 + \mathbf{Z}/3$.

Proof. (1) is equivalent to that $m(F_i) = n(F_i) + 1$ for $2 \leq i \leq t$.
(2) follows from the same argument as in Lemma 17.7 (1).
(3) By the same argument as in Lemma 17.7 (2) the restriction to I of the projection $S^*/S = P^*/P \oplus Q(G)^*/Q(G) \to P^*/P \cong \mathbf{Z}/9 \oplus \mathbf{Z}/3$ is injective. Q.E.D.

Lemma 30.7. *If $\rho = r + 8$ and $a = 2$, then*

$$\nu(A) + 2\nu(D) + 2\nu(E) \leq 18 - r.$$

Here, $\nu(T)$ denotes the number of components of G of type T.

Proof. By $\nu(T)$ we also denote the number of singular fibers of Φ of type T. Set $t_1 = \nu(I)$. By Proposition 30.6 (1) we know $t_1 = \nu(A) - \nu(III) - \nu(IV) + \nu(I_1)$

and $t - t_1 = 1 + \nu(D) + \nu(E) + \nu(II) + \nu(III) + \nu(IV)$. Substituting these and the assumptions $\rho = r + 8$, $a = 2$ into the equality $\rho = 26 + a - 2t + t_1$, we have $18 - r = \nu(A) + 2\nu(D) + 2\nu(E) + 2\nu(II) + \nu(III) + \nu(IV) + \nu(I_1)$. Q.E.D.

By the results so far we can assume that the I1-condition fails for $G \in PC(U_{1,0})$ in what follows. In the following several sections we assume this condition when it is necessary. In particular, the number r of vertices of G is equal to 12, 11, or 10.

Moreover, we divide our case into 3 subcases.

[[1]] G has a component of type D.

[[2]] G has a component of type E but has no component of type D.

[[3]] All components of G are of type A.

The case [[i]] is discussed in following Section $30 + i$.

31
Graphs with a component of type D, the case of $U_{1,0}$

In this section we consider the case [[1]] in the last part of Section 30. Note that if we consider a graph different from D_{11}, then we can assume $a = 2$ by Lemma 30.4 and that the claims in Proposition 30.6 and Lemma 30.7 hold.

Lemma 31.1. *Let G be a Dynkin graph with components of type A, D or E only. We assume that G has 12 vertices and has a component of type D.*

(1) The following 2 conditions are equivalent:

$\langle a \rangle$ The root lattice $Q = Q(G)$ of type G satisfies $l((Q^*/Q)_2) \leq 2$.

$\langle b \rangle$ G belongs to one of the following 4 classes MPD12, EXD12, MID12, NID12:

MPD12:	(6 items.)		
	D_{12},	$D_9 + A_3$,	$E_8 + D_4$,
	$D_7 + D_5$,	$D_7 + A_4 + A_1$,	$A_6 + D_6$.
EXD12:	(1 item.)	$E_6 + D_4 + A_2$.	
MID12:	(3 items.)		
	$A_8 + D_4$,	$D_4 + 2A_4$,	$D_4 + 4A_2$.
NID12:	(21 items.)		
	$D_{11} + A_1$,	$D_{10} + A_2$,	$D_9 + A_2 + A_1$,
	$D_8 + A_4$,	$D_8 + 2A_2$,	$D_7 + A_5$,
	$D_7 + A_3 + A_2$,	$D_7 + 2A_2 + A_1$,	$A_6 + D_5 + A_1$,
	$A_6 + D_4 + A_2$,	$D_6 + E_6$,	$D_6 + A_4 + A_2$,
	$D_6 + 3A_2$,	$E_6 + D_5 + A_1$,	$A_5 + D_5 + A_2$,
	$2D_5 + A_2$,	$D_5 + A_4 + A_3$,	$D_5 + A_4 + A_2 + A_1$,
	$D_5 + A_3 + 2A_2$,	$D_5 + 3A_2 + A_1$,	$D_4 + A_4 + 2A_2$.

(2) If $G \in PC(U_{1,0})$, then G satisfies the above equivalent conditions in (1).

Proof. Claim (1) is easy. Claim (2) follows from Proposition 30.6 (3). Q.E.D.

Lemma 31.2. (NPD12) *Any one of the 6 graphs in NPD12 has no component of type A_8 and can be made from the basic graph $A_8 + G_2$ by 2 tie transformations. Thus, it belongs to $PC(U_{1,0})$.*

Proof. Below we show an example of the graph made from $A_8 + G_2$ by the first tie transformation. Applying a tie transformation again we get the desired graph.

$$\begin{array}{lcllcl} D_{12} & \longleftarrow & A_{11}, & D_9 + A_3 & \longleftarrow & A_8 + A_3, \\ E_8 + D_4 & \longleftarrow & D_9 + G_2, & D_7 + D_5 & \longleftarrow & D_9 + G_2, \\ D_7 + A_4 + A_1 & \longleftarrow & D_9 + G_2, & A_6 + D_6 & \longleftarrow & D_9 + G_2. \end{array}$$

Q.E.D.

Remark. Also the member $A_8 + D_4$ of MID12 can be made from $A_8 + G_2$ by 2 tie transformation. However, the component A_8 remains as an obstruction component.

Lemma 31.3. (MID12) *Any member of MID12 does not belong to $PC(U_{1,0})$.*

Proof. At the beginning we consider a more general situation. By G we denote the graph with r vertices to be considered. Assume that there exist a primitive embedding $P \hookrightarrow \Lambda_3$ and a full embedding $Q(G) \hookrightarrow \Lambda_3/P$ such that any component of type A_8 is not an obstruction component. We would like to deduce a contradiction.

Let $S = P \oplus Q(G) \hookrightarrow \Lambda_3$ denote the induced embedding. We denote by $\widetilde{S}$ the primitive hull of S in Λ_3, and by T the orthogonal complement of S in Λ_3. T has signature $(12 - r, 2)$. Set $I = \widetilde{S}/S$. By Proposition 30.6 (3) $I = I_3$. By $d(L)$ we denote the discriminant of a lattice L.

The case of $G = A_8 + D_4$.

$S^*/S \cong \mathbf{Z}/9 \oplus \mathbf{Z}/9 \oplus \mathbf{Z}/3 \oplus (\mathbf{Z}/2 + \mathbf{Z}/2)$. The first component of the right-hand side corresponds to P, the second to A_8, the third also to P, and the fourth and the fifth to D_4.

The discriminant quadratic form on $Q(A_8)^*/Q(A_8) \cong \mathbf{Z}/9$ can be written $q(y) \equiv 8y^2/9 \bmod 2\mathbf{Z}$ for $y \in \mathbf{Z}/9$. Notice that if we choose $2 \in \mathbf{Z}/9$ as the generator, it can be also written in the form $q(y) \equiv 32y^2/9 \equiv -4y^2/9 \bmod 2\mathbf{Z}$. Thus, the discriminant quadratic form of S can be written

$$q_S(\overline{\alpha}) \equiv 4(x^2 - y^2)/9 - 2z^2/3 + a_1^2 + a_1a_2 + a_2^2 \bmod 2\mathbf{Z}$$

for $\overline{\alpha} = (x, y, z, a_1, a_2) \in \mathbf{Z}/9 \oplus \mathbf{Z}/9 \oplus \mathbf{Z}/3 \oplus (\mathbf{Z}/2 + \mathbf{Z}/2)$.

Since $3 = l((S^*/S)_3) > \operatorname{rank} T = 2$, $I_3 \neq 0$. A non-zero element $\overline{\alpha} \in I_3 \subset (S^*/S)_3 \cong \mathbf{Z}/9 \oplus \mathbf{Z}/9 \oplus \mathbf{Z}/3$ satisfies $q_S(\overline{\alpha}) \equiv 0$, and $\overline{\alpha} = (\pm1, \pm1, 0)$, $(\pm2, \pm2, 0)$, $(\pm4, \pm4, 0)$, $(\pm3, 0, 0)$, $(0, \pm3, 0)$ or $(\pm3, \pm3, 0)$.

If $\overline{\alpha} = (\pm x, \pm x, 0)$ with $x = 1$, 2 or 4, then the restriction to I_3 of the projection $(S^*/S)_3 \to (Q(G)^*/Q(G))_3 \cong Q(A_8)^*/Q(A_8)$ is surjective. This implies that A_8 is an obstruction component. It contradicts the assumption.

If $\overline{\alpha} = (\pm3, 0, 0)$, then the embedding $P \hookrightarrow \Lambda_3$ is not primitive, a contradiction.

If $\overline{\alpha} = (0, \pm3, 0)$, then $Q(A_8) \subset Q(E_8) \subset \widetilde{Q(G)} \subset \Lambda_3/P$ and it contradicts fullness.

Therefore, $\overline{\alpha} = (\pm3, \pm3, 0)$. Exchanging the sign of the generators of the components, we can assume $\overline{\alpha} = (3, 3, 0)$. I_3 is the group of order 3 generated by $(3, 3, 0)$. Then, we have $(\widetilde{S}^*/\widetilde{S})_3 \cong (I^{\perp}/I)_3 \cong \mathbf{Z}/3 + \mathbf{Z}/3 + \mathbf{Z}/3$; thus, $2 = \operatorname{rank} T \geq l((\widetilde{S}^*/\widetilde{S})_3) = 3$, which is a contradiction.

In conclusion, $A_8 + D_4 \notin PC(U_{1,0})$.

The case $G = D_4 + 2A_4$.

We have $d(T) = |d(T)| = |d(\widetilde{S})| = |d(P)||d(G)|/m^2 = 3^3 \times 2^2 \times 5^2/m^2$ with $m = [\widetilde{S} : S] = \#(I)$. Since $I = I_3$, $5 \not| m$.

On the other hand, $(\widetilde{S}^*/\widetilde{S})_5 \cong (S^*/S)_5 \cong \mathbf{Z}/5 \oplus \mathbf{Z}/5$, and the discriminant quadratic form on it can be written $q_5(x, y) \equiv 4(x^2 + y^2)/5 \bmod 2\mathbf{Z}$ for $(x, y) \in \mathbf{Z}/5 \oplus \mathbf{Z}/5$. Thus, $T \otimes \mathbf{Z}_5$ is isomorphic to the 5-adic lattice of rank 2 defined by the 2×2 diagonal matrix whose diagonal entries are $-2^2 \times 5$ and $-2^2 \times 5$. Thus, $d(T) \equiv 2^4 \times 5^2 \bmod \mathbf{Z}_5^{*2}$. Comparing 2 expressions of $d(T)$, we have $3 \in \mathbf{Z}_5^{*2}$, which implies that $z^2 \equiv 3 \pmod 5$ for some integer z, a contradiction.

In conclusion, $D_4 + 2A_4 \notin PC(U_{1,0})$.

The case $G = D_4 + 4A_2$.

$(S^*/S)_3 \cong \mathbf{Z}/9 \oplus (\mathbf{Z}/3)^{\oplus 5}$. The first component and the second of the right-hand side correspond to P, and the remaining 4 components $\mathbf{Z}/3$ correspond to A_2 respectively. The discriminant quadratic form is given by $q(\overline{\alpha}) \equiv 4x^2/9 - 2y^2/3 + 2(z_1^2 + z_2^2 + z_3^2 + z_4^2)/3 \bmod 2\mathbf{Z}$ for $\overline{\alpha} = (x, y, z_1, z_2, z_3, z_4) \in \mathbf{Z}/9 \oplus (\mathbf{Z}/3)^{\oplus 5}$. Since $2 \geq \operatorname{rank} T \geq l((S^*/S)_3) - 2l(I_3) = 6 - 2l(I_3)$, we have $l(I_3) \geq 2$. I_3 is not cyclic.

Any non-zero element in I_3 is isotropic, and an isotropic element in $(S^*/S)_3$ is one of the following:

$$\begin{array}{ll} \overline{\beta}_1 = (0,0,\{\pm1,\pm1,\pm1,0\}), & \overline{\delta}_1 = (0,\pm1,\pm1,\pm1,\pm1,\pm1), \\ \overline{\beta}_2 = (0,\pm1,\{\pm1,0,0,0\}), & \overline{\delta}_2 = (\pm3,0,\{\pm1,\pm1,\pm1,0\}), \\ \overline{\beta}_3 = (\pm3,\pm1,\{\pm1,0,0,0\}), & \overline{\delta}_3 = (\pm2,\pm1,\pm1,\pm1,\pm1,\pm1), \\ \overline{\beta}_4 = (\pm3,0,0,0,0,). & \end{array}$$

Recall that $\{x_1, x_2, \ldots, x_k\}$ stands for $x_{\sigma(1)}, x_{\sigma(2)}, \ldots, x_{\sigma(k)}$ for some permutation σ.

If $\overline{\beta}_1 \in I_3$, then $Q(3A_2) \subset Q(E_6) \subset \widetilde{Q(G)} \subset \Lambda_3/P$ and it contradicts fullness.

If $\overline{\beta}_2 \in I_3$ or $\overline{\beta}_3 \in I_3$, then $Q(A_2) \subset Q(G_2) \subset \widetilde{Q(G)} \subset \Lambda_3/P$ and it also contradicts fullness.

If $\overline{\beta}_4 \in I_3$, then the embedding $P \hookrightarrow \Lambda_3$ is not primitive, a contradiction.

Therefore, any non-zero element in I_3 has the form of $\overline{\delta}_1$, $\overline{\delta}_2$ or $\overline{\delta}_3$.

Consider the restriction $I_3 \to \mathbf{Z}/9$ to I_3 of the projection to the first component of $(S^*/S)_3$. Since I_3 is not cyclic its kernel contains a non-zero element. Thus, $\overline{\delta}_1 \in I_3$. Exchanging the sign of the generators of the components of $(S^*/S)_3$ we can assume $(0,1,1,1,1,1) \in I_3$.

Considering the projection to the second component we know $\pm\overline{\delta}_2 \in I_3$. Exchanging the order of A_2-components we can assume $(\pm3,0,1,\pm1,\pm1,0) \in I_3$. Thus, in particular, $(\pm3,0,1,\pm1,\pm1,0) - (0,1,1,1,1,1) = (\pm3,-1,0,*,*,-1) \in I_3$. This coincides with neither $\overline{\delta}_1$, $\overline{\delta}_2$ nor $\overline{\delta}_3$. We get a contradiction. Thus, $D_4 + 4A_2 \notin PC(U_{1,0})$.

Q.E.D.

The class EXD12 is an exception in Theorem 0.8 and will be studied in Section 34. The class NID12 will be discussed later in this section. (Lemma 31.10.)

Next, we treat the case $r = 11$.

Lemma 31.4. *Let G be a Dynkin graph with 11 vertices and with components A, D or E only. We assume that G contains a component of type D. Set $G = G' + G_D$, where G_D is the sum of components of type D. The following 3 conditions are equivalent:*
(1) $Q(G)$ does not satisfy the I1-condition, i.e., for some prime number p $\epsilon_p(Q(G)) \neq (3, d(Q(G)))_p$.
(2) For $p = 3$ or $p = 5$ the following $(*_p)$ holds:

$$(*_p) \qquad \epsilon_p(Q(G')) \neq (3, d(Q(G')))_p.$$

(3) G belongs to one of the following 5 classes MPD11, NPD11, EXD11, MID11, NID11:

MPD11:	(1 item.)	$A_5 + D_4 + A_2$.	
NPD11:	(12 items.)		
	$D_{10} + A_1$,	$A_7 + D_4$,	$E_7 + D_4$,
	$D_6 + A_5$,	$D_7 + A_4$,	$D_7 + A_3 + A_1$,
	$A_6 + D_4 + A_1$,	$E_6 + D_4 + A_1$,	$2D_5 + A_1$,
	$D_5 + A_4 + 2A_1$,	$A_4 + D_4 + A_3$,	$A_4 + D_4 + A_2 + A_1$.
EXD11:	(1 item.)	$D_4 + 3A_2 + A_1$.	
MID11:	(8 items.)	$D_8 + 3A_1$,	$D_6 + D_4 + A_1$,
	$D_6 + 5A_1$,	$D_5 + A_3 + 3A_1$,	$2D_4 + 3A_1$,
	$D_4 + 2A_3 + A_1$,	$D_4 + A_3 + 2A_2$,	$D_4 + 7A_1$.
NID11:	(5 items.)	$D_7 + 2A_2$,	$E_6 + D_5$,
	$A_5 + D_5 + A_1$,	$D_5 + 3A_2$,	$D_5 + 2A_2 + 2A_1$.

Proof. (1)$\Leftrightarrow$(2) Since $\epsilon_p(Q(G)) = \epsilon_p(Q(G'))$ and $(3, d(Q(G)))_p = (3, d(Q(G')))_p$, (1) is equivalent to that $(*_p)$ holds for some prime number p.

Note here that if $(*_p)$ holds for $p = 2$, then $(*_p)$ holds for some odd prime number p because of the product formula. We can assume moreover $p \geq 3$.

Thus, it suffices to show that for $p \geq 7$ $(*_p)$ never holds. Note that G' has at most 7 vertices.

Assume $p \geq 11$. By Lemma 7.5 (3) $p \not| d(Q(G'))$. By Lemma 7.5 (4) $\epsilon_p(Q(G')) = (3, d(Q(G')))_p = +1$.

Assume $p = 7$. If $7 \not| d(Q(G'))$, then, by the same reason as above, $(*_7)$ never holds. If $7 \mid d(Q(G'))$, then G' contains a component A_6 and $G' = A_6 + A_1$ or A_6. In both cases we have $\epsilon_7(Q(G')) = (3, d(Q(G')))_7 = -1$.

(2)$\Leftrightarrow$(3) First, we determine all G' satisfying (2). We write $d = d(Q(G'))$ and $\epsilon_p = \epsilon_p(Q(G'))$ for simplicity.

Consider the case where $(*_5)$ holds. If $5 \not| d$, then $\epsilon_5 = (3, d)_5 = +1$. Thus, $5 \mid d$. We can write $G' = A_4 + G''$. Set $d'' = d(Q(G''))$. G'' has at most 3 vertices, and $5 \not| d''$. $\epsilon_5 = (5, d'')_5$. $(3, d)_5 = (3, 5)_5 = -1$. Thus, $(*_5) \Leftrightarrow d'' \equiv \pm 1 \pmod 5$. We have 7 possibilities A_3, $A_2 + A_1$, $3A_1$, A_2, $2A_1$, A_1 and the empty graph, but only 4 of them satisfy the last condition. We have $G' = A_4 + A_3$, $A_4 + A_2 + A_1$, $A_4 + 2A_1$ or A_4.

Next, we consider the case where $(*_3)$ holds and $3 \mid d$. In particular, G' contains E_6, A_5 or A_2 as a component. If G' contains E_6, A_5 or $3A_2$, then we have only 7 possibilities and all of them satisfy $(*_3)$.

Assume that we can write $G' = 2A_2 + G''$ with $3 \not| d'' = d(Q(G''))$. Since $\epsilon_3 = -1$ and $(3,d)_3 = (3,d'')_3$, we know $(*_3) \Leftrightarrow d'' \equiv 1 \pmod 3$. We can conclude that $G'' = A_3$, $2A_1$ or the empty graph. If $G' = A_2 + G''$ with $3 \not| d'' = d(Q(G''))$, then $\epsilon_3 = -(3,d'')_3 = (3,d)_3$ and $(*_3)$ never holds.

Finally, we consider the case where $(*_3)$ holds and $3 \not| d$. In this case, $\epsilon_3 = +1$. Thus, $(*_3) \Leftrightarrow d \equiv 2 \pmod 3$. We know that G' is one of the following 13 possibilities: A_7, E_7, $A_6 + A_1$, $A_4 + A_3$, $2A_3 + A_1$, $7A_1$, $A_4 + 2A_1$, $A_3 + 3A_1$, $5A_1$, A_4, $A_3 + A_1$, $3A_1$, A_1.

It is easy to determine the corresponding G. Q.E.D.

Lemma 31.5. (MPD11) *The Dynkin graph $A_5 + D_4 + A_2$ contains no component of type A_8 and can be made from the basic graph $A_8 + G_2$ by 2 tie transformations. In particular, $A_5 + D_4 + A_2 \in PC(U_{1,0})$.*

Proof. Indeed, we can make $A_8 + A_3$ from $A_8 + G_2$. From $A_8 + A_3$ we can make $A_5 + D_4 + A_2$. Q.E.D.

Lemma 31.6. (NPD11) *Any Dynkin graph G in NPD11 is a subgraph of a graph belonging to MPD12. Thus, G can be made from $A_8 + G_2$ by 2 transformations and $G \in PC(U_{1,0})$.*

Lemma 31.7. (MID11) *The 8 graphs in MID11 do not belong to $PC(U_{1,0})$.*

Proof. We assume the same situation as in the verification of Lemma 31.3, and we use the same notations.

First, we consider one of the 7 graphs except $D_4 + A_3 + 2A_2$ as G. In this case, $Q = Q(G)$ satisfies $l((Q^*/Q)_2) \geq 5$. On the other hand, $(\widetilde{S}^*/\widetilde{S})_2 \cong (S^*/S)_2$ since $I_2 = 0$. Thus, $3 = \operatorname{rank} T \geq l((\widetilde{S}^*/\widetilde{S})_2) = l((S^*/S)_2) \geq 5$. We have a contradiction.

Consider the case $G = D_4 + A_3 + 2A_2$. In this case, $(S^*/S)_3 \cong \mathbf{Z}/9 \oplus \mathbf{Z}/3 \oplus \mathbf{Z}/3 \oplus \mathbf{Z}/3$. The first and second components correspond to P, and the third and fourth components correspond to A_2 respectively. The discriminant form can be written $q(\overline{\alpha}) \equiv 4x^2/9 - 2y^2/6 + 2(z_1^2 + z_2^2)/3 \bmod 2\mathbf{Z}$ for $\overline{\alpha} = (x, y, z_1, z_2) \in \mathbf{Z}/9 \oplus \mathbf{Z}/3 \oplus \mathbf{Z}/3 \oplus \mathbf{Z}/3$. Since $l(S^*/S)_3 = 4 > 3 = \operatorname{rank} T$, we have $I = I_3 \neq 0$.

On the other hand, an isotropic element in $(S^*/S)_3$ is one of the following: $\overline{\beta}_1 = (0, \pm 1, \{\pm 1, 0\})$, $\overline{\beta}_2 = (\pm 3, \pm 1, \{\pm 1, 0\})$, $\overline{\beta}_3 = (\pm 3, 0, 0, 0)$. I_3 contains one of them. If $\overline{\beta}_1 \in I_3$ or $\overline{\beta}_2 \in I_3$, then we have $Q(A_2) \subset Q(G_2) \subset \widetilde{Q(G)} \subset \Lambda_3/P$, which contradicts being full. If $\overline{\beta}_3 \in I_3$, then the embedding $P \hookrightarrow \Lambda_3$ is not primitive. We get a contradiction in any case. Q.E.D.

Lemma 31.8. $D_5 + 2A_2 + A_1 \notin PC(U_{1,0})$.

Proof. Set $G = D_5 + 2A_2 + A_1$.

Again, we assume the same situation as in the verification of Lemma 31.3, and we use the same notations.

Now, in this case, $(S^*/S)_3$ and the discriminant quadratic form on it are same as those in the above case $D_4 + A_3 + 2A_2$. Thus, by the same reasoning we can conclude

$I = I_3 = 0$. We have $\widetilde{S} = S$. In particular, $d(T) = 2^3 \times 3^5$ and $T \otimes \mathbf{Z}_3$ is isomorphic to the 3-adic lattice defined by the 4×4 diagonal matrix whose diagonal entries are $-2^2 \times 3^2$, 2×3, -2×3 and -2×3. Thus, $d(T) \equiv -2^5 \times 3^5 \bmod \mathbf{Z}_3^{*2}$. Comparing 2 expressions of $d(T)$ we have $-1 \in \mathbf{Z}_3^{*2}$, which implies that $z^2 \equiv -1 \pmod 3$ for some integer z, a contradiction. Q.E.D.

Lemma 31.9. (NID11) *Any graph G in NID11 contains $D_5 + 2A_2 + A_1$ as a subgraph; thus, $G \notin PC(U_{1,0})$.*

Lemma 31.10. (NID12) *Any graph in the 21 members of NID12 contains as a subgraph either $D_5 + 2A_2 + A_1$ or the member $D_4 + A_3 + 2A_2$ of MID11. Thus, $G \notin PC(U_{1,0})$.*

The unique member $D_4 + 3A_2 + A_1$ of EXD11 is an exception in Theorem 0.8 and will be discussed in Section 34.

We proceed to the case $r = 10$.

Lemma 31.11. *Let G be a Dynkin graph with 10 vertices and with components of type A, D or E only. Assume that G has a component of type D. We write $G = G' + G_D$ where G_D is the sum of components of type D. The following conditions are equivalent:*
(1) $Q(G)$ does not satisfy the I1-condition.
(2) $d = d(Q(G'))$ and $\epsilon_p = \epsilon_p(Q(G'))$ satisfies the following $(**_p)$ for $p = 3$ or 2:

$$(**_p) \qquad\qquad 3d \in \mathbf{Q}_p^{*2} \text{ and } \epsilon_p \neq (-1,3)_p.$$

(3) $(**_3)$ holds.
(4) $(**_2)$ holds.
(5) $G' = E_6$, $A_5 + A_1$ or $3A_2$.
(6) G belongs to the following class NPD10:

$$NPD10: \quad \text{(3 items.)} \quad E_6 + D_4, \quad A_5 + D_4 + A_1, \quad D_4 + 3A_2.$$

Proof. (1)⇔(2) Since $d(Q(G)) = 4^m d$ for some m and $\epsilon_p = \epsilon_p(Q(G))$ (1) is equivalent to that $(**_p)$ holds for some prime number p. Thus, it suffices to show that for $p \geq 5$ $(**_p)$ never holds.

Assume that for $p \geq 5$ $(**_p)$ holds. Since $(-1, 3)_p = 1$, we have $\epsilon_p = -1$. By Lemma 7.5 (4) $p \mid d$. Moreover, by the assumption $3d \in \mathbf{Q}_p^{*2}$, we have $p^2 \mid d$. By Lemma 7.5 (3) we have $8 \leq \operatorname{rank} Q(G') \leq 6$, a contradiction.
(2)⇒(3) Trivial.
(3)⇒(5) Since $3d \in \mathbf{Q}_3^{*2}$ we have $3 \mid d$. G' contains E_6, A_5 or A_2 as a component. If $G' = A_5$, then $3d = 2 \times 3^2 \notin \mathbf{Q}_3^{*2}$. If $G' = 2A_2 + G''$ with $3 \nmid d'' = d(Q(G''))$, then $3d = 3^3 d'' \notin \mathbf{Q}_3^{*2}$. Assume $G' = A_2 + G''$ with $3 \nmid d'' = d(Q(G''))$. Since $3d = 3^2 d'' \in \mathbf{Q}_3^{*2}$, we have $d'' \equiv 1 \pmod 3$. On the other hand, since $1 = \epsilon_3 = -(3, d'')_3$, we have $d'' \equiv 2 \pmod 3$, a contradiction.
(2)⇒(4) Trivial.

(4)⇒(5) We apply Lemma 7.7. We divide the case into 5 subcases.
⟨1⟩ G' contains a component E_6.
$G' = E_6$ since G' has at most 6 vertices.
⟨2⟩ G' contains a component A_6.
$G' = A_6$. Since $e(A_6) = (0,1,0) \in (\mathbf{Z}/2)^3$ and $e(3) = (0,1,1)$, we have $3d \notin \mathbf{Q}_2^{*2}$, a contradiction.
⟨3⟩ G' contains a component A_5.
$e(A_5) = (1,1,1)$. Considering the first component of $e(G')$ we have $G' = A_5 + A_1$.
⟨4⟩ G' contains a component A_4.
Considering the second component of $e(G')$ we have $G' = A_4 + A_2$. However, in this case, $e(G') = (0,1,0) \neq (0,1,1)$; thus, $3d \notin \mathbf{Q}_2^{*2}$, a contradiction.
⟨5⟩ All components of G' are of type A_3, A_2 or A_1.
By the first component of $e(G')$ we know that the number of A_1-components of G' is even. By the second component we know that the number of A_2-components in G' is odd. We can write $G' = kA_3 + (2l+1)A_2 + 2mA_1$. Since $3k + 2(2l+1) + 2m \leq 6$, we know $G' = 3A_2$, $A_3 + A_2$, $A_2 + 4A_1$, $A_2 + 2A_1$, or A_2. Calculating ϵ_2 for each case we know $\epsilon_2 = -1 = (-1,3)_2$ if $G' \neq 3A_2$.
(5)⇒(3) This implication can be checked easily by calculation.
(5)⇔(6) Trivial. Q.E.D.

Lemma 31.12. (NPD10) *Any one G of the 3 graphs in NPD10 is a subgraph of either $E_6 + D_4 + A_1$ in NPD11 or $A_5 + D_4 + A_2$ in MPD11. In particular, we can make G from the basic graph $A_8 + G_2$ by 2 transformations and $G \in PC(U_{1,0})$.*

We complete this section here. We have shown Proposition 30.1 under the assumption [[1]] in the last part of Section 30.

32 Graphs with a component of type E, the case of $U_{1,0}$

In this section we consider the case [[2]], i.e., the case where $G \in PC(U_{1,0})$ has a component of type E but has no component of type D. We assume that the I1-condition fails for G, if necessary. By Lemma 30.4 the claims in Proposition 30.6 and in Lemma 30.7 hold in this case.

Lemma 32.1. *Let G be a Dynkin graph with 12 vertices such that every component is of type A or E only. We assume that G has a component of type E.*
(1) The following 2 conditions are equivalent:
⟨a⟩ The root lattice $Q = Q(G)$ of type G satisfies $l((Q^*/Q)_2) \leq 2$.
⟨b⟩ G belongs to the following class NIE12:

NIE12: (17 items.)

$E_8 + A_4$,	$E_8 + A_3 + A_1$,	$E_8 + 2A_2$,
$E_8 + A_2 + 2A_1$,	$E_7 + A_5$,	$E_7 + A_4 + A_1$,
$E_7 + A_3 + A_2$,	$E_7 + 2A_2 + A_1$,	$A_6 + E_6$,

$$\begin{array}{lll} 2E_6, & E_6+A_5+A_1, & E_6+A_4+A_2, \\ E_6+2A_3, & E_6+A_4+2A_1, & E_6+A_3+A_2+A_1, \\ E_6+3A_2, & E_6+2A_2+2A_1. & \end{array}$$

(2) If $G \in PC(U_{1,0})$, then G satisfies the above equivalent conditions in (1).

Proof. Claim (1) is easy. Claim (2) follows from Proposition 30.6 (3). Q.E.D.

All graphs in NIE12 do not belong to $PC(U_{1,0})$. We will see this later. (Lemma 32.5.)

Lemma 32.2. *Let G be a Dynkin graph with 11 vertices. G has a component of type E and every component of G is either of type A or of type E if and only if G belongs to one of the following 3 classes:*

$$\begin{array}{llll} \textit{MPE11:} & \text{(6 items.)} & & \\ & E_8+A_3, & E_8+3A_1, & E_7+A_4, \\ & E_7+A_3+A_1, & E_6+A_5, & E_6+A_3+A_2. \\ \textit{MIE11:} & \text{(8 items.)} & & \\ & E_8+A_2+A_1, & E_7+2A_2, & E_7+A_2+2A_1, \\ & E_7+4A_1, & E_6+A_4+A_1, & E_6+A_3+2A_1, \\ & E_6+2A_2+A_1, & E_6+5A_1. & \\ \textit{EXE11:} & \text{(1 item.)} & E_6+A_2+3A_1. & \end{array}$$

Lemma 32.3. (MPE11) *Any member G in MPE11 can be made from one of the 4 basic graphs A_8+G_2, E_7+G_2, $E_8+A_2(3)$, $E_6+A_2+A_1(3)$ by 2 tie transformations. In particular, $G \in PC(U_{1,0})$.*

Proof. Below we show examples of the process to make the desired graph G. Every arrow indicates a tie transformation.

$$\begin{array}{lllll} E_8+A_3 & \longleftarrow & D_9+G_2 & \longleftarrow & A_8+G_2 \\ E_9+3A_1 & \longleftarrow & D_9+G_2 & \longleftarrow & A_8+G_2 \\ E_7+A_4 & \longleftarrow & E_7+A_3 & \longleftarrow & E_7+G_2 \\ E_7+A_3+A_1 & \longleftarrow & E_7+A_3 & \longleftarrow & E_7+G_2 \\ E_6+A_5 & \longleftarrow & E_8+G_2 & \longleftarrow & E_8+A_2(3) \\ E_6+A_3+A_2 & \longleftarrow & E_6+A_2+G_2 & \longleftarrow & E_6+A_2+A_2(3) \end{array}$$

Since G contains no obstruction component A_8, we have $G \in PC(U_{1,0})$. Q.E.D.

Lemma 32.4. (MIE11) *Any member G in MIE11 do not belong to $PC(U_{1,0})$.*

Proof. We deduce a contradiction assuming the existence of a primitive embedding $P \hookrightarrow \Lambda_3$ and the existence of a full embedding $Q = Q(G) \hookrightarrow \Lambda_3/P$. By $S = P \oplus Q \hookrightarrow \Lambda_3$ we denote the induced embedding. T stands for the orthogonal complement of S in Λ_3. T has signature (1, 2). We denote $I = \widetilde{S}/S$ where $\widetilde{S}$ is the primitive hull of S in Λ_3. By Proposition 30.6 (3) $I = I_3$.

• $G = E_7 + 4A_1$ or $E_6 + 5A_1$.

We have $(\widetilde{S}^*/\widetilde{S})_2 \cong (S^*/S)_2 \cong (Q^*/Q)_2$ since $I_2 = 0$. Thus, $l((\widetilde{S}^*/\widetilde{S})_2) = l((Q^*/Q)_2) = 5$ in these 2 cases. However, $3 = \operatorname{rank} T \geq l((\widetilde{S}^*/\widetilde{S})_2)$. We have a contradiction.

• $G = E_7 + 2A_2$.

For this case we can apply the method explained in Lemma 31.7 for $D_4 + A_3 + 2A_2$. Since $3 = \operatorname{rank} T < l((S^*/S)_3) = 4$, we have $I_3 \neq 0$. On the other hand, we can show that if an isotropic element in $(S^*/S)_3$ belongs to I, then $P \hookrightarrow \Lambda_3$ is not primitive or $Q \hookrightarrow \Lambda_3/P$ is not full. Thus, $I_3 = 0$, a contradiction.

• $G = E_8 + A_2 + A_1$ or $E_7 + A_2 + 2A_1$.

In both cases we can write $G = A_2 + G'$ with $d' = d(Q(G')) \equiv 2 \pmod 3$. To these cases we can apply the method explained in Lemma 31.8 for $D_5 + 2A_2 + A_1$.

If an isotropic element in $(S^*/S)_3$ belongs to I_3, then we can show that either $P \hookrightarrow \Lambda_3$ is not primitive or $Q \hookrightarrow \Lambda_3/P$ is not full. Thus, $I = I_3 = 0$ and $\widetilde{S} = S$. Calculating the discriminant of T in 2 different methods we can deduce that $d' \equiv 1 \pmod 3$, a contradiction.

• $G = E_6 + A_4 + A_1$ or $E_6 + A_3 + 2A_1$.

In both cases we can write $G = E_6 + G'$ with $d' = d(Q(G')) \equiv 1 \pmod 3$. Also, to these cases we can apply the method explained in Lemma 31.8 for $D_5 + 2A_2 + A_1$. We can deduce $d' \equiv 2 \pmod 3$, a contradiction.

• $G = E_6 + 2A_2 + A_1$.

In this case, $S^*/S = (S^*/S)_3 \oplus \mathbf{Z}/2$ and $(S^*/S)_3 \cong \mathbf{Z}/9 \oplus (\mathbf{Z}/3)^{\oplus 4}$. The first component of the right-hand side of the last equality and the second correspond to P, the third to E_6, and the fourth and the fifth correspond to one of 2 A_2 respectively. The discriminant quadratic form on $(S^*/S)_3$ can be written $q(\overline{\alpha}) \equiv 4x^2/9 - 2(y^2+z^2)/3 + 2(w_1^2+w_2^2)/3 \bmod 2\mathbf{Z}$ for $\overline{\alpha} = (x, y, z, w_1, w_2) \in \mathbf{Z}/9 \oplus (\mathbf{Z}/3)^{\oplus 4}$. We have $I = I_3 \neq 0$, since $3 = \operatorname{rank} T < l((S^*/S)_3) = 5$.

On the other hand, any isotropic element in $(S^*/S)_3$ is one of the following:

$$\overline{\beta}_1 = (0, \pm 1, 0, \{\pm 1, 0\}), \qquad \overline{\beta}_2 = (0, 0, \pm 1, \{\pm 1, 0\}),$$
$$\overline{\beta}_3 = (\pm 3, 0, 0, 0, 0), \qquad \overline{\beta}_4 = (\pm 3, \pm 1, 0, \{\pm 1, 0\}),$$
$$\overline{\gamma} = (\pm 3, 0, \pm 1, \{\pm 1, 0\}).$$

If $\overline{\beta}_1$ or $\overline{\beta}_4 \in I_3$, then we have $Q(A_2) \subset Q(G_2) \subset \widetilde{Q(G)} \subset \Lambda_3/P$, which contradicts the fullness. If $\overline{\beta}_2 \in I_2$, then we have $Q(E_6 + A_2) \subset Q(E_8) \subset \widetilde{Q(G)} \subset \Lambda_3/P$, which contradicts the fullness, too. If $\overline{\beta}_3 \in I_3$, then $P \hookrightarrow \Lambda_3$ is not primitive, a contradiction. In conclusion, I_3 is a cyclic group of order 3 generated by $\overline{\gamma}$. Exchanging the sign of generators of components of $(S^*/S)_3$ and exchanging the order of 2 of A_2 components we can assume $\overline{\gamma} = (3, 0, 1, 1, 0) \in I_3$. Set

$$\overline{\delta}_1 = (0, 1, 0, 0, 0), \qquad \overline{\delta}_2 = (0, 0, 0, 0, 1), \qquad \overline{\delta}_3 = (1, 0, 1, 0, 0).$$

Let J be the subgroup generated by $\overline{\delta}_i$'s. We know $J + I_3 = I_3^{\perp}$ and $J \cap I_3 = 0$ in $(S^*/S)_3$. Thus, $\widetilde{S}^*/\widetilde{S} \cong I^{\perp}/I \cong \mathbf{Z}/9 \oplus \mathbf{Z}/3 \oplus \mathbf{Z}/3 \oplus \mathbf{Z}/2$. Therefore, we have $d(T) = |d(\widetilde{S})| = 2 \times 3^4$.

On the other hand, since $\overline{\delta}_i$'s are mutually orthogonal and since $q_3(\overline{\delta}_1) \equiv -2/3$ $q_3(\overline{\delta}_2) \equiv 2/3$ and $q_3(\overline{\delta}_3) \equiv -2/9$, the discriminant form on J can be written $-2x^2/9 + 2y^2/3 - 2z^2/3$ for $(x, y, z) \in \mathbf{Z}/9 \oplus \mathbf{Z}/3 \oplus \mathbf{Z}/3$. The discriminant quadratic form of $T \otimes \mathbf{Z}_3$ is equal to (-1) times that on $(\widetilde{S}^*/\widetilde{S})_3 \cong J$. It implies $d(T) \equiv -2^3 \times 3^4 \bmod \mathbf{Z}_3^{*2}$. Comparing the 2 expressions of $d(T)$ we have $-1 \in \mathbf{Z}_3^{*2}$, which implies that $z^2 \equiv -1 \pmod 3$ for some integer z, a contradiction. Q.E.D.

The unique member $E_6 + A_2 + 3A_1$ of EXE11 is an exception in Theorem 0.8, and will be studied in Section 34.

Lemma 32.5. (NIE12) *Any member G in NIE12 has a subgraph G' belonging to MIE11. Thus, $G \notin PC(U_{1,0})$.*

We proceed to the case $r = 10$.

Lemma 32.6. *Let G be a Dynkin graph with 10 vertices. We assume that G has a component of type E and every component of G is either of type E or of type A. $Q = Q(G)$ does not satisfy the I1-condition if, and only if, $G = E_6 + 4A_1$.*

Proof. Set $G = G_E + G'$, where G_E is the sum of components of type E. G' has at most 4 vertices. We denote $d = d(Q) = d(Q(G))$ and $d' = d(Q(G'))$. Since the I1-condition fails, the following $(**_p)$ holds for some prime number p:

$$(**_p) \qquad 3d \in \mathbf{Q}_p^{*2} \text{ and } \epsilon_p(Q) \neq (-1, 3)_p.$$

Assume that $(**_p)$ holds for $p \geq 5$. We have $p \mid d$ since $\epsilon_p(Q) = -1$. Moreover, we have $p^2 \mid d$, since $3d \in \mathbf{Q}_p^{*2}$. Thus, we conclude $p^2 \mid d'$, since $p \nmid d(Q(G_E))$. By Lemma 7.5 (3) we have $8 \leq \operatorname{rank} Q(G') \leq 4$, a contradiction.

Therefore, $(**_p)$ holds for $p = 3$ or 2.

Assume that $(**_3)$ holds. Since $3d \in \mathbf{Q}_3^{*2}$, the total number of components of G of type A_2 or E_6 is odd. We have 6 possibilities satisfying this condition. Checking $(**_3)$ for these 6 we can conclude $G = E_6 + 4A_1$.

Assume that $(**_2)$ holds. By Lemma 7.7 we know that $3d \in \mathbf{Q}_2^{*2}$ if, and only if, $G = E_8 + A_2$, $E_7 + A_2 + A_1$, $E_6 + 2A_2$ or $E_6 + 4A_1$. Computing ϵ_2 we get the conclusion.

The converse is now obvious. Q.E.D.

Lemma 32.7. *The graph $G = E_6 + 4A_1$ can be made from the basic graph $A_8 + G_2$ by 2 tie transformations. Thus, in particular, $E_6 + 4A_1 \in PC(U_{1,0})$.*

Proof. Indeed, from $A_8 + G_2$ we can make $D_9 + G_2$ by a tie transformation. $E_6 + 4A_1$ can be made from $D_9 + G_2$ by a tie transformation. Q.E.D.

In this section we have shown Proposition 30.1 under the assumption [[2]] in the last part of Section 30.

33
Combinations of graphs of type A in the case $U_{1,0}$

In this section we treat the case [[3]], i.e., the case where G has components of type A only. As in the previous sections we assume that the I1-condition fails for G, when this assumption is necessary. By Lemma 30.4, if G is not a subgraph of $A_{11}+A_1$ the claims in Proposition 30.6 and Lemma 30.7 hold.

Lemma 33.1. $PC(U_{1,0}) \subset PC(W_{1,0})$.

Proof. Recall that $P = P(U_{1,0}) = P(2,3,3,3)$ is a lattice of rank 8 with a basis e_0, ..., e_7. Let $\overline{P}$ be the sublattice generated by e_0, ..., e_6. $\overline{P}$ is isomorphic to the lattice P defined for $W_{1,0}$, i.e., $P(2,2,3,3)$.

If $G \in PC(U_{1,0})$, then we have a lattice embedding $P \oplus Q(G) \hookrightarrow \Lambda_3$ satisfying Looijenga's conditions (L1) and (L2). We can check easily that the induced embedding $\overline{P} \oplus Q(G) \hookrightarrow \Lambda_3$ also satisfies (L1) and (L2). Thus, $G \in PC(W_{1,0})$. Q.E.D.

First, we consider the case where the number r of vertices in G is equal to 12. By Lemma 30.7 G has at most 6 components if $G \in PC(U_{1,0})$. Thus, we can consider only graphs with the form $G = A_{k_1} + \cdots + A_{k_6}$, $k_1 \geq k_2 \geq \cdots \geq k_6 \geq 0$, $\sum_{i=1}^{6} k_i = 12$. (A_0 stands for the empty graph.) Such graphs correspond to divisions of 12 into a sum of 6 non-negative integers. As we considered in the case $J_{3,0}$, $W_{1,0}$, there are 58 such divisions. By the description for $W_{1,0}$ and by Lemma 33.1 it is enough to treat only the following 29:

[1] 12	[2] 11+1	[3] 10+2	[4] 9+3
[5] 8+4	[6] 7+5	[7] 6+6	[8] 10+1+1
[9] 9+2+1	[10] 8+3+1	[11] 8+2+2	[12] 7+4+1
[13] 7+3+2	[14] 6+5+1	[15] 6+4+2	
[17] 5+5+2	[18] 5+4+3		[20] 9+1+1+1
[21] 8+2+1+1	[22] 7+3+1+1		[24] 6+4+1+1
[25] 6+3+2+1		[27] 5+5+1+1	
	[30] 5+3+2+2	[31] 4+4+3+1	
[36] 7+2+1+1+1	[37] 6+3+1+1+1	[42] 4+4+2+1+1	[46] 3+3+3+2+1
[51] 5+2+2+1+1+1			

We divide these 29 graphs into the 3 classes MPA12, MIA12, NIA12.

MPA12: (3 items.) [1] [2] [4]
NIA12: (5 items.) [5] [7] [9] [17] [31]
NIA12: (21 items.)
[3] [6] [8] [10] [11] [12] [13] [14] [15] [18] [20]
[21] [22] [24] [25] [27] [30] [36] [42] [46] [51]

Lemma 33.2. (MPA12) *Any graph G of 3 members in MPA12 can be made from $A_8 + G_2$ by 2 tie transformations, and it has no component of type A_8. In particular, $G \in PC(U_{1,0})$.*

Proof. Indeed we can make A_{11} or $A_9 + G_2$ from $A_8 + G_2$ by a tie transformation. From A_{11} we can make [1] A_{12} or [2] $A_{11} + A_1$ by a tie transformation. [4] $A_9 + A_5$ can be made from $A_9 + G_2$ by a tie transformation. Q.E.D.

Lemma 33.3. (MIA12) *Any graph G in MIA12 does not belong to $PC(U_{1,0})$.*

Proof. Let G be the graph under consideration. Assuming that there exist a primitive embedding $P \hookrightarrow \Lambda_3$ and a full embedding $Q = Q(G) \hookrightarrow \Lambda_3/P$ such that any component of type A_8 is not an obstruction component, we will deduce a contradiction.

$S = P \oplus Q(G) \hookrightarrow \Lambda_3$ denotes the induced embedding. T stands for the orthogonal complement of S in Λ_3. T has signature $(12 - r, 2)$. We denote $I = \widetilde{S}/S$ where $\widetilde{S}$ is the primitive hull of S in Λ_3. By Lemma 30.6 (3) $I = I_3$, since G is not a subgraph of $A_{11} + A_1$ in our case. Any non-zero element in I_3 is an isotropic element in $(S^*/S)_3$.

[5] $A_8 + A_4$.

In this case, we have to be careful about the obstruction component A_8. $(S^*/S)_3 \cong \mathbf{Z}/9 \oplus \mathbf{Z}/9 \oplus \mathbf{Z}/3$. The first component $\mathbf{Z}/9$ of the right-hand side and the third $\mathbf{Z}/3$ correspond to P, and the second $\mathbf{Z}/9$ to A_8. Recall that the discriminant form of $Q(A_8)$ is given by $x \in \mathbf{Z}/9 \mapsto 8x^2/9 \bmod 2\mathbf{Z}$. However, if we take $2 \in \mathbf{Z}/9$ as the generator, this is given by $y \in \mathbf{Z}/9 \mapsto 8 \times 4y^2/9 \equiv -4y^2/9 \bmod 2\mathbf{Z}$. Thus, the discriminant quadratic form on $(S^*/S)_3$ is given by $q_3(\overline{\alpha}) \equiv 4(x^2 - y^2)/9 - 2z^2/3 \bmod 2\mathbf{Z}$ for $\overline{\alpha} = (x, y, z) \in \mathbf{Z}/9 \oplus \mathbf{Z}/9 \oplus \mathbf{Z}/3$.

On the other hand, we have $I_3 \neq 0$ since $2 = \operatorname{rank} T \geq l((S^*/S)_3) - 2l(I_3) = 3 - 2l(I_3)$. Thus, some isotropic element belongs to I_3. Any isotropic element in $(S^*/S)_3$ is one of the following:

$$\begin{array}{lll} \overline{\beta}_1 = (\pm 3, 0, 0), & \overline{\beta}_2 = (0, \pm 3, 0), & \\ \overline{\gamma}_1 = (\pm m, \pm m, 0), & \overline{\gamma}_2 = (\pm m, \pm 2m, \pm 1), & (\text{m=1, 2 or 4.}) \\ \overline{\delta} = (\pm 3, \pm 3, 0). & & \end{array}$$

If $\overline{\beta}_1 \in I_3$, then the embedding $P \hookrightarrow \Lambda_3$ is not primitive. If $\overline{\beta}_2 \in I_3$, then $Q(A_8) \subset Q(E_8) \subset \widetilde{Q(G)} \subset \Lambda_3/P$, which contradicts the fullness. If $\overline{\gamma}_1$ or $\overline{\gamma}_2 \in I_3$, then the restriction to I_3 of the projection $(S^*/S)_3 \to (Q(G)^*/Q(G))_3 = Q(A_8)^*/Q(A_8)$ is surjective and A_8 is an obstruction component, which contradicts the assumption. In conclusion, $I = I_3$ is a cyclic group of order 3 generated by $\overline{\delta}$. Thus, $(\widetilde{S}^*/\widetilde{S})_3 \cong I_3^{\perp}/I_3 \cong (\mathbf{Z}/3)^3$. It implies $2 = \operatorname{rank} T \geq l((\widetilde{S}^*/\widetilde{S})_3) = 3$, a contradiction.

[7] $2A_6$.

Setting $\#I = \#I_3 = 3^k$, we know $d(T) = |d(\widetilde{S})| = 3^{3-2k} \times 7^2$.

On the other hand, the discriminant quadratic form on $(S^*/S)_7 \cong \mathbf{Z}/7 \oplus \mathbf{Z}/7$ is given by $6(x^2+y^2)/7 \bmod 2\mathbf{Z}$ for $(x, y) \in \mathbf{Z}/7 \oplus \mathbf{Z}/7$. It implies that $T \otimes \mathbf{Z}_7$ is isomorphic to the 7-adic lattice defined by the 2×2 diagonal matrix whose diagonal entries are -6×7 and -6×7. Thus, $d(T) \equiv 2^2 \times 3^2 \times 7^2 \bmod \mathbf{Z}_7^{*2}$. By the 2 expressions of $d(T)$ we have $3 \in \mathbf{Z}_7^{*2}$, which implies that $z^2 \equiv 3 \pmod 7$ for some integer z, a contradiction.

[9] $A_9+A_2+A_1$. We can apply the method explained in Lemma 31.7 in the case $D_4+A_3+2A_2$ to this case. First, we have $I_3 \neq 0$ by the inequality $\operatorname{rank} T \geq l((S^*/S)_3) - 2l(I_3)$. On the other hand, solving the equation $q(\overline{\alpha}) \equiv 0$, we can show that if an isotropic

element in $(S^*/S)_3$ belongs to I_3, then either $Q \hookrightarrow \Lambda_3/P$ is not full or $P \hookrightarrow \Lambda_3$ is not primitive. Thus, $I_3 = 0$, a contradiction.

[17] $2A_5 + A_2$.

We can apply the method in Lemma 31.3 for $D_4 + 4A_2$ to this case. First, we know that I_3 is not cyclic by the inequality rank $T \geq l((S^*/S)_3) - 2l(I_3)$. Let ISO be the set of isotropic elements in $(S^*/S)_3$. We can determine ISO explicitly. Let ISO' be the collection of elements $\overline{\alpha} \in ISO$ such that if $\overline{\alpha} \in I_3$, then either $P \hookrightarrow \Lambda_3$ is not primitive or $Q \hookrightarrow \Lambda_3/P$ is not full. We consider the complement $ISO'' = ISO - ISO'$. Considering projections from $(S^*/S)_3$ to a direct summand, we know that some elements $\overline{\gamma}_1, \overline{\gamma}_2$ belong to I_3. However, we can choose such elements with $\overline{\gamma}_1 + \overline{\gamma}_2 \notin ISO''$, and we get a contradiction.

[31] $2A_4 + A_3 + A_1$.

The method explained in Lemma 31.8 for $D_5 + 2A_2 + A_1$ is applicable to this case. Solving the congruent equation $q(\overline{\alpha}) \equiv 0$ on $(S^*/S)_3$ we can show $I = I_3 = 0$ under the assumption of being primitive and being full. Thus, $\widetilde{S} = S$ and $d(T) = |d(S)| = 5^2 \times 3^3 \times 2^3$. On the other hand, by the 3-adic method we have $d(T) \equiv -2^3 \times 3^3 \bmod \mathbf{Z}_3^{*2}$. Thus, we get $-1 \in \mathbf{Z}_3^{*2}$, a contradiction. Q.E.D.

We will discuss the class NIA12 later. (Lemma 33.8.) We proceed to the case $r = 11$. By Lemma 30.7 our Dynkin graph corresponds to a division of 11 into a sum of 7 non-negative integers. We have 49 such divisions as follows:

⟨1⟩ 11	⟨2⟩ 10+1	⟨3⟩ 9+2
⟨4⟩ 8+3	⟨5⟩ 7+4	⟨6⟩ 6+5
⟨7⟩ 9+1+1	⟨8⟩ 8+2+1	⟨9⟩ 7+3+1
⟨10⟩ 7+2+2	⟨11⟩ 6+4+1	⟨12⟩ 6+3+2
⟨13⟩ 5+5+1	⟨14⟩ 5+4+2	⟨15⟩ 5+3+3
⟨16⟩ 4+4+3	⟨17⟩ 8+1+1+1	⟨18⟩ 7+2+1+1
⟨19⟩ 6+3+1+1	⟨20⟩ 6+2+2+1	⟨21⟩ 5+4+1+1
⟨22⟩ 5+3+2+1	⟨23⟩ 5+2+2+2	⟨24⟩ 4+4+2+1
⟨25⟩ 4+3+3+1	⟨26⟩ 4+3+2+2	⟨27⟩ 3+3+3+2
⟨28⟩ 7+1+1+1+1	⟨29⟩ 6+2+1+1+1	⟨30⟩ 5+3+1+1+1
⟨31⟩ 5+2+2+1+1	⟨32⟩ 4+4+1+1+1	⟨33⟩ 4+3+2+1+1
⟨34⟩ 4+2+2+2+1	⟨35⟩ 3+3+3+1+1	⟨36⟩ 3+3+2+2+1
⟨37⟩ 3+2+2+2+2	⟨38⟩ 6+1+1+1+1+1	⟨39⟩ 5+2+1+1+1+1
⟨40⟩ 4+3+1+1+1+1	⟨41⟩ 4+2+2+1+1+1	⟨42⟩ 3+3+2+1+1+1
⟨43⟩ 3+2+2+2+1+1	⟨44⟩ 2+2+2+2+2+2+1	⟨45⟩ 5+1+1+1+1+1+1
⟨46⟩ 4+2+1+1+1+1+1	⟨47⟩ 3+3+1+1+1+1+1	⟨48⟩ 3+2+2+1+1+1+1
⟨49⟩ 2+2+2+2+1+1+1		

We divide these into 4 classes MPA11, NPA11, AEA11, MIA11.

MPA11: (8 items.)

⟨4⟩ ⟨8⟩ ⟨14⟩ ⟨19⟩ ⟨22⟩ ⟨23⟩ ⟨25⟩ ⟨32⟩

NPA11: (12 items.)

⟨1⟩ ⟨2⟩ ⟨3⟩ ⟨5⟩ ⟨6⟩ ⟨7⟩ ⟨9⟩ ⟨11⟩ ⟨12⟩ ⟨13⟩ ⟨15⟩ ⟨16⟩

AEA11: (9 items.)
$\langle 18\rangle$ $\langle 20\rangle$ $\langle 24\rangle$ $\langle 27\rangle$ $\langle 31\rangle$ $\langle 33\rangle$ $\langle 36\rangle$ $\langle 37\rangle$ $\langle 44\rangle$
MIA11: (20 items.)
$\langle 10\rangle$ $\langle 17\rangle$ $\langle 21\rangle$ $\langle 26\rangle$ $\langle 28\rangle$ $\langle 29\rangle$ $\langle 30\rangle$ $\langle 34\rangle$ $\langle 35\rangle$ $\langle 38\rangle$
$\langle 39\rangle$ $\langle 40\rangle$ $\langle 41\rangle$ $\langle 42\rangle$ $\langle 43\rangle$ $\langle 45\rangle$ $\langle 46\rangle$ $\langle 47\rangle$ $\langle 48\rangle$ $\langle 49\rangle$

Lemma 33.4. (MPA11) *Every graph G in MPA11 can be make from the basic graph $A_8 + G_2$ or $E_6 + A_2 + A_2(3)$ by 2 combinations of elementary transformations and tie transformation. We can make G without an obstruction component A_8. In particular, $G \in PC(U_{1,0})$.*

Proof. In the following we show an example of the process for each graph. Except the top left 2 arrows every arrow indicates a tie transformation:

$$
\begin{array}{llllll}
\langle 4\rangle & A_8 + A_3 & \overset{elementary}{\longleftarrow} & A_8 + A_3 & \longleftarrow & A_8 + G_2 \\
\langle 8\rangle & A_8 + A_2 + A_1 & \overset{elementary}{\longleftarrow} & A_8 + G_2 + A_1 & \longleftarrow & A_8 + G_2 \\
\langle 14\rangle & A_5 + A_4 + A_2 & \longleftarrow & A_8 + A_3 & \longleftarrow & A_8 + G_2 \\
\langle 19\rangle & A_6 + A_3 + 2A_1 & \longleftarrow & A_8 + A_3 & \longleftarrow & A_8 + G_2 \\
\langle 22\rangle & A_5 + A_3 + A_2 + A_1 & \longleftarrow & A_8 + A_3 & \longleftarrow & A_8 + G_2 \\
\langle 23\rangle & A_5 + 3A_2 & \longleftarrow & E_6 + A_2 + G_2 & \longleftarrow & E_6 + A_2 + A_2(3) \\
\langle 25\rangle & A_4 + 2A_3 + A_1 & \longleftarrow & A_8 + A_3 & \longleftarrow & A_8 + G_2 \\
\langle 32\rangle & 2A_4 + 3A_1 & \longleftarrow & D_9 + G_2 & \longleftarrow & A_8 + G_2
\end{array}
$$

Lemma 33.5. (NPA11) *Every graph G in NPA11 does not contain a component A_8 and is a subgraph of one of the 3 members A_{12}, $A_{11}+A_1$, A_9+A_3 of MPA12. Therefore, G can be made from the basic graph A_8+G_2 by 2 of elementary or tie transformations, and in particular, G belongs to $PC(U_{1,0})$.*

Lemma 33.6. (AEA11) *For every graph G in AEA11 the root lattice $Q = Q(G)$ satisfies the I1-condition. Thus, we need not consider them further.*

Proof. By the I1-condition for every prime number p following $(*_p)$ holds:

$$(*_p) \qquad \epsilon_p(Q) = (3,\, d(Q))_p.$$

By the product formula we need not consider the case $p = 2$. For every graph G we can check this condition easily. Q.E.D.

Lemma 33.7. (MIA11) *Any graph G in MIA11 does not belong to $PC(U_{1,0})$.*

Proof. It is easily checked that G is not a subgraph of $A_{11} + A_1$. Thus, we can assume $a = 2$.

For the following 11 graphs the corresponding division of 11 contains 5 odd numbers:

$\langle 28\rangle$, $\langle 30\rangle$, $\langle 35\rangle$, $\langle 38\rangle$, $\langle 39\rangle$, $\langle 40\rangle$, $\langle 42\rangle$, $\langle 45\rangle$, $\langle 46\rangle$, $\langle 47\rangle$, $\langle 48\rangle$

This implies $l((S^*/S)_2) = l((Q(G)^*/Q(G))_2) \geq 5$. By Lemma 30.6 (3) $I_2 = 0$ and $l((\widetilde{S}^*/\widetilde{S})_2) = l((S^*/S)_2)$. Thus, we get $3 = \operatorname{rank} T \geq l((\widetilde{S}^*/\widetilde{S})_2) \geq 5$, a contradiction.

For the following 2 cases, by the description in $W_{1,0}$, $G \notin PC(W_{1,0})$:

$$\langle 34\rangle,\ \langle 43\rangle$$

By Lemma 33.1 we have $G \notin PC(U_{1,0})$.

Following 7 cases are remaining:

$$\langle 10\rangle,\ \langle 17\rangle,\ \langle 21\rangle,\ \langle 26\rangle,\ \langle 29\rangle,\ \langle 41\rangle,\ \langle 49\rangle$$

We assume the situations and notations at the beginning of the verification of Lemma 33.3.

$\langle 10\rangle$ $A_7 + 2A_2$, $\langle 26\rangle$ $A_4 + A_3 + 2A_2$, $\langle 41\rangle$ $A_4 + 2A_2 + 3A_1$.

To these 3 cases we can apply the method explained in Lemma 31.7 for $D_4 + A_3 + 2A_2$. First, by the inequality $\operatorname{rank} T \geq l((S^*/S)_3 - 2l(I_3)$ we can conclude $I_3 \neq 0$. On the other hand, solving the congruent equation $q(\overline{\alpha}) \equiv 0$ on $(S^*/S)_3$ we can show that if an isotropic element belongs to I_3, then $P \hookrightarrow \Lambda_3$ is not primitive or $Q \hookrightarrow \Lambda_3/P$ is not full, which contradicts the assumption. Thus, $I_3 = 0$, a contradiction.

$\langle 49\rangle$ $4A_2 + 3A_1$.

We can apply the method in Lemma 31.3 for $D_4 + 4A_2$ to this case. We can go along the line explained in Lemma 33.3 in the case [17] $2A_5 + A_2$.

$\langle 21\rangle$ $A_5 + A_4 + 2A_1$, $\langle 29\rangle$ $A_6 + A_2 + 3A_1$.

To these 2 cases the method in Lemma 31.8 for $D_5 + 2A_2 + A_1$ can be applied. We get a contradiction -1 or $2 \in \mathbf{Z}_3^{*2}$.

$\langle 17\rangle$ $A_8 + 3A_1$.

The method in this case is very similar to that in Lemma 32.4 for $E_6 + 2A_2 + A_1$. However, an obstruction component A_8 is concerning.

Now, we have $(S^*/S)_3 \cong \mathbf{Z}/9 \oplus \mathbf{Z}/9 \oplus \mathbf{Z}/3$. The first component of the right-hand side and the third correspond to P and the second to A_8. The discriminant quadratic form is given by $q_3(\overline{\alpha}) \equiv 4(x^2-y^2)/9 - 2z^2/3 \bmod 2\mathbf{Z}$ for $\overline{\alpha} = (x, y, z) \in \mathbf{Z}/9 \oplus \mathbf{Z}/9 \oplus \mathbf{Z}/3$.

First, assume $I = I_3 = 0$. $d(T) = |d(S)| = 3^5 \times 2^3$. On the other hand, since the discriminant form of $T \otimes \mathbf{Z}_3$ is equal to $-q_3$ we have $d(T) \equiv -2^5 \times 3^5 \bmod \mathbf{Z}_3^{*2}$. By the 2 expressions we have $-1 \in \mathbf{Z}_3^{*2}$, a contradiction. We conclude $I = I_3 \neq 0$.

An isotropic element in $(S^*/S)_3$ is one of the following:

$$\begin{array}{lll} \overline{\beta}_1 = (0, \pm 3, 0), & \overline{\beta}_2 = (\pm 3, 0, 0), & \\ \overline{\gamma}_1 = (\pm m, \pm m, 0), & \overline{\gamma}_2 = (\pm m, \pm 2m, \pm 1), & (m = 1,\ 2 \text{ or } 4.) \\ \overline{\delta} = (\pm 3, \pm 3, 0). & & \end{array}$$

I_3 contains one of them.

If $\overline{\beta}_1 \in I_3$, then $Q(A_8) \subset Q(E_8) \subset \widetilde{Q(G)} \subset \Lambda_3/P$, which contradicts the fullness. If $\overline{\beta}_2 \in I_3$, then the embedding $P \hookrightarrow \Lambda_3$ is not primitive, a contradiction. If $\overline{\gamma}_1$ or $\overline{\gamma}_2$ belongs to I_3, then the restriction to I_3 of the projection $(S^*/S)_3 \to (Q(G)^*/Q(G))_3 = (Q(A_8)^*/Q(A_8))_3$ is surjective. It implies that A_8 is an obstruction component, a contradiction. Thus, I_3 coincide with a cyclic group of order 3 generated by $\overline{\delta}$. Exchanging the sign of the generators of $(S^*/S)_3$ we can assume $\overline{\delta} = (3, 3, 0) \in I_3$. Set $\overline{\epsilon}_1 = (1, -2, 0)$, $\overline{\epsilon}_2 = (2, -1, 0)$ and $\overline{\epsilon}_3 = (0, 0, 1)$. $\overline{\epsilon}_i$'s are mutually orthogonal and generate $I_3^{\perp}$. $q_3(\bar{\epsilon}_1) \equiv 2/3$, $q_3(\bar{\epsilon}_2) \equiv -2/3 \bmod 2\mathbf{Z}$. Thus, the quadratic form on $(\widetilde{S}^*/\widetilde{S})_3 \cong (I^{\perp}/I)_3 \cong \mathbf{Z}/3 \oplus \mathbf{Z}/3 \oplus \mathbf{Z}/3$ is given by $\tilde{q}_3(\bar{x}) \equiv 2a^2/3 - 2(b^2+c^2)/3 \bmod 2\mathbf{Z}$ for $\bar{x} = (a, b, c) \in \mathbf{Z}/3 \oplus \mathbf{Z}/3 \oplus \mathbf{Z}/3$. We have $d(T) = |d(\widetilde{S})| = 2^3 \times 3^3 \equiv -2^3 \times 3^3 \bmod \mathbf{Z}_3^{*2}$. It implies $-1 \in \mathbf{Z}_3^{*2}$, a contradiction. Q.E.D.

Lemma 33.8. (NIA12) *Every graph G in NIA12 contains a subgraph G' belonging to MIA11. Thus, $G \notin PC(U_{1,0})$.*

The last remaining case is the case $r = 10$.

Lemma 33.9. *Let G be a Dynkin graph with 10 vertices. We assume that every component of G is of type A. The root lattice $Q = Q(G)$ of type G does not satisfy the I1-condition if, and only if, $G = A_5 + 5A_1$, $3A_2 + 4A_1$ or $10A_1$.*

Proof. We write $d = d(Q)$ and $\epsilon_p = \epsilon_p(Q)$ for simplicity. Recall that the I1-condition fails if, and only if, the following $(**_p)$ holds for some prime number p:

$$(**_p) \qquad 3d \in \mathbf{Q}_p^{*2} \text{ and } \epsilon_p \neq (-1,\, 3)_p.$$

Assume that $(**_p)$ holds for $p \geq 7$. $\epsilon_p = -1$. Thus, $p \mid d$ by Lemma 7.5 (4). Moreover, since $3d \in \mathbf{Q}_p^{*2}$, $p^2 \mid d$. By Lemma 7.5 (3) we have $12 \leq \operatorname{rank} Q = 10$, a contradiction.

Assume that $(**_5)$ holds. By the same reason as above we can write $G = 2A_4 + G'$ where $G' = A_2$ or $2A_1$. However, then, $\epsilon_5 = (5,\, 5)_5 = +1 = (-1,\, 3)_5$, which contradicts the assumption.

Assume that $(**_3)$ holds. Since $3d \in \mathbf{Q}_3^{*2}$ the total number of components of type A_5 or A_2 is odd. Since $(-1,\, 3)_3 = -1$, $\epsilon_3 = +1$ under the assumption. If $G = A_5 + 2A_2 + A_1$, then $\epsilon_3 = -1$. Thus, $G \neq A_5 + 2A_2 + A_1$. Thus, if G contains a component A_5, then we can write $G = A_5 + G'$ with $3 \not| \, d' = d(Q(G'))$. $3d = 2 \times 3^2 d' \in \mathbf{Q}_3^{*2} \Leftrightarrow d' \equiv 2 \pmod 3$. $+1 = \epsilon_3 = -(3,\, d')_3 \Leftrightarrow d' \equiv 2 \pmod 3$. By the last condition we know $G' = 5A_1$ and $G = A_5 + 5A_1$. Below we assume that G has no component A_5.

If $G = 5A_2$, then $\epsilon_3 = -1$, a contradiction. If $G = 3A_2 + G'$ with $3 \not| \, d' = d(Q(G'))$, then $3d = 3^4 d' \in \mathbf{Q}_3^{*2} \Leftrightarrow d' \equiv 1 \pmod 3$, and $+1 = \epsilon_3 = (3,\, d')_3 \Leftrightarrow d' \equiv 1 \pmod 3$. By this condition we know $G' = 4A_1$ and $G = 3A_2 + 4A_1$. If $G = A_8 + A_2$, then $\epsilon_3 = -1$, a contradiction. If we can write $G = A_2 + G'$ with $3 \not| \, d' = d(Q(G'))$, then $3d = 3^2 d' \in \mathbf{Q}_3^{*2} \Leftrightarrow d' \equiv 1 \pmod 3$, and $+1 = \epsilon_3 = -(3,\, d')_3 \Leftrightarrow d' \equiv 2 \pmod 3$, which are a contradiction. The above exhausts all possibilities.

Next, assume $(**_2)$ holds. By Lemma 7.7 we have $3d \in \mathbf{Q}_2^{*2} \Leftrightarrow G = A_{10}$, $A_8 + A_2$, $A_7 + A_2 + A_1$, $A_6 + A_4$, $A_5 + 2A_2 + A_1$, $A_5 + 5A_1$, $2A_4 + A_2$, $2A_3 + A_2 + 2A_1$, $5A_2$, $3A_2 + 4A_1$, $A_2 + 8A_1$, or $10A_1$. Calculating ϵ_2 for each possibility we get the conclusion. Q.E.D.

We divide the above 3 graphs into 2 classes MIA10 and EXA10.

MIA10:	(2 items.)	$A_5 + 5A_1$,	$10A_1$.
EXA10:	(1 item.)	$3A_2 + 4A_1$.	

Lemma 33.10. (MIA10) *Any graph G in MIA10 does not belong to $PC(U_{1,0})$.*

Proof. Assume $G \in PC(U_{1,0})$. We use the same notations at the beginning of the verification of Lemma 33.3. The division of 10 corresponding to $A_5 + 5A_1$ or $10A_1$ contains 6 or more odd integers. Thus, $l((S^*/S)_2) = l((Q^*/Q)_2) \geq 6$. Since $I_2 = 0$ by Lemma 30.6 (3), $l((\widetilde{S}^*/\widetilde{S})_2) = l((S^*/S)_2)$. Thus, we have $4 = \mathrm{rank}\, T \geq l((\widetilde{S}^*/\widetilde{S})_2) \geq 6$, a contradiction. Q.E.D.

The unique member $3A_2 + 4A_1$ of EXA10 is an exception in Theorem 0.8. We will treat this in the next section.

In this section we have shown Proposition 30.1 under the assumption [[3]] at the last part of Section 30. By results in Section 31, Section 32 and Section 33, Proposition 30.1 has been established.

34
The exceptions in the case $U_{1,0}$

Only the remaining part of our verification of Theorem 0.8 is the argument on the 4 exceptions $E_6 + D_4 + A_2$, $E_6 + A_2 + 3A_1$, $D_4 + 3A_2 + A_1$ and $3A_2 + 4A_1$.

Lemma 34.1. *(1) $E_6 + A_2 + 3A_1$ and $D_4 + 3A_2 + A_1$ are subgraphs of $E_6 + D_4 + A_2$. (2) $3A_2 + 4A_1$ is a common subgraph of $E_6 + A_2 + 3A_1$ and $D_4 + 3A_2 + A_1$.*

If we can show the following proposition, we can complete the verification of Theorem 0.8:

Proposition 34.2. *If $G = E_6 + D_4 + A_2$, $E_6 + A_2 + 3A_1$, $D_4 + 3A_2 + A_1$ or $3A_2 + 4A_1$, then there exists a full embedding $Q(G) \hookrightarrow \Lambda_3/P$ when we fix a primitive embedding $P \hookrightarrow \Lambda_3$. Thus, in particular, $G \in PC(U_{1,0})$.*

Proof. By Lemma 34.1, if the above proposition holds for $G = E_6 + D_4 + A_2$, then it holds also for the other 3 cases. Thus, we give the proof only for $G = E_6 + D_4 + A_2$. We fix a primitive embedding $P \hookrightarrow \Lambda_3$.

Let $\alpha_1, \ldots, \alpha_6$ and β_1, β_2 be the root bases of the root lattice $Q(E_6)$ and $Q(A_2)$ as in the following figure:

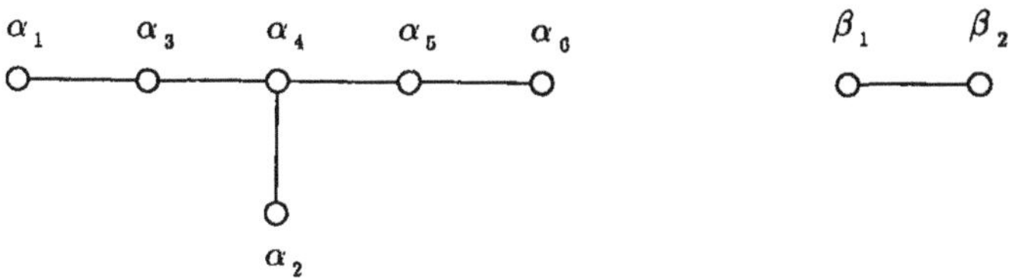

Set $\omega = (4\alpha_1 + 3\alpha_2 + 5\alpha_3 + 6\alpha_4 + 4\alpha_5 + 2\alpha_6)/3$ and $\chi = (2\beta_1 + \beta_2)/3$. We have $\omega \cdot \alpha_1 = 1$, $\omega \cdot \alpha_i = 0$ $(2 \leq i \leq 6)$, $\chi \cdot \beta_1 = 1$ and $\chi \cdot \beta_2 = 0$. Thus, $\omega \in Q(E_6)^*$, $\chi \in Q(A_2)^*$, $\omega^2 = 4/3$, and $\chi^2 = 2/3$.

Set, moreover, $u_1 = 3e_0 + 2(e_2 + e_3 + e_4) + e_5 + e_6 + e_7 \in P$. We can check $u_1^2 = 0$, $u_1 \cdot e_1 = 3$ and $u_1 \cdot e_i = 0$ $(2 \le i \le 7)$. Thus, $\nu = u_1/3 \in P^*$, $\nu^2 = 0$ and $\nu \notin P$.

Set $S = P \oplus Q(G)$ and $\hat{S} = S + \mathbf{Z}(\nu + \omega + \chi)$. $\hat{S}$ is an overlattice of S with index 3. We have $(\nu + \omega + \chi)^2 = \nu^2 + \omega^2 + \chi^2 = 2$.

We would like to compute the discriminant quadratic form of $\hat{S}$ here. $(S^*/S)_3 \cong \mathbf{Z}/9 \oplus \mathbf{Z}/3 \oplus \mathbf{Z}/3 \oplus \mathbf{Z}/3$. The first component of the right-hand side and the second correspond to P, the third to E_6, and the fourth to A_2. The discriminant quadratic form on it can be written $q_3(\overline{\alpha}) \equiv 4x^2/9 - 2(y^2+z^2)/3 + 2w^2/3 \bmod 2\mathbf{Z}$ for $\overline{\alpha} = (x, y, z, w) \in \mathbf{Z}/9 \oplus \mathbf{Z}/3 \oplus \mathbf{Z}/3 \oplus \mathbf{Z}/3$. Set $I = \hat{S}/S$. We know $I = I_3 \subset (S^*/S)_3$ and I is a cyclic group of order 3 generated by $\overline{\beta} = (3, 0, 1, 1)$. Setting $\overline{\gamma}_1 = (1, 0, -1, 0)$ and $\overline{\gamma}_2 = (0, 1, 0, 0)$, we know that $\overline{\beta}$, $\overline{\gamma}_1$ and $\overline{\gamma}_2$ generate $I^{\perp} \cap (S^*/S)_3$ and they are mutually orthogonal. Furthermore, $\langle \overline{\gamma}_1, \overline{\gamma}_2 \rangle \cap \langle \overline{\beta} \rangle = 0$. (By $\langle * \rangle$ we denote the subgroup generated by $*$.) Thus, we have $\hat{S}^*/\hat{S} \cong \mathbf{Z}/9 \oplus \mathbf{Z}/3 \oplus (\mathbf{Z}/2 + \mathbf{Z}/2)$. Since $q_3(\overline{\gamma}_1) \equiv -2/9$ and $q_3(\overline{\gamma}_2) \equiv -2/3$, the discriminant form of $\hat{S}$ can be written $q_{\hat{S}}(\overline{\alpha}) \equiv -2x^2/9 - 2y^2/3 + a^2 + ab + b^2 \bmod 2\mathbf{Z}$ for $\overline{\alpha} = (x, y, a, b) \in \mathbf{Z}/9 \oplus \mathbf{Z}/3 \oplus (\mathbf{Z}/2 + \mathbf{Z}/2)$. Note here that if we take $4 \in \mathbf{Z}/9$ as the generator, it can be written also $q_{\hat{S}}(\overline{\alpha}) \equiv 4x^2/9 - 2y^2/3 + a^2 + ab + b^2 \bmod 2\mathbf{Z}$.

Let $T = \mathbf{Z}\alpha + \mathbf{Z}\beta$ be the negative definite even lattice defined by $\alpha^2 = \beta^2 = -12$ and $\alpha \cdot \beta = -6$. $d(T) = 2^2 \cdot 3^3$. Let η, $\mu \in T^*$ be the dual basis of α, β.

$$\eta = -(\alpha/3^2) + (\beta/2 \cdot 3^2), \qquad \mu = (\alpha/2 \cdot 3^2) - (\beta/3^2).$$

Below we write $\overline{\xi} = \xi \bmod T \in T^*/T$ for $\xi \in T^*$. Set $\overline{\gamma} = 2\overline{\eta}$, $\overline{\delta} = 2\overline{\eta} + 4\overline{\mu}$, $\overline{\epsilon}_1 = 3^2\overline{\eta}$ and $\overline{\epsilon}_2 = 3^2\overline{\mu}$. We have $\langle \overline{\gamma} \rangle \cong \mathbf{Z}/9$, $\langle \overline{\delta} \rangle \cong \mathbf{Z}/3$, $\langle \overline{\epsilon}_1, \overline{\epsilon}_2 \rangle \cong \mathbf{Z}/2 + \mathbf{Z}/2$, and $T^*/T = \langle \overline{\gamma} \rangle \oplus \langle \overline{\delta} \rangle \oplus \langle \overline{\epsilon}_1, \overline{\epsilon}_2 \rangle$. The discriminant quadratic form satisfies $q_T(x\overline{\gamma} + y\overline{\delta} + a\overline{\epsilon}_1 + b\overline{\epsilon}_2) \equiv -4x^2/9 + 2y^2/3 - a^2 - ab - b^2 \bmod 2\mathbf{Z}$.

Therefore, there is a group isomorphism $r : \hat{S}^*/\hat{S} \xrightarrow{\sim} T^*/T$ with $q_T r = -q_{\hat{S}}$.

By Lemma 5.5 (4) we have an embedding $\hat{S} \oplus T \hookrightarrow \Lambda_3$ into an even unimodular lattice Λ_3 with signature (19, 3) such that $\hat{S}$ and T are orthogonal complements of each other in Λ_3.

We consider the composed embedding $S = P \oplus Q(G) \hookrightarrow \hat{S} \hookrightarrow \Lambda_3$. By definition the primitive hull of S coincides with $\hat{S}$. The induced embedding $P \hookrightarrow \Lambda_3$ is primitive.

It turns out here that the induced embedding $Q = Q(G) \hookrightarrow \Lambda_3/P$ is full. We see this below. By the construction $\widetilde{Q} = \hat{S}/P = Q + \mathbf{Z}(\nu + \omega + \chi \bmod P)$ in Λ_3/P. Let F be the orthogonal complement of P in Λ_3. By the natural embedding we can regard $F \subset \Lambda_3/P$. $\Lambda_3/P \cong F^*$. We have a group isomorphism $s : P^*/P \to F^*/F$ whose graph coincides with $\Lambda_3/(P \oplus F)$. We regard $\lambda = \nu + \omega + \chi \bmod P \in \Lambda_3/P \cong F^*$. Then, since $\nu + \omega + \chi \in \Lambda_3$, we have $\lambda \bmod F = s(\nu \bmod P)$. Since $P^*/P \ni \nu \bmod P \neq 0$, $\lambda \notin F$. Thus, $\widetilde{Q} \cap F = Q$. Since $\widetilde{Q}$ is an even lattice, every root in it is a long root and belongs to F. Therefore, we can conclude that Q is full in Λ_3/P. Q.E.D.

By Proposition 34.2 Theorem 0.8 has been established.

Now, we cannot omit any of 4 exceptions in Theorem 0.8. The following proposition shows this fact:

Proposition 34.3. *Let G be one of $E_6 + D_4 + A_2$, $E_6 + A_2 + 3A_1$, $D_4 + 3A_2 + A_1$, and $3A_2 + 4A_1$.*
(1) We fix a primitive embedding $P \hookrightarrow \Lambda_2$. For any positive definite full root submodule $L \subset \Lambda_2/P$ we cannot make G from the Dynkin graph of L by one elementary or tie transformation.
(2) We fix a primitive embedding $\hat{P} \hookrightarrow \Lambda_2$. For any positive definite full submodule $L \subset \Lambda_2/\hat{P}$ we cannot make G from the Dynkin graph of L by one dual elementary transformation.

Proof. (1) Assume the contrary. Then, there exist a full embedding $Q(G) \hookrightarrow \Lambda_3/P$ and a primitive isotropic element $u \in \Lambda_3/P$ in a nice position with respect to $Q(G)$. Set $G' = 3A_2 + 4A_1$. By Lemma 34.1 G' is a subgraph of G, and if we consider the induced embedding $Q(G') \hookrightarrow Q(G) \hookrightarrow \Lambda_3/P$, u is in a nice position with respect to $Q(G')$. Thus, it suffices to show the above proposition when $G = G'$. We consider only the case $G = 3A_2 + 4A_1$ below.

By Lemma 33.9 $Q = Q(G)$ does not satisfy the I1-condition. Thus, u is not orthogonal to $Q(G)$. There exist a root basis $\Delta \subset Q(G)$ and a long root $\theta \in \Delta$ such that $u \cdot \theta = 1$ and $u \cdot \alpha = 0$ for any $\alpha \in \Delta$ with $\alpha \neq \theta$. Moreover, $u \in F$, where F denotes the orthogonal complement of P in Λ_3. We regard $F \subset \Lambda_3/P$ through the natural inclusion.

Let $S = P \oplus Q(G) \hookrightarrow \Lambda_3$ be the induced embedding. By T we denote the orthogonal complement of S in Λ_3. We set $I = \widetilde{S}/S$ where $\widetilde{S}$ denotes the primitive hull of S in Λ_3. Since $3A_2 + 4A_1$ is not a subgraph of $A_{11} + A_1$, we can assume $a = 2$ by Proposition 30.4. By Proposition 30.6 (3) we have $I = I_3$. We denote the natural lift $u \in F \subset \Lambda_3$ of u again by u. u is orthogonal to P. Since $u \in \Lambda_3 \subset \widetilde{S}^* \oplus T^*$ we can write $u = \sigma + \tau$ with $\sigma \in \widetilde{S}^*$ and $\tau \in T^*$. σ is orthogonal to P, $\sigma \cdot \theta = 1$, and $\sigma \cdot \alpha = 0$ for $\alpha \in \Delta$ with $\alpha \neq \theta$. Thus, σ coincides with the fundamental weight ω_θ associated with the pair (θ, Δ). $\sigma = \omega_\theta$. We have 2 cases.

⟨1⟩ θ belongs to a component of Δ of type A_2.
⟨2⟩ θ belongs to a component of Δ of type A_1.

On the other hand, $(S^*/S)_3 \cong \mathbf{Z}/9 \oplus \mathbf{Z}/3 \oplus \mathbf{Z}/3 \oplus \mathbf{Z}/3 \oplus \mathbf{Z}/3$. The first component of the right-hand side and the second correspond to P, and the third, the fourth, and the fifth correspond to an A_2-component respectively. The discriminant quadratic form is given by $q_3(\overline{\alpha}) \equiv 4x^2/9 - 2y^2/3 + 2(z_1^2 + z_2^2 + z_3^2)/3 \bmod 2\mathbf{Z}$ for $\overline{\alpha} = (x, y, z_1, z_2, z_3) \in \mathbf{Z}/9 \oplus (\mathbf{Z}/3)^{\oplus 4}$. An isotropic element in $(S^*/S)_3$ is one of the following:

$$\begin{aligned} &\overline{\beta}_1 = (0,0,\pm1,\pm1,\pm1), && \overline{\beta}_2 = (0,\pm1,\{\pm1,0,0\}), \\ &\overline{\beta}_3 = (\pm3,\pm1,\{\pm1,0,0\}), && \overline{\beta}_4 = (\pm3,0,0,0,0), \\ &\overline{\delta} = (\pm3,0,\pm1,\pm1,\pm1). && \end{aligned}$$

If $\overline{\beta}_1 \in I_3$, then $Q(3A_2) \subset Q(E_6) \subset \widetilde{Q(G)} \subset \Lambda_3/P$. If $\overline{\beta}_2 \in I_3$ or $\overline{\beta}_3 \in I_3$, then $Q(A_2) \subset Q(G_2) \subset \widetilde{Q(G)} \subset \Lambda_3/P$. Both cases contradict fullness. If $\overline{\beta}_4 \in I_3$, then P is not primitive in Λ_3, a contradiction. Therefore, we can conclude that $I = I_3$ is the group of order 3 generated by $\overline{\delta}$. $I = I_3 = \langle\overline{\delta}\rangle$. Exchanging the sign of the generators of $(S^*/S)_3$, we can assume $\overline{\delta} = (3,0,1,1,1)$.

Now, $\sigma \in \widetilde{S}^* \subset S^*$ if, and only if, $b(\overline{\sigma}, \overline{\delta}) \equiv 0 \bmod \mathbf{Z}$ where b is the discriminant bilinear form and $\overline{\sigma} = \sigma \bmod S \in S^*/S$. In the case $\langle 1\rangle$ $\overline{\sigma} = \overline{\omega_\theta} = \omega_\theta \bmod S = (0,0,\{\pm1,0,0\})$; thus, $0 \equiv b(\overline{\sigma}, \overline{\delta}) \equiv \pm 2/3 \bmod \mathbf{Z}$, a contradiction. The case $\langle 1\rangle$ never takes place.

Assume that the case $\langle 2\rangle$ takes place. We can write $\sigma = \omega_\theta = \theta/2$. Since $2\tau = 2u - \theta$ is orthogonal to S and belongs to Λ_3, we have $2\tau \in T$. Since 2 is invertible in $\mathbf{Z}_3$ we have, moreover, $\tau \in T \otimes \mathbf{Z}_3$. On the other hand, $\tau^2 + (\theta/2)^2 = u^2 = 0$, $(\theta/2)^2 = 1/2$; thus, $\tau^2 = -1/2$.

Now, we consider the following elements in $(S^*/S)_3$:

$$\overline{\epsilon}_1 = (1,0,1,0,0), \qquad \overline{\epsilon}_2 = (3,0,1,-1,0), \qquad \overline{\epsilon}_3 = (0,1,0,0,0).$$

We see $\langle \overline{\delta}, \overline{\epsilon}_1, \overline{\epsilon}_2, \overline{\epsilon}_3\rangle = I_3^\perp$ and $\langle \overline{\epsilon}_1, \overline{\epsilon}_2, \overline{\epsilon}_3\rangle \cap \langle\overline{\delta}\rangle = 0$. Thus, there are isomorphisms $(\widetilde{S}^*/\widetilde{S})_3 \cong (I^\perp/I)_3 \cong \langle \overline{\epsilon}_1, \overline{\epsilon}_2, \overline{\epsilon}_3\rangle$ preserving the discriminant forms. Note that $\overline{\epsilon}_i$'s are mutually orthogonal and $q(\overline{\epsilon}_1) \equiv -8/9$, $q(\overline{\epsilon}_2) \equiv q(\overline{\epsilon}_3) \equiv -2/3$. The discriminant quadratic form on $(\widetilde{S}^*/\widetilde{S})_3$ is given by $\widetilde{q}(x,y,z) \equiv -8x^2/9 - 2(y^2+z^2)/3 \bmod 2\mathbf{Z}$ for $(x,y,z) \in \mathbf{Z}/9 \oplus \mathbf{Z}/3 \oplus \mathbf{Z}/3 \cong (\widetilde{S}^*/\widetilde{S})_3$.

Let $U = \mathbf{Z}_3^4$ be the 3-adic lattice of rank 4 with the quadratic form given by $\xi^2 = 2^3 \cdot 3^2 \xi_1^2 + 2\cdot 3(\xi_2^2 + \xi_3^2) + 2\xi_4^2$ for $\xi = (\xi_1, \xi_2, \xi_3, \xi_4) \in \mathbf{Z}_3^4$. It is easy to see that U and $T \otimes \mathbf{Z}_3$ have the same discriminant quadratic form and the discriminant in $\mathbf{Z}_3^*/\mathbf{Z}_3^{*2}$. Thus, they are isomorphic by Lemma 12.2. Let $\xi \in U$ be the element corresponding to $\tau \in T \otimes \mathbf{Z}_3$ under this isomorphism. We have $(2\xi_4)^2 \equiv -1 \bmod 3\mathbf{Z}_3$, since $\xi^2 = -1/2$. It implies that $z^2 \equiv -1 \pmod 3$ for some integer z, which is a contradiction. The case $\langle 2\rangle$ does not take place, either.

(2) By the same reason as in (1) we can assume $G = 3A_2 + 4A_1$. The claim follows from that $Q(G)$ does not satisfy the I1-condition. Q.E.D.

References

1. Arnold, V.: Local normal forms of functions. Invent. Math. **35**, 87–109 (1976)
2. Arnold, V.: Singularity theory. Lond. Math. Soc. Lecture note series **53**. Cambridge: Cambridge Univ. Press 1981.
3. Bourbaki, N.: Groupes et algèbre de Lie. Chaps. 4–6. Paris: Hermann 1968
4. Cassels, J. W. S.: Rational quadratic forms. London: Academic Press 1978
5. Conway, J. H., Sloane, N. J. A.: Leech roots and Vinberg groups. Proc. R. Soc. Lond. A **384**, 233–258 (1982)
6. Durfee, A. H.: Fifteen characterization of rational double points and simple critical points. L'enseignement. Math. II **25**, 131–163 (1979)
7. Dynkin, E. B.: Semisimple subalgebras of semisimple Lie groups. Mat. Sb. N. S. **30 (72)**, 349–462 (1952) (English translation: Americam Math. Soc. Translations **6**, 111–244 (1957))
8. Kodaira, K.: On compact analytic surfaces II, III. Ann. of Math. **77**, 563–626 (1963), **78**, 1–40 (1963)
9. Looijenga, E.: The smoothing components of a triangle singularity. II. Math. Ann. **269**, 357–387 (1984)
10. Milnor, J., Husemoller, D.: Symmetric bilinear forms. Berlin Heidelberg New York: Springer 1973
11. Nikulin, V. V.: Integral symmetric bilinear forms and some of their applications. Mat. USSR Izv. **43** No. 1, (1979) (English translation: Math. USSR Izv. **14** No. 1, 103–167 (1980))
12. Saito, K.: Algebraic surfaces for regular systems of weights. In: Algebraic geometry and commutative algebra, 517–614. Tokyo: Kinokuniya 1987
13. Serre, J.-P.: Cours d'arithmètique. Paris: Presses Univ. France 1970
14. Shioda, T.: On elliptic modular surfaces. J. Math. Soc. Japan **24** No. 1, 20–59 (1972)
15. Urabe, T.: Dynkin graphs and combinations of singularities on quartic surfaces. Proc. Japan Acad. Ser. A **61**, 266–269 (1985)
16. Urabe, T.: Elementary transformations of Dynkin graphs and singularities on quartic surfaces. Invent. Math. **87**, 549–572 (1987)
17. Urabe, T.: Dynkin graphs and combinations of singularities on plane sextic curves. In: Randell, R. (ed.) Singularities. Proceedings, the University of Iowa 1986. (Contemporary Math. **90**, 295–316) Providence, Rhode Island: American Math. Soc. 1989
18. Urabe, T.: Tie transformations of Dynkin graphs and singularities on quartic surfaces. Invent. Math. **100**, 207–230 (1990)
19. Vinberg, È. B.: Discrete groups generated by reflections in Lobačevskiĭ spaces. Mat. Sb. **72** No. 3, 471–488 (1967) (English translation: Math. USSR Sb. **1** No. 3, 429–444 (1967))
20. Vinberg, È. B.: On groups of unit elements of certain quadratic forms. Mat. Sb. **87** No. 1, 18–36 (1972) (English translation: Math. USSR Sb. **16** No. 1, 17–35 (1972))
21. Vinberg, È. B.: On unimodular integral quadratic forms. Funkt. Analiz Prilozh **6** No. 2, 24–31 (1972) (English translation: Funct. Analysis Appl. **6**, 105–111 (1972))
22. Vinberg, È. B.: Some arithmetical discrete groups in Lobačevskiĭ spaces. In: Discrete subgroups of Lie groups and applications to moduli, 323–348. Oxford: Oxford Univ. Press 1975
23. Vinberg, È. B., Kaplinskaja, I. M.: On the groups $O_{18,1}(\mathbf{Z})$ and $O_{19,1}(\mathbf{Z})$. Dokl. Akad. Nauk SSSR **238** No. 6 (1978) (English translation: Soviet Math. **19** No. 1, 194–197 (1978))
24. Vinberg, È. B.: The two most algebraic K3 surfaces. Math. Ann. **265**, 1–21 (1983)
25. Wall, C. T. C.: Exceptional deformations of quadrilateral singularities and singular K3 surfaces. Bull. London Math. Soc. **19**, 174–176 (1987)

Appendix
Singularities of type A, D, E on plane sextic curves

In Urabe [17] we have developed the theory on possible combinations of singularities of type A, D or E on plane sextic curves. However, when we wrote [17], we did not introduce the concept of type BC Dynkin graphs and the concept of Coxeter-Vinberg graphs. Thus, here we give the revision of the theorems in [17] using type BC Dynkin graphs and add more precise theorems following from the study of Coxeter-Vinberg graphs. As explained in [17], we associate the curve singularity locally isomorphic to the one defined by the following local equation with a connected type A, D or E Dynkin graph:

$$A_k: \quad x^2 + y^{k+1} = 0 \quad (k \geq 0), \qquad D_l: \quad x^2y + y^{l-1} = 0 \quad (l \geq 4),$$
$$E_6: \quad x^3 + y^4 = 0, \qquad E_7: \quad x^3 + xy^3 = 0, \qquad E_8: \quad x^3 + y^5 = 0.$$

Theorem A.1. *Let G be a Dynkin graph. By r we denote the number of vertices in G. Then, the following conditions (a) and (b) are equivalent:*

(a) There exists a sextic curve in the projective space of dimension 2 whose combination of singularities just agrees with G, and, moreover, one of the following conditions $\langle 1\rangle$, $\langle 2\rangle$, $\langle 3\rangle$, $\langle 4\rangle$ holds for the root (quasi-)lattice $Q = Q(G)$ of type G. By $d(Q)$ we denote the discriminant of Q:
$\langle 1\rangle$ $r = 17$, $2d(Q)$ is a square number, and for every prime number p, $\epsilon_p(Q) = 1$.
$\langle 2\rangle$ $r = 16$, and for every prime number p, $\epsilon_p(Q) = (-2,\, d(Q))_p$.
$\langle 3\rangle$ $r = 15$, and for every prime number p, $-2d(Q) \notin \mathbf{Q}_p^{*2}$ or $\epsilon_p(Q) = (-1,\, -1)_p$.
$\langle 4\rangle$ $r \leq 14$.

(b) G contains no vertex corresponding to a short root, and G can be obtained from one of the following 4 basic Dynkin graphs by elementary transformations repeated twice:

The basic graphs: A_{17}, $D_{16} + BC_1$, $D_{10} + E_7$, $2E_8 + BC_1$.

Remarks. The symbol $\epsilon_p(Q) = \pm 1$ denotes the Hasse symbol. The symbol $(a,\, b)_p = \pm 1$ is the Hilbert symbol. (See Section 7.)

To go further we would like to draw the Coxeter-Vinberg graph of the quasi-lattice $\Lambda_2/\mathbf{Z}\lambda$ with $\lambda^2 = -2$. (Λ_2 is the even unimodular lattice of signature (18, 2).)

We fix an embedding $\mathbf{Z}\lambda \hookrightarrow \Lambda_2$ where λ is a vector with $\lambda^2 = -2$. Such an embedding is unique up to integral orthogonal transformations of Λ_2. (Lemma 12.4.) By F we denote the orthogonal complement of $\mathbf{Z}\lambda$ in Λ_2. It is identified with its image under the canonical surjective homomorphism $\Lambda_2 \to \Lambda_2/\mathbf{Z}\lambda$. We have $\Lambda_2/\mathbf{Z}\lambda \cong F^*$ as quasi-lattices. The discriminant quadratic form of F is given by $x \in \mathbf{Z}/2 \cong F^*/F \mapsto x^2/2 \bmod 2\mathbf{Z}$.

Let Λ_1 be the even unimodular lattice with signature (17, 1) and w_0 be a vector with $w_0^2 = +2$. Both F and $\Lambda_1 \oplus \mathbf{Z}w_0$ are even indefinite lattices, and they have the same signature and the same discriminant quadratic form. Thus, $F \cong \Lambda_1 \oplus \mathbf{Z}w_0$. It is easy to check that $\Lambda_2/\mathbf{Z}\lambda \cong F^* \cong \Lambda_1 \oplus \mathbf{Z}(w_0/2)$ is a regular root module whose root has length $\sqrt{2}$ or $1/\sqrt{2}$.

We can give more concrete presentation for Λ_1. Let $K = \sum_{i=0}^{17} \mathbf{Z}v_i$ be the lattice satisfying $v_0^2 = -1$, $v_i^2 = +1$ for $(1 \le i \le 17)$ and $v_i \cdot v_j = 0$ for $i \ne j$. Set $\chi = \sum_{i=0}^{17} v_i \in K$. $\chi^2 = 16$. Set $M = \{y \in K \mid y \cdot \chi \equiv 0 \pmod 2\} = \{\sum_{i=0}^{17} y_i v_i \in K \mid \sum_{i=0}^{17} y_i$ is an even integer.$\}$. Then, we have $\Lambda_1 \cong M + \mathbf{Z}(\chi/2)$. We get the following:

$$\Lambda_2/\mathbf{Z}\lambda \cong F^* \cong [M + \mathbf{Z}(\chi/2)] \oplus \mathbf{Z}(w_0/2).$$

We apply Vinberg's algorithm to this presentation. The vector v_0 is chosen as the controlling vector. As the root basis orthogonal to v_0 we can choose:

$$\begin{aligned} e_i &= -v_i + v_{i+1} \quad (1 \le i \le 16), \\ e_{17} &= -v_{16} + v_{17}, \qquad e_{18} = w_0/2. \end{aligned}$$

$e_i^2 = 2$ for $1 \le i \le 17$ and $e_{18}^2 = 1/2$. By the algorithm we get the following vectors successively:

$$\begin{aligned} e_{19} &= v_0 + v_1 + v_2 + v_3, \\ e_{20} &= v_0 + v_1 - w_0, \\ e_{21} &= 3v_0/2 + (v_1 + \cdots + v_{16})/2 - v_{17}/2, \\ e_{22} &= 3v_0 + v_1 + \cdots + v_9 - w_0. \end{aligned}$$

We consider the set $\Delta^* = \{e_i \mid 1 \le i \le 22,\ i \ne 18\}$ of long roots among the above 22 vectors. Drawing the graph for Δ^* we know that it has the automorphism group isomorphic to the symmetric group S_3 of degree 3, and $f = e_{17} + e_{19} + e_{22}$ and $u = \sum_{i=1}^{16} e_i + e_{20} + e_{21}$ are invariant vectors. Since $f^2 = 6$, $u^2 = 0$ and $f \cdot u = -3$, the invariant vector $f + 2u$ satisfies $(f + 2u)^2 = -6 < 0$. Note, moreover, that Δ^* generates $F \cong \Lambda_1 \oplus \mathbf{Z}w_0$. By the same arguments as that in the latter half of Section 21, we know that the Coxeter-Vinberg graph of $\Lambda_2/\mathbf{Z}\lambda$ has the action of S_3 and that vectors conjugate to e_{18} under the action of S_3 are fundamental roots. We get the following additional 2 fundamental roots:

$$\begin{aligned} e_{23} &= 4v_0 + (v_1 + \cdots + v_{12}) - 3w_0/2 \\ e_{24} &= 9v_0/2 + 3(v_1 + \cdots + v_6)/2 + (v_7 + \cdots + v_{17})/2 - 3w_0/2. \end{aligned}$$

Drawing the graph for these 24 vectors we get the graph on the next page. (We can actually see that the subgraph corresponding to Δ^* has an action of S_3.)

We can check the condition $\langle a \rangle$ in Proposition 8.5 easily. Indeed, the extended Dynkin graph with the maximal rank 17 is of type A_{17}, $D_{16} + BC_1$, $D_{10} + E_7$ or $2E_8 + BC_1$, and the type of such a subgraph uniquely determines the conjugacy class in the entire graph. To check the condition $\langle b \rangle$ in Proposition 8.5 we apply Lemma 8.6. Only the 3 subgraphs with a dotted edge and with only 2 vertices are Lannér subgraphs. Let Γ be one of 3 Lannér subgraphs. The subgraph Ξ defined in Lemma 8.6 is a Dynkin subgraph of type $A_{11} + E_6$. Thus, the sum of the numbers of vertices in Γ and Ξ is equal to $19 = \operatorname{rank} \Lambda_2/\mathbf{Z}\lambda$. Thus, also the condition $\langle b \rangle$ is satisfied. The graph on the next page is the Coxeter-Vinberg graph of $\Lambda_2/\mathbf{Z}\lambda$ with $\lambda^2 = -2$.

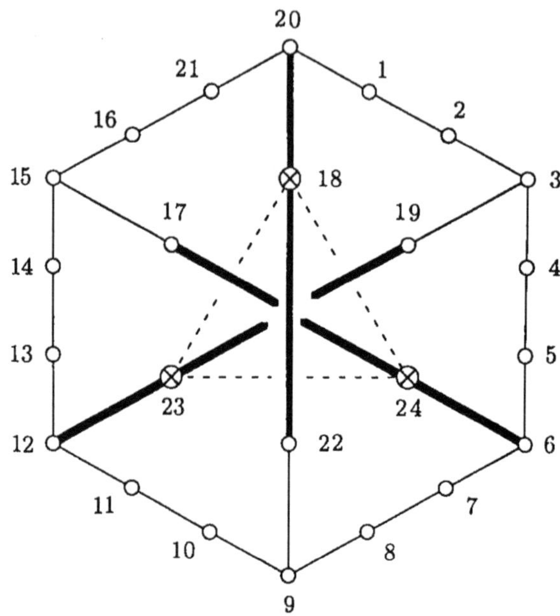

Lemma A.2. *The following 2 conditions are equivalent:*

(1) G is a maximal Dynkin subgraph of the above Coxeter-Vinberg graph. Let $\Delta = \{e_i \mid i \in I\}$ *be the set of roots corresponding to the set of vertices of G. There is no isotropic vector* $u \in \Lambda_2/\mathbf{Z}\lambda$ *in a nice position with respect to the root lattice generated by* Δ.

(2) G is one of the following 13 graphs:

$A_{15}+A_3$, $A_{13}+A_4+A_1$, $A_{12}+A_6$, $A_{11}+A_5+2A_1$,
$A_{10}+A_7+A_1$, $2A_9$, $A_9+A_6+A_3$, $A_9+2A_4+A_1$,
$2A_7+A_3+A_1$, $A_7+A_6+A_4+A_1$, $A_7+A_5+A_4+2A_1$, $3A_6$,
$3A_5+3A_1$.

Proof. (1)⇒(2) By Lemma 8.7 G has 18 vertices.

Case 1. G contains the vertex 18, 23 or 24.

By the symmetry of the graph we can assume that G contains the vertex 18. Then, G does not contain the vertex 20, 22, 23, and 24, since they are connected with 18 by a bold edge or a dotted edge. Consider the vector $u_1 = -(e_{22}+2e_{18})$. This is isotropic.

It satisfies

$$u_1 \cdot e_i = 0, \quad 1 \leq i \leq 24, \quad i \neq 9,\ 20,\ 23,\ 24$$
$$u_1 \cdot e_9 = 1, \quad u_1 \cdot e_{20} = 2, \quad u_1 \cdot e_{23} = u_1 \cdot e_{24} = 3.$$

Thus, u_1 is orthogonal to $\Delta - \{e_9\}$, $u_1 \cdot e_9 = 1$ and e_9 is a long root. Thus, u_1 is in a nice position with respect to the root basis Δ, which contradicts the assumption. This case never takes place.

Case 2. G contains neither the vertex 18, 23 nor 24.

If $e_{20} \notin \Delta$, then u_1 is in a nice position with respect to Δ, a contradiction. Thus, G contains the vertex 20. By symmetry, also the vertex 6 and 12 belong to G.

Consider here the vector $u_2 = -(e_{12}+2e_{13}+3e_{14}+4e_{15}+3e_{16}+2e_{17}+e_{20}+2e_{21})$. This is isotropic.

$$u_2 \cdot e_i = 0, \quad 1 \leq i \leq 24, \quad i \neq 1,\ 11,\ 18,\ 23,\ 24$$
$$u_2 \cdot e_1 = u_2 \cdot e_{11} = 1, \quad u_2 \cdot e_{18} = u_2 \cdot e_{23} = 1, \quad u_2 \cdot e_{24} = 2.$$

We consider e_1 and e_{11}. Both are long roots. If one of them does not belong to Δ, then u_2 is in a nice position with respect to Δ. Thus, G contains the vertex 1 and 11. By symmetry G contains also the vertex 7, 21; 5 and 13.

We here consider the following 3 subgraphs of type D_4:

G_1^*: Vertices 2, 3, 4, 19.
G_2^*: Vertices 8, 9, 10, 22.
G_3^*: Vertices 14, 15, 16, 17.

G is obtained by erasing 3 vertices belonging to $G_1^* + G_2^* + G_3^*$ in addition to the above mentioned vertices 18, 23, 24 from the above Coxeter-Vinberg graph.

Assume that for some i with $1 \leq i \leq 3$ G_i^* contains none of these 3 vertices to be erased. Then, G contains a subgraph isomorphic to the extended Dynkin graph of type E_7 and containing G_i^*, a contradiction. Thus, erasing one vertex from each G_i^* for $i = 1, 2, 3$ from the above Coxeter-Vinberg graph after erasing the vertex 18, 23 and 24, we get graph G.

Consider further the vector $u_3 = -(\sum_{i=1}^{16} e_i + e_{20} + e_{21})$. Also, this is isotropic.

$$u_3 \cdot e_i = 0, \quad 1 \leq i \leq 24, \quad i \neq 17,\ 18,\ 19,\ 22,\ 23,\ 24$$
$$u_3 \cdot e_{17} = u_3 \cdot e_{19} = u_3 \cdot e_{22} = 1, \quad u_3 \cdot e_{18} = u_3 \cdot e_{23} = u_3 \cdot e_{24} = 1.$$

Vectors e_{17}, e_{19}, e_{22} are long roots. Thus, if at most one of the vertex 17, 19 and 22 belongs to G, then u_3 is in a nice position with respect to Δ. Thus, one of the following 2 cases takes place:

(2-1) All the vertices 17, 19 and 22 belong to G.
(2-2) Only one of the vertices 17, 19, 22 does not belong to G.

Case (2-1). If all the vertices 9, 3 and 15 do not belong to G, then $G = 3A_5 + 3A_1$.

If 2 of the vertices 9, 3, 15 do not belong to G, then $G = A_7 + A_5 + A_4 + 2A_1$.

If only one of the vertices 9, 3, 15 do not belong to G, then $G = A_9 + 2A_4 + A_1$, $2A_7 + A_3 + A_1$, or $A_7 + A_6 + A_4 + A_1$.

If none of the vertices 9, 3, 15 belongs to G, then $G = 3A_6$ or $A_9 + A_6 + A_3$.

Case (2-2). By symmetry we can assume that the vertex 22 does not belong to G. If the vertex i and the vertex j do not belong to G we get the following graph depending on i and j:

i, j	G
2, 16	$A_{15}+A_3$
2, 15	$A_{13}+A_4+A_1$
2, 14	$A_{12}+A_6$
3, 15	$A_{11}+A_5+2A_1$
3, 14	$A_{10}+A_7+A_1$
4, 14	$2A_9$

By symmetry we need not consider other cases.
(2)⇒(1). By the above description obviously G is a maximal Dynkin subgraph. If the root lattice has an isotropic vector in a nice position, then G can be obtained from one of the basic graphs A_{17}, $D_{16}+BC_1$, $D_{10}+E_7$, $2E_8+BC_1$ by one tie transformation. However, since every component of G is of type A, it is easy to see that this is not the case. Q.E.D.

It is not difficult to write the condition $\mathrm{I}(Q\oplus\mathbf{Z}\lambda,\ \Lambda_3)$ explicitly. (See Section 7.) We get the following:

Theorem A.3. *Let G be a Dynkin graph with r vertices. Then, the following conditions (A) and (B) are equivalent:*

(A) There exists a sextic curve in the projective space of dimension 2 whose combination of singularities just agrees with G, and, moreover, one of the following conditions $\langle 1\rangle$, $\langle 2\rangle$, $\langle 3\rangle$ holds for the root (quasi-)lattice $Q=Q(G)$ of type G. By $d(Q)$ we denote the discriminant of Q:

$\langle 1\rangle$ $r=18$, and for every prime number p, $\epsilon_p(Q)=(2,\ d(Q))_p$.

$\langle 2\rangle$ $r=17$, and for every prime number p, $2d(Q)\notin \mathbf{Q}_p^{*2}$ or $\epsilon_p(Q)=1$.

$\langle 3\rangle$ $r\leq 16$.

(B) G contains no vertex corresponding to a short root and either the following (B-1) or (B-2) holds:

(B-1) G can be obtained from one of the following 13 sub-basic Dynkin graphs by one elementary transformation:

The sub-basic Dynkin graphs

$A_{15}+A_3$,	$A_{13}+A_4+A_1$,	$A_{12}+A_6$,	$A_{11}+A_5+2A_1$,
$A_{10}+A_7+A_1$,	$2A_9$,	$A_9+A_6+A_3$,	$A_9+2A_4+A_1$,
$2A_7+A_3+A_1$,	$A_7+A_6+A_4+A_1$,	$A_7+A_5+A_4+2A_1$,	$3A_6$,
$3A_5+3A_1$.			

(B-2) G can be obtained from one of the 4 basic Dynkin graphs in Theorem A.1 by one of the following 3 kinds of procedures:

The procedures:

$\langle 1\rangle$ elementary transformations repeated twice

$\langle 2\rangle$ an elementary transformation following after a tie transformation

$\langle 3\rangle$ a tie transformation following after an elementary transformation.

In our case of sextic curves we need not consider obstruction components.

Theorem A.4. *(1) Assume that a Dynkin graph G without a vertex corresponding to a short root can be made from one of the basic graphs in Theorem A.1 by tie transformations repeated twice. Then, there exists a sextic curve in the projective space of dimension 2 whose combination of singularities just agrees with G.*
(2) Assume that a Dynkin graph G without a vertex corresponding to a short root can be made from one of the sub-basic graphs in Theorem A.3 by one tie transformation. Then, there exists a sextic curve in the projective space of dimension 2 whose combination of singularities just agrees with G.

Now, it comes into question whether every Dynkin graph describing the singularities on a plane sextic curve can be made by the procedures described in Theorem A.3 and Theorem A.4. Perhaps there exist several exceptions to this claim. To determine the exceptional graphs is the theme of the future study. We complete this book here.

Lecture Notes in Mathematics

For information about Vols. 1–1364
please contact your bookseller or Springer-Verlag

Vol. 1365: M. Giaquinta (Ed.), Topics in Calculus of Variations. Seminar, 1987. X, 196 pages. 1989.

Vol. 1366: N. Levitt, Grassmannians and Gauss Maps in PL-Topology. V, 203 pages. 1989.

Vol. 1367: M. Knebusch, Weakly Semialgebraic Spaces. XX, 376 pages. 1989.

Vol. 1368: R. Hübl, Traces of Differential Forms and Hochschild Homology. III, 111 pages. 1989.

Vol. 1369: B. Jiang, Ch.-K. Peng, Z. Hou (Eds.), Differential Geometry and Topology. Proceedings, 1986–87. VI, 366 pages. 1989.

Vol. 1370: G. Carlsson, R.L. Cohen, H.R. Miller, D.C. Ravenel (Eds.), Algebraic Topology. Proceedings, 1986. IX, 456 pages. 1989.

Vol. 1371: S. Glaz, Commutative Coherent Rings. XI, 347 pages. 1989.

Vol. 1372: J. Azéma, P.A. Meyer, M. Yor (Eds.), Séminaire de Probabilités XXIII. Proceedings. IV, 583 pages. 1989.

Vol. 1373: G. Benkart, J.M. Osborn (Eds.), Lie Algebras. Madison 1987. Proceedings. V, 145 pages. 1989.

Vol. 1374: R.C. Kirby, The Topology of 4-Manifolds. VI, 108 pages. 1989.

Vol. 1375: K. Kawakubo (Ed.), Transformation Groups. Proceedings, 1987. VIII, 394 pages, 1989.

Vol. 1376: J. Lindenstrauss, V.D. Milman (Eds.), Geometric Aspects of Functional Analysis. Seminar (GAFA) 1987–88. VII, 288 pages. 1989.

Vol. 1377: J.F. Pierce, Singularity Theory, Rod Theory, and Symmetry-Breaking Loads. IV, 177 pages. 1989.

Vol. 1378: R.S. Rumely, Capacity Theory on Algebraic Curves. III, 437 pages. 1989.

Vol. 1379: H. Heyer (Ed.), Probability Measures on Groups IX. Proceedings, 1988. VIII, 437 pages. 1989.

Vol. 1380: H.P. Schlickewei, E. Wirsing (Eds.), Number Theory, Ulm 1987. Proceedings. V, 266 pages. 1989.

Vol. 1381: J.-O. Strömberg, A. Torchinsky, Weighted Hardy Spaces. V, 193 pages. 1989.

Vol. 1382: H. Reiter, Metaplectic Groups and Segal Algebras. XI, 128 pages. 1989.

Vol. 1383: D.V. Chudnovsky, G.V. Chudnovsky, H. Cohn, M.B. Nathanson (Eds.), Number Theory, New York 1985–88. Seminar. V, 256 pages. 1989.

Vol. 1384: J. Garcia-Cuerva (Ed.), Harmonic Analysis and Partial Differential Equations. Proceedings, 1987. VII, 213 pages. 1989.

Vol. 1385: A.M. Anile, Y. Choquet-Bruhat (Eds.), Relativistic Fluid Dynamics. Seminar, 1987. V, 308 pages. 1989.

Vol. 1386: A. Bellen, C.W. Gear, E. Russo (Eds.), Numerical Methods for Ordinary Differential Equations. Proceedings, 1987. VII, 136 pages. 1989.

Vol. 1387: M. Petkovi´c, Iterative Methods for Simultaneous Inclusion of Polynomial Zeros. X, 263 pages. 1989.

Vol. 1388: J. Shinoda, T.A. Slaman, T. Tugué (Eds.), Mathematical Logic and Applications. Proceedings, 1987. V, 223 pages. 1989.

Vol. 1000: Second Edition. H. Hopf, Differential Geometry in the Large. VII, 184 pages. 1989.

Vol. 1389: E. Ballico, C. Ciliberto (Eds.), Algebraic Curves and Projective Geometry. Proceedings, 1988. V, 288 pages. 1989.

Vol. 1390: G. Da Prato, L. Tubaro (Eds.), Stochastic Partial Differential Equations and Applications II. Proceedings, 1988. VI, 258 pages. 1989.

Vol. 1391: S. Cambanis, A. Weron (Eds.), Probability Theory on Vector Spaces IV. Proceedings, 1987. VIII, 424 pages. 1989.

Vol. 1392: R. Silhol, Real Algebraic Surfaces. X, 215 pages. 1989.

Vol. 1393: N. Bouleau, D. Feyel, F. Hirsch, G. Mokobodzki (Eds.), Séminaire de Théorie du Potentiel Paris, No. 9. Proceedings. VI, 265 pages. 1989.

Vol. 1394: T.L. Gill, W.W. Zachary (Eds.), Nonlinear Semigroups, Partial Differential Equations and Attractors. Proceedings, 1987. IX, 233 pages. 1989.

Vol. 1395: K. Alladi (Ed.), Number Theory, Madras 1987. Proceedings. VII, 234 pages. 1989.

Vol. 1396: L. Accardi, W. von Waldenfels (Eds.), Quantum Probability and Applications IV. Proceedings, 1987. VI, 355 pages. 1989.

Vol. 1397: P.R. Turner (Ed.), Numerical Analysis and Parallel Processing. Seminar, 1987. VI, 264 pages. 1989.

Vol. 1398: A.C. Kim, B.H. Neumann (Eds.), Groups – Korea 1988. Proceedings. V, 189 pages. 1989.

Vol. 1399: W.-P. Barth, H. Lange (Eds.), Arithmetic of Complex Manifolds. Proceedings, 1988. V, 171 pages. 1989.

Vol. 1400: U. Jannsen. Mixed Motives and Algebraic K-Theory. XIII, 246 pages. 1990.

Vol. 1401: J. Steprans, S. Watson (Eds.), Set Theory and its Applications. Proceedings, 1987. V, 227 pages. 1989.

Vol. 1402: C. Carasso, P. Charrier, B. Hanouzet, J.-L. Joly (Eds.), Nonlinear Hyperbolic Problems. Proceedings, 1988. V, 249 pages. 1989.

Vol. 1403: B. Simeone (Ed.), Combinatorial Optimization. Seminar, 1986. V, 314 pages. 1989.

Vol. 1404: M.-P. Malliavin (Ed.), Séminaire d´Algèbre Paul Dubreil et Marie-Paul Malliavin. Proceedings, 1987–1988. IV, 410 pages. 1989.

Vol. 1405: S. Dolecki (Ed.), Optimization. Proceedings, 1988. V, 223 pages. 1989. Vol. 1406: L. Jacobsen (Ed.),

Analytic Theory of Continued Fractions III. Proceedings, 1988. VI, 142 pages. 1989.

Vol. 1407: W. Pohlers, Proof Theory. VI, 213 pages. 1989.

Vol. 1408: W. Lück, Transformation Groups and Algebraic K-Theory. XII, 443 pages. 1989.

Vol. 1409: E. Hairer, Ch. Lubich, M. Roche. The Numerical Solution of Differential-Algebraic Systems by Runge-Kutta Methods. VII, 139 pages. 1989.

Vol. 1410: F.J. Carreras, O. Gil-Medrano, A.M. Naveira (Eds.), Differential Geometry. Proceedings, 1988. V, 308 pages. 1989.

Vol. 1411: B. Jiang (Ed.), Topological Fixed Point Theory and Applications. Proceedings. 1988. VI, 203 pages. 1989.

Vol. 1412: V.V. Kalashnikov, V.M. Zolotarev (Eds.), Stability Problems for Stochastic Models. Proceedings, 1987. X, 380 pages. 1989.

Vol. 1413: S. Wright, Uniqueness of the Injective III_1 Factor. III, 108 pages. 1989.

Vol. 1414: E. Ramirez de Arellano (Ed.), Algebraic Geometry and Complex Analysis. Proceedings, 1987. VI, 180 pages. 1989.

Vol. 1415: M. Langevin, M. Waldschmidt (Eds.), Cinquante Ans de Polynômes. Fifty Years of Polynomials. Proceedings, 1988. IX, 235 pages.1990.

Vol. 1416: C. Albert (Ed.), Géométrie Symplectique et Mécanique. Proceedings, 1988. V, 289 pages. 1990.

Vol. 1417: A.J. Sommese, A. Biancofiore, E.L. Livorni (Eds.), Algebraic Geometry. Proceedings, 1988. V, 320 pages. 1990.

Vol. 1418: M. Mimura (Ed.), Homotopy Theory and Related Topics. Proceedings, 1988. V, 241 pages. 1990.

Vol. 1419: P.S. Bullen, P.Y. Lee, J.L. Mawhin, P. Muldowney, W.F. Pfeffer (Eds.), New Integrals. Proceedings, 1988. V, 202 pages. 1990.

Vol. 1420: M. Galbiati, A. Tognoli (Eds.), Real Analytic Geometry. Proceedings, 1988. IV, 366 pages. 1990.

Vol. 1421: H.A. Biagioni, A Nonlinear Theory of Generalized Functions, XII, 214 pages. 1990.

Vol. 1422: V. Villani (Ed.), Complex Geometry and Analysis. Proceedings, 1988. V, 109 pages. 1990.

Vol. 1423: S.O. Kochman, Stable Homotopy Groups of Spheres: A Computer-Assisted Approach. VIII, 330 pages. 1990.

Vol. 1424: F.E. Burstall, J.H. Rawnsley, Twistor Theory for Riemannian Symmetric Spaces. III, 112 pages. 1990.

Vol. 1425: R.A. Piccinini (Ed.), Groups of Self-Equivalences and Related Topics. Proceedings, 1988. V, 214 pages. 1990.

Vol. 1426: J. Azéma, P.A. Meyer, M. Yor (Eds.), Séminaire de Probabilités XXIV, 1988/89. V, 490 pages. 1990.

Vol. 1427: A. Ancona, D. Geman, N. Ikeda, École d'Eté de Probabilités de Saint Flour XVIII, 1988. Ed.: P.L. Hennequin. VII, 330 pages. 1990.

Vol. 1428: K. Erdmann, Blocks of Tame Representation Type and Related Algebras. XV. 312 pages. 1990.

Vol. 1429: S. Homer, A. Nerode, R.A. Platek, G.E. Sacks, A. Scedrov, Logic and Computer Science. Seminar, 1988. Editor: P. Odifreddi. V, 162 pages. 1990.

Vol. 1430: W. Bruns, A. Simis (Eds.), Commutative Algebra. Proceedings. 1988. V, 160 pages. 1990.

Vol. 1431: J.G. Heywood, K. Masuda, R. Rautmann, V.A. Solonnikov (Eds.), The Navier-Stokes Equations – Theory and Numerical Methods. Proceedings, 1988. VII, 238 pages. 1990.

Vol. 1432: K. Ambos-Spies, G.H. Müller, G.E. Sacks (Eds.), Recursion Theory Week. Proceedings, 1989. VI, 393 pages. 1990.

Vol. 1433: S. Lang, W. Cherry, Topics in Nevanlinna Theory. II, 174 pages.1990.

Vol. 1434: K. Nagasaka, E. Fouvry (Eds.), Analytic Number Theory. Proceedings, 1988. VI, 218 pages. 1990.

Vol. 1435: St. Ruscheweyh, E.B. Saff, L.C. Salinas, R.S. Varga (Eds.), Computational Methods and Function Theory. Proceedings, 1989. VI, 211 pages. 1990.

Vol. 1436: S. Xambó-Descamps (Ed.), Enumerative Geometry. Proceedings, 1987. V, 303 pages. 1990.

Vol. 1437: H. Inassaridze (Ed.), K-theory and Homological Algebra. Seminar, 1987–88. V, 313 pages. 1990.

Vol. 1438: P.G. Lemarié (Ed.) Les Ondelettes en 1989. Seminar. IV, 212 pages. 1990.

Vol. 1439: E. Bujalance, J.J. Etayo, J.M. Gamboa, G. Gromadzki. Automorphism Groups of Compact Bordered Klein Surfaces: A Combinatorial Approach. XIII, 201 pages. 1990.

Vol. 1440: P. Latiolais (Ed.), Topology and Combinatorial Groups Theory. Seminar, 1985–1988. VI, 207 pages. 1990.

Vol. 1441: M. Coornaert, T. Delzant, A. Papadopoulos. Géométrie et théorie des groupes. X, 165 pages. 1990.

Vol. 1442: L. Accardi, M. von Waldenfels (Eds.), Quantum Probability and Applications V. Proceedings, 1988. VI, 413 pages. 1990.

Vol. 1443: K.H. Dovermann, R. Schultz, Equivariant Surgery Theories and Their Periodicity Properties. VI, 227 pages. 1990.

Vol. 1444: H. Korezlioglu, A.S. Ustunel (Eds.), Stochastic Analysis and Related Topics VI. Proceedings, 1988. V, 268 pages. 1990.

Vol. 1445: F. Schulz, Regularity Theory for Quasilinear Elliptic Systems and – Monge Ampère Equations in Two Dimensions. XV, 123 pages. 1990.

Vol. 1446: Methods of Nonconvex Analysis. Seminar, 1989. Editor: A. Cellina. V, 206 pages. 1990.

Vol. 1447: J.-G. Labesse, J. Schwermer (Eds), Cohomology of Arithmetic Groups and Automorphic Forms. Proceedings, 1989. V, 358 pages. 1990.

Vol. 1448: S.K. Jain, S.R. López-Permouth (Eds.), Non-Commutative Ring Theory. Proceedings, 1989. V, 166 pages. 1990.

Vol. 1449: W. Odyniec, G. Lewicki, Minimal Projections in Banach Spaces. VIII, 168 pages. 1990.

Vol. 1450: H. Fujita, T. Ikebe, S.T. Kuroda (Eds.), Functional-Analytic Methods for Partial Differential Equations. Proceedings, 1989. VII, 252 pages. 1990.

Vol. 1451: L. Alvarez-Gaumé, E. Arbarello, C. De Concini, N.J. Hitchin, Global Geometry and Mathematical Physics. Montecatini Terme 1988. Seminar. Editors: M. Francaviglia, F. Gherardelli. IX, 197 pages. 1990.

Vol. 1452: E. Hlawka, R.F. Tichy (Eds.), Number-Theoretic Analysis. Seminar, 1988–89. V, 220 pages. 1990.

Vol. 1453: Yu.G. Borisovich, Yu.E. Gliklikh (Eds.), Global Analysis – Studies and Applications IV. V, 320 pages. 1990.

Vol. 1454: F. Baldassari, S. Bosch, B. Dwork (Eds.), p-adic Analysis. Proceedings, 1989. V, 382 pages. 1990.

Vol. 1455: J.-P. Françoise, R. Roussarie (Eds.), Bifurcations of Planar Vector Fields. Proceedings, 1989. VI, 396 pages. 1990.

Vol. 1456: L.G. Kovács (Ed.), Groups – Canberra 1989. Proceedings. XII, 198 pages. 1990.

Vol. 1457: O. Axelsson, L.Yu. Kolotilina (Eds.), Preconditioned Conjugate Gradient Methods. Proceedings, 1989. V, 196 pages. 1990.

Vol. 1458: R. Schaaf, Global Solution Branches of Two Point Boundary Value Problems. XIX, 141 pages. 1990.

Vol. 1459: D. Tiba, Optimal Control of Nonsmooth Distributed Parameter Systems. VII, 159 pages. 1990.

Vol. 1460: G. Toscani, V. Boffi, S. Rionero (Eds.), Mathematical Aspects of Fluid Plasma Dynamics. Proceedings, 1988. V, 221 pages. 1991.

Vol. 1461: R. Gorenflo, S. Vessella, Abel Integral Equations. VII, 215 pages. 1991.

Vol. 1462: D. Mond, J. Montaldi (Eds.), Singularity Theory and its Applications. Warwick 1989, Part I. VIII, 405 pages. 1991.

Vol. 1463: R. Roberts, I. Stewart (Eds.), Singularity Theory and its Applications. Warwick 1989, Part II. VIII, 322 pages. 1991.

Vol. 1464: D. L. Burkholder, E. Pardoux, A. Sznitman, Ecole d'Eté de Probabilités de Saint- Flour XIX-1989. Editor: P. L. Hennequin. VI, 256 pages. 1991.

Vol. 1465: G. David, Wavelets and Singular Integrals on Curves and Surfaces. X, 107 pages. 1991.

Vol. 1466: W. Banaszczyk, Additive Subgroups of Topological Vector Spaces. VII, 178 pages. 1991.

Vol. 1467: W. M. Schmidt, Diophantine Approximations and Diophantine Equations. VIII, 217 pages. 1991.

Vol. 1468: J. Noguchi, T. Ohsawa (Eds.), Prospects in Complex Geometry. Proceedings, 1989. VII, 421 pages. 1991.

Vol. 1469: J. Lindenstrauss, V. D. Milman (Eds.), Geometric Aspects of Functional Analysis. Seminar 1989-90. XI, 191 pages. 1991.

Vol. 1470: E. Odell, H. Rosenthal (Eds.), Functional Analysis. Proceedings, 1987-89. VII, 199 pages. 1991.

Vol. 1471: A. A. Panchishkin, Non-Archimedean L-Functions of Siegel and Hilbert Modular Forms. VII, 157 pages. 1991.

Vol. 1472: T. T. Nielsen, Bose Algebras: The Complex and Real Wave Representations. V, 132 pages. 1991.

Vol. 1473: Y. Hino, S. Murakami, T. Naito, Functional Differential Equations with Infinite Delay. X, 317 pages. 1991.

Vol. 1474: S. Jackowski, B. Oliver, K. Pawałowski (Eds.), Algebraic Topology, Poznań 1989. Proceedings. VIII, 397 pages. 1991.

Vol. 1475: S. Busenberg, M. Martelli (Eds.), Delay Differential Equations and Dynamical Systems. Proceedings, 1990. VIII, 249 pages. 1991.

Vol. 1476: M. Bekkali, Topics in Set Theory. VII, 120 pages. 1991.

Vol. 1477: R. Jajte, Strong Limit Theorems in Noncommutative L_2-Spaces. X, 113 pages. 1991.

Vol. 1478: M.-P. Malliavin (Ed.), Topics in Invariant Theory. Seminar 1989-1990. VI, 272 pages. 1991.

Vol. 1479: S. Bloch, I. Dolgachev, W. Fulton (Eds.), Algebraic Geometry. Proceedings, 1989. VII, 300 pages. 1991.

Vol. 1480: F. Dumortier, R. Roussarie, J. Sotomayor, H. Żoładek, Bifurcations of Planar Vector Fields: Nilpotent Singularities and Abelian Integrals. VIII, 226 pages. 1991.

Vol. 1481: D. Ferus, U. Pinkall, U. Simon, B. Wegner (Eds.), Global Differential Geometry and Global Analysis. Proceedings, 1991. VIII, 283 pages. 1991.

Vol. 1482: J. Chabrowski, The Dirichlet Problem with L^2-Boundary Data for Elliptic Linear Equations. VI, 173 pages. 1991.

Vol. 1483: E. Reithmeier, Periodic Solutions of Nonlinear Dynamical Systems. VI, 171 pages. 1991.

Vol. 1484: H. Delfs, Homology of Locally Semialgebraic Spaces. IX, 136 pages. 1991.

Vol. 1485: J. Azéma, P. A. Meyer, M. Yor (Eds.), Séminaire de Probabilités XXV. VIII, 440 pages. 1991.

Vol. 1486: L. Arnold, H. Crauel, J.-P. Eckmann (Eds.), Lyapunov Exponents. Proceedings, 1990. VIII, 365 pages. 1991.

Vol. 1487: E. Freitag, Singular Modular Forms and Theta Relations. VI, 172 pages. 1991.

Vol. 1488: A. Carboni, M. C. Pedicchio, G. Rosolini (Eds.), Category Theory. Proceedings, 1990. VII, 494 pages. 1991.

Vol. 1489: A. Mielke, Hamiltonian and Lagrangian Flows on Center Manifolds. X, 140 pages. 1991.

Vol. 1490: K. Metsch, Linear Spaces with Few Lines. XIII, 196 pages. 1991.

Vol. 1491: E. Lluis-Puebla, J.-L. Loday, H. Gillet, C. Soulé, V. Snaith, Higher Algebraic K-Theory: an overview. IX, 164 pages. 1992.

Vol. 1492: K. R. Wicks, Fractals and Hyperspaces. VIII, 168 pages. 1991.

Vol. 1493: E. Benoît (Ed.), Dynamic Bifurcations. Proceedings, Luminy 1990. VII, 219 pages. 1991.

Vol. 1494: M.-T. Cheng, X.-W. Zhou, D.-G. Deng (Eds.), Harmonic Analysis. Proceedings, 1988. IX, 226 pages. 1991.

Vol. 1495: J. M. Bony, G. Grubb, L. Hörmander, H. Komatsu, J. Sjöstrand, Microlocal Analysis and Applications. Montecatini Terme, 1989. Editors: L. Cattabriga, L. Rodino. VII, 349 pages. 1991.

Vol. 1496: C. Foias, B. Francis, J. W. Helton, H. Kwakernaak, J. B. Pearson, H_∞-Control Theory. Como, 1990. Editors: E. Mosca, L. Pandolfi. VII, 336 pages. 1991.

Vol. 1497: G. T. Herman, A. K. Louis, F. Natterer (Eds.), Mathematical Methods in Tomography. Proceedings 1990. X, 268 pages. 1991.

Vol. 1498: R. Lang, Spectral Theory of Random Schrödinger Operators. X, 125 pages. 1991.

Vol. 1499: K. Taira, Boundary Value Problems and Markov Processes. IX, 132 pages. 1991.

Vol. 1500: J.-P. Serre, Lie Algebras and Lie Groups. VII, 168 pages. 1992.

Vol. 1501: A. De Masi, E. Presutti, Mathematical Methods for Hydrodynamic Limits. IX, 196 pages. 1991.

Vol. 1502: C. Simpson, Asymptotic Behavior of Monodromy. V, 139 pages. 1991.

Vol. 1503: S. Shokranian, The Selberg-Arthur Trace Formula (Lectures by J. Arthur). VII, 97 pages. 1991.

Vol. 1504: J. Cheeger, M. Gromov, C. Okonek, P. Pansu, Geometric Topology: Recent Developments. Editors: P. de Bartolomeis, F. Tricerri. VII, 197 pages. 1991.

Vol. 1505: K. Kajitani, T. Nishitani, The Hyperbolic Cauchy Problem. VII, 168 pages. 1991.

Vol. 1506: A. Buium, Differential Algebraic Groups of Finite Dimension. XV, 145 pages. 1992.

Vol. 1507: K. Hulek, T. Peternell, M. Schneider, F.-O. Schreyer (Eds.), Complex Algebraic Varieties. Proceedings, 1990. VII, 179 pages. 1992.

Vol. 1508: M. Vuorinen (Ed.), Quasiconformal Space Mappings. A Collection of Surveys 1960-1990. IX, 148 pages. 1992.

Vol. 1509: J. Aguadé, M. Castellet, F. R. Cohen (Eds.), Algebraic Topology - Homotopy and Group Cohomology. Proceedings, 1990. X, 330 pages. 1992.

Vol. 1510: P. P. Kulish (Ed.), Quantum Groups. Proceedings, 1990. XII, 398 pages. 1992.

Vol. 1511: B. S. Yadav, D. Singh (Eds.), Functional Analysis and Operator Theory. Proceedings, 1990. VIII, 223 pages. 1992.

Vol. 1512: L. M. Adleman, M.-D. A. Huang, Primality Testing and Abelian Varieties Over Finite Fields. VII, 142 pages. 1992.

Vol. 1513: L. S. Block, W. A. Coppel, Dynamics in One Dimension. VIII, 249 pages. 1992.

Vol. 1514: U. Krengel, K. Richter, V. Warstat (Eds.), Ergodic Theory and Related Topics III, Proceedings, 1990. VIII, 236 pages. 1992.

Vol. 1515: E. Ballico, F. Catanese, C. Ciliberto (Eds.), Classification of Irregular Varieties. Proceedings, 1990. VII, 149 pages. 1992.

Vol. 1516: R. A. Lorentz, Multivariate Birkhoff Interpolation. IX, 192 pages. 1992.

Vol. 1517: K. Keimel, W. Roth, Ordered Cones and Approximation. VI, 134 pages. 1992.

Vol. 1518: H. Stichtenoth, M. A. Tsfasman (Eds.), Coding Theory and Algebraic Geometry. Proceedings, 1991. VIII, 223 pages. 1992.

Vol. 1519: M. W. Short, The Primitive Soluble Permutation Groups of Degree less than 256. IX, 145 pages. 1992.

Vol. 1520: Yu. G. Borisovich, Yu. E. Gliklikh (Eds.), Global Analysis – Studies and Applications V. VII, 284 pages. 1992.

Vol. 1521: S. Busenberg, B. Forte, H. K. Kuiken, Mathematical Modelling of Industrial Process. Bari, 1990. Editors: V. Capasso, A. Fasano. VII, 162 pages. 1992.

Vol. 1522: J.-M. Delort, F. B. I. Transformation. VII, 101 pages. 1992.

Vol. 1523: W. Xue, Rings with Morita Duality. X, 168 pages. 1992.

Vol. 1524: M. Coste, L. Mahé, M.-F. Roy (Eds.), Real Algebraic Geometry. Proceedings, 1991. VIII, 418 pages. 1992.

Vol. 1525: C. Casacuberta, M. Castellet (Eds.), Mathematical Research Today and Tomorrow. VII, 112 pages. 1992.

Vol. 1526: J. Azéma, P. A. Meyer, M. Yor (Eds.), Séminaire de Probabilités XXVI. X, 633 pages. 1992.

Vol. 1527: M. I. Freidlin, J.-F. Le Gall, Ecole d'Eté de Probabilités de Saint-Flour XX – 1990. Editor: P. L. Hennequin. VIII, 244 pages. 1992.

Vol. 1528: G. Isac, Complementarity Problems. VI, 297 pages. 1992.

Vol. 1529: J. van Neerven, The Adjoint of a Semigroup of Linear Operators. X, 195 pages. 1992.

Vol. 1530: J. G. Heywood, K. Masuda, R. Rautmann, S. A. Solonnikov (Eds.), The Navier-Stokes Equations II – Theory and Numerical Methods. IX, 322 pages. 1992.

Vol. 1531: M. Stoer, Design of Survivable Networks. IV, 206 pages. 1992.

Vol. 1532: J. F. Colombeau, Multiplication of Distributions. X, 184 pages. 1992.

Vol. 1533: P. Jipsen, H. Rose, Varieties of Lattices. X, 162 pages. 1992.

Vol. 1534: C. Greither, Cyclic Galois Extensions of Commutative Rings. X, 145 pages. 1992.

Vol. 1535: A. B. Evans, Orthomorphism Graphs of Groups. VIII, 114 pages. 1992.

Vol. 1536: M. K. Kwong, A. Zettl, Norm Inequalities for Derivatives and Differences. VII, 150 pages. 1992.

Vol. 1537: P. Fitzpatrick, M. Martelli, J. Mawhin, R. Nussbaum, Topological Methods for Ordinary Differential Equations. Montecatini Terme, 1991. Editors: M. Furi, P. Zecca. VII, 218 pages. 1993.

Vol. 1538: P.-A. Meyer, Quantum Probability for Probabilists. X, 287 pages. 1993.

Vol. 1539: M. Coornaert, A. Papadopoulos, Symbolic Dynamics and Hyperbolic Groups. VIII, 138 pages. 1993.

Vol. 1540: H. Komatsu (Ed.), Functional Analysis and Related Topics, 1991. Proceedings. XXI, 413 pages. 1993.

Vol. 1541: D. A. Dawson, B. Maisonneuve, J. Spencer, Ecole d' Eté de Probabilités de Saint-Flour XXI - 1991. Editor: P. L. Hennequin. VIII, 356 pages. 1993.

Vol. 1542: J.Fröhlich, Th.Kerler, Quantum Groups, Quantum Categories and Quantum Field Theory. VII, 431 pages. 1993.

Vol. 1543: A. L. Dontchev, T. Zolezzi, Well-Posed Optimization Problems. XII, 421 pages. 1993.

Vol. 1544: M.Schürmann, White Noise on Bialgebras. VII, 146 pages. 1993.

Vol. 1545: J. W. Morgan, K. G. O'Grady. Differential Topology of Complex Surfaces. VIII, 224 pages. 1993.

Vol. 1546: V. V. Kalashnikov, V. M. Zolotarev (Eds.), Stability Problems for Stochastic Models. Proceedings, 1991. VIII, 229 pages. 1993.

Vol. 1547: P. Harmand, D. Werner, W. Werner, M-ideals in Banach Spaces and Banach Algebras. VIII, 387 pages. 1993.

Vol. 1548: T. Urabe, Dynkin Graphs and Quadrilateral Singularities. VI, 233 pages. 1993.

GPSR Compliance
The European Union's (EU) General Product Safety Regulation (GPSR) is a set of rules that requires consumer products to be safe and our obligations to ensure this.

If you have any concerns about our products, you can contact us on

ProductSafety@springernature.com

In case Publisher is established outside the EU, the EU authorized representative is:

Springer Nature Customer Service Center GmbH
Europaplatz 3
69115 Heidelberg, Germany

www.ingramcontent.com/pod-product-compliance
Ingram Content Group UK Ltd.
Pitfield, Milton Keynes, MK11 3LW, UK
UKHW021828190726
13853UKWH00003B/1255

* 9 7 8 3 6 6 2 2 1 2 2 4 0 *